Graphs, Morphisms and Statistical Physics

DIMACS
Series in Discrete Mathematics and Theoretical Computer Science

Volume 63

Graphs, Morphisms and Statistical Physics

DIMACS Workshop
Graphs, Morphisms and Statistical Physics
March 19–21, 2001
DIMACS Center

J. Nešetřil
P. Winkler
Editors

Center for Discrete Mathematics
and Theoretical Computer Science
A consortium of Rutgers University, Princeton University,
AT&T Labs–Research, Bell Labs (Lucent Technologies),
NEC Laboratories America, and Telcordia Technologies
(with partners at Avaya Labs, IBM Research, and Microsoft Research)

American Mathematical Society

This DIMACS volume presents the proceedings from the workshop Graphs, Morphisms and Statistical Physics held at the DIMACS Center, Rutgers University, March 19–21, 2001.

2000 *Mathematics Subject Classification.* Primary 05-06, 05A16, 05C15, 05C60, 60-06, 60J10, 60K35, 68R10, 82B20.

Library of Congress Cataloging-in-Publication Data

DIMACS Workshop Graphs, Morphisms and Statistical Physics (2001 : DIMACS Center)
   Graphs, morphisms, and statistical physics : DIMACS Workshop Graphs, Morphisms and Statistical Physics, March 19–21, 2001, DIMACS Center / J. Nesetril, P. Winkler, editors.
       p. cm. — (DIMACS series in discrete mathematics and theoretical computer science ; v. 63)
   Includes bibliographical references.
   ISBN 0-8218-3551-3 (acid-free paper)
   1. Graph theory–Congresses.   2. Morphisms (Mathematics)–Congresses.   3. Statistical physics–Congresses.   I. Nesetril, Jaroslav.   II. Winkler, P. (Peter), 1946–   III. Title.   IV. Series.

QC20.7.G7D56   2004
530.15′15–dc22   2003063743

Contents

Foreword

A workshop on Graphs, Morphisms, and Statistical Physics was held at Rutgers University on March 19-21, 2001. We would like to express our appreciation to Jaroslav Nešetřil and Peter Winkler for their efforts to organize and plan this successful conference.

The workshop was part of a unique partnership between the DIMACS Center, headquarted in New Jersey, USA, and the DIMATIA Center, headquartered in Prague, Czech Republic. The partnership has included a series of workshops, an exchange of visitors, a joint undergraduate research (REU) program, and joint research projects.

Issues at the confluence of graph theory, computer science, probability, and statistical physics have come together in recent years to create exciting new research agendas. Areas of common interest include connections between random graphs and percolation, between slow mixing and phase transition, and between graph morphisms and hard-constraint models.

This volume is not simply a workshop proceedings. It includes papers from researchers whose work added to the coverage of the volume and also includes descriptions of work that was stimulated by the workshop particularly and the partnership between DIMACS and DIMATIA generally.

DIMACS gratefully acknowledges the generous support that makes these programs possible. In particular, the National Science Foundation supported this workshop, matched by a similar grant from its sister agency in the Czech Republic. In addition, the New Jersey Commission on Science and Technology, DIMACS' partners at Rutgers, Princeton, AT&T Labs-Research, Bell Labs, NEC Laboratories America, and Telcordia Technologies, and its affiliated partners at Avaya Labs, HP Labs, IBM Research, and Microsoft Research have generously supported the partnership between DIMACS and DIMATIA.

We are particularly grateful to DIMATIA and also to the Institute of Theoretical Computer Science (ITI) at Charles University for their support of our many joint projects and in particular for contributing so vitally to the success of the workshop and this volume.

Fred S. Roberts
Director

Robert Tarjan
Co-Director for Princeton

Preface

The intersection of combinatorics and statistical physics has been an area of great activity over the past few years, fertilized by an exchange not only of techniques but of objectives. Spurred by computing theorists interested in approximation algorithms, statistical physicists and discrete mathematicians have overcome language problems and found a wealth of common ground in probabilistic combinatorics.

Close connections between percolation and random graphs, between graph morphisms and hard-constraint models, and between slow mixing and phase transition, have led to new results and new perspectives. These connections can help in understanding typical, as opposed to extremal, behavior of combinatorial phenomena such as graph coloring and homomorphisms.

Any "nearest neighbor" system of statistical physics can be interpreted as a space of graph morphisms—for example, morphisms to the two-node graph with one edge and one loop correspond to the "hard-core lattice gas model" and to random independent sets. Given the set of morphisms from a (possibly infinite) graph G to a graph H, when do we see long range order? When does changing the morphism site by site (heat bath) yield rapid mixing, or even eventual mixing? The special case of proper colorings (corresponding to the anti-ferromagnetic Potts model at 0 temperature) is especially interesting to graph theorists but more general notions of graph colorings may also yield some intriguing new questions.

Inspired by these issues, and encouraged by Fred Roberts and other colleagues, we organized a "DIMACS/DIMATIA Workshop on Graphs, Morphisms and Statistical Physics" which took place at Rutgers University, March 19–21, 2001. The workshop was attended by 53 scientists from the U.S. and abroad, and the present volume is an outgrowth of this meeting.

Some of the topics we cover here are: percolation, random colorings, homomorphisms from and to a fixed graph, mixing, combinatorial phase transitions, threshold phenomena, scaling windows, and some purely combinatorial aspects of graph coloring.

In addition to the participants we asked several other colleagues whose research complemented the scope of the volume. The volume also contains several contributions which were directly influenced by the fruitful atmosphere of this meeting. We thank all participants for their time and effort which made the success of this meeting possible.

This workshop was a joint activity of DIMACS, New Jersey and DIMATIA, Prague and it was supported by both of these centers. Most of the organization of the volume was done by Robert Šámal (DIMATIA–ITI) and without his effort the volume would hardly be possible.

The final stage of preparation of the volume was supported by Institute of Theoretical Computer Science (ITI) at Charles University, Prague (under grant LN00A056).

Jaroslav Nešetřil
Peter Winkler

List of delivered talks

Stefan Boettcher	Efficient Local Search Near Phase Transitions
Béla Bollobás	Large Subgraphs of Random Graphs
Christian Borgs	Slow Mixing for Random H-Colorings on the Hypercubic Lattice
Graham Brightwell	Proper Colorings of Regular Trees
Jennifer Chayes	A Phase Transition in the Optimum Partitioning Problem
Josep Díaz	Counting H-Morphisms
Peter Gács	Compatible Sequences and a Slow Winkler Percolation
Gabriel Istrate	Computational Complexity and Phase Transition?
Pavol Hell	List Homomorphisms
Irene Hueter	Threshold Phenomena for Interacting or Constraint Particles
Jacob Katriel	On the Solution of a Spin Model
Roman Kotecký	Potts Antiferromagnet: The Case of Entropic Contours?
Hanno Lefmann	Cliques and Independent Sets in Graphs and Hypergraphs and Applications
Martin Loebl	The Pfaffian Approach to the Ising Problem and the Dimer Problem in 2-Dimensional and 3-Dimensional Lattices
Elchanan Mossel	Information Flow on Trees
Jaroslav Nešetřil	Homomorphisms: Structure, Extremal and Random
Kihong Park	Denial of Service Attacks and the Internet Power Law
Yuval Peres	Rapid Mixing for Glauber Dynamics Without Uniqueness of Gibbs States
Gert Sabidussi	Large Independent Sets and 3-Colourings of 4-Regular Hamiltonian Graphs
Benny Sudakov	A Sharp Threshold for Network Reliability
Claude Tardif	The Chromatic Number of the Product of Two Graphs is at Least Half the Minimum of the Fractional Chromatic Numbers of the Factors
David Wilson	Critical Resonance for Nonintersecting Lattice Paths
Peter Winkler	Graph Homomorphisms and Long Range Order

List of participants

Sorin Alexe
Rutgers University
RUTCOR, 640 Bartholomew Road, Piscataway, NJ 08854
salexe@rutcor.rutgers.edu

Eric Allender
Rutgers University
Computer Science Department, 110 Frelinghuysen Road, Piscataway, NJ 08854
allender@cs.rutgers.edu

Stefan Boettcher
Emory University
Physics Department, 1510 Clifton Road, Atlanta, GA 30322
sboettc@emory.edu

Béla Bollobás
University of Memphis
Mathematics, 243B Dunn Hall, 3725 Norriswood, Memphis, TN 38152
bollobas@msci.memphis.edu

Christian H. Borgs
Microsoft Research
Theory Group, 1 Microsoft Way, Redmond WA, 98052
borgs@microsoft.com

Endré Boros
Rutgers University
RUTCOR, 640 Bartholomew Road, Piscataway, NJ 08854
boros@rutcor.rutgers.edu

Graham Richard Brightwell
London School of Economics
Mathematics, Houghton Street, London WC2A 2AE, U.K.
G.R.Brightwell@lse.ac.uk

Jennifer T. Chayes
Microsoft Research
Theory Group, 1 Microsoft Way, Redmond, WA 98052
jchayes@microsoft.com

Raissa M. D'Souza
Bell Laboratories
Fundamental Mathematics, 600 Mountain Avenue, Murray Hill, NJ 07974
raissa@bell-labs.com

Josep Díaz
Universitat Politecnica Catalunya
LSI, Jordi Girona Salgado 1-3, Barcelona, 08034, Spain
diaz@lsi.upc.es

Nancy Eaton
University of Rhode Island
Mathematics, 9 Greenhouse Road, Suite 3, Kingston, RI 02881
eaton@math.uri.edu

Glenn E. Faubert
University of Rhode Island
Mathematics, Greenhouse Road, Kingston, RI 02881
gfaubert@math.uri.edu

Peter Gács
Boston University
Computer Science, 111 Cummington Street, Boston, MA 02215
gacs@bu.edu

David Galvin
Rutgers University
Department of Mathematics, 110 Frelinghuysen Road, Piscataway, NJ 08854
dgalvin@math.rutgers.edu

David Gamarnik
IBM - TJ Watson Research Center
Math Sciences, PO Box 218, Yorktown Heights, NY
gamarnik@watson.ibm.com

Peter L. Hammer
Rutgers University
RUTCOR, 640 Bartholomew Road, Piscataway, NJ 08854
hammer@rutcor.rutgers.edu

Stephen Hartke
Rutgers University
Department of Mathematics, 110 Frelinghuysen Road, Piscataway, NJ 08854
hartke@math.rutgers.edu

Pavol Hell
Simon Fraser University
Computing Science, Burnaby, BC V5A1S6, Canada
pavol@cs.sfu.ca

Irene Hueter
University of Florida
Mathematics, 358 Little Hall, PO Box 118105, Gainesville, FL 32611-8105
hueter@math.ufl.edu

Gabriel Istrate
Los Alamos National Laboratory
Center for Nonlinear Sciences and D-2 Division, MS M997, Los Alamos, NM 87545
istrate@lanl.gov

Mel Janowitz
Rutgers University
DIMACS, 96 Frelinghuysen Road, Piscataway, NJ 08854-8018
melj@dimacs.rutgers.edu

Jeff Kahn
Rutgers University
Mathematics Department, 110 Frelinghuysen Road, Piscataway, NJ 08854
jkahn@math.rutgers.edu

Jacob Katriel
Rutgers University
Chemistry, Piscataway, NJ
jkatriel@tx.technion.ac.il

Tanya Khovanova
Lotus Management
321 Witherspoon Street, Princeton, NJ 08540
tanyakh@yahoo.com

Alexy Khrabrov
NEC Research Institute
alexy@research.nj.nec.com

Roman Kotecký
Charles University
Center for Theoretical Study, Jilska 1, 110 00 Praha 1, Czech Republic
kotecky@cucc.ruk.cuni.cz

Michal Koucký
Rutgers University
Computer Science Department, 110 Frelinghuysen Road, Piscataway, NJ 08854
mkoucky@paul.rutgers.edu

Klay Thomas Kruczek
Rutgers Univeristy
Mathematics, 3112 Cypress Court, Monmouth Junction, NJ 08852
kruczek@math.rutgers.edu

Hanno Lefmann
TU-Chemnitz
Fakultaet fuer Informatik, Strasse der Nationen 62, Chemnitz, D-09107
Germany
lefmann@informatik.tu-chemnitz.de

Martin Loebl
Georgia Institute of Technology
School of Mathematics, Skiles 116A, Atlanta, GA 30332-0160
loebl@math.gatech.edu

Michael W. Mahoney
Yale University
Computer Science, 400 West 119th Street, Apt. 11A, New York, NY 10027
michael.mahoney@yale.edu

Russell Martin
Georgia Institute of Technology
Mathematics, Atlanta, GA 30332-0160
martin@math.gatech.edu

Elchanan Mossel
Microsoft Research
1 One Microsoft Way, Redmond, WA 98052
mossel@microsoft.com

Shobhana Murali
Georgia Institute of Technology
School of Mathematics, Atlanta, GA 30332-0160
murali@math.gatech.edu

Serban Miron Nacu
University of California, Berkeley
Statistics, 367 Evans Hall #3860, Berkeley, CA 94720-3860
serban@stat.berkeley.edu

Jaroslav Nešetřil
DIMATIA-ITI
Charles University, Malostranske nam. 25, 11800 Praha 1, Czech Republic
nesetril@kam.ms.mff.cuni.cz

Kihong Park
Purdue University
Computer Science, 1398 Computer Science Building, West Lafayette, IN 47907
park@cs.purdue.edu

Alair Pereira do Lago
DIMACS/Univ. of Sao Paulo
96 Frelinghuysen Road, Piscataway, NJ 08854
alair@dimacs.rutgers.edu

Yuval Peres
University of California, Berkelcy
Statistics and Mathematics, Evans Hall, Berkeley, CA 94720
peres@stat.berkeley.edu

Dana Randall
Georgia Tech
College of Computing, Atlanta, GA 30332-0128
randall@cc.gatech.edu

Fred S. Roberts
Rutgers University
DIMACS, 96 Frelinghuysen Road, Piscataway, NJ 08854-8018
froberts@dimacs.rutgers.edu

Gert Sabidussi
Universite de Montreal
Mathematics, CP 6128/Centre-ville, Montreal (Qc) H3C 3J7, Canada
sab@dms.umontreal.ca

Michael Saks
Rutgers University
Mathematics Department
saks@math.rutgers.edu

Clifford Smyth
Rutgers University
Department of Mathematics, 110 Frelinghuysen Road, Piscataway, NJ 08854
csmyth@math.rutgers.edu

Jeff Steif
Georgia Institute of Technology
Mathematics, Skiles Hall, Atlanta, GA 30332-0160
steif@math.gatech.edu

Amy Stern
Rutgers University
Department of Mathematics, 110 Frelinghuysen Road, Piscataway, NJ 08854
aestern@math.rutgers.edu

Benny Sudakov
Princeton University
Mathematics, Fine Hall, Washington Road, Princeton, NJ 08544
bsudakov@math.princeton.edu

Claude Tardif
University of Regina
Department of Mathematics and Statistics, University of Regina, Regina, SK S4S-
0A2, Canada
tardif@math.uregina.ca

Pavel Valtr
Charles University, Prague
Department of Applied Mathematics
valtr@kam.ms.mff.cuni.cz

Eric Vigoda
University of Edinburgh
Computer Science, JCMB/King's Buildings, Edinburgh EH9 3JZ, United Kingdom
vigoda@dcs.ed.ac.uk

David B. Wilson
Microsoft Research
One Microsoft Way, Redmond, WA 98052
dbwilson@microsoft.com

Peter Winkler
Bell Labs
700 Mountain Avenue, Murray Hill, NJ 07974
pw@lucent.com

Yi Zhao
Rutgers University
Department of Mathematics, 110 Frelinghuysen Road, Piscataway, NJ 08854
yzhao@math.rugers.edu

DIMACS Series in Discrete Mathematics
and Theoretical Computer Science
Volume **63**, 2004

Efficient Local Search near Phase Transitions in Combinatorial Optimization

Stefan Boettcher

ABSTRACT. We introduce Extremal Optimization, a new heuristic for finding high-quality solutions to hard optimization problems. For many physical
and combinatorial optimization problems studied, Extremal Optimization appears to be particularly successful near phase transitions, which are believed
to harbor some of the hardest instances. We give an overview over the im-
plementations for a variety of NP-hard optimization problems. We present
numerical results for partitioning on some large graphs from testbeds and the
ground states of spin glasses. In addition, we will demonstrate the capabilities
of the method in determining costs and backbones near phase transitions, in
particular, for graph bipartitioning and for random graph coloring.

1. Introduction

Many natural systems have, without any centralized organizing facility, developed into complex structures that optimize their use of resources in sophisticated
ways [**3**]. Biological evolution has formed efficient and strongly interdependent networks in which resources rarely go to waste. Even the morphology of inanimate
landscapes exhibits patterns that seem to serve a purpose, such as the efficient
drainage of water [**40, 14**].

Natural systems that exhibit self-organizing qualities often possess a common
feature: a large number of strongly coupled entities with similar properties. Hence,
at some coarse level they permit a statistical description. An external resource
(sunlight in the case of evolution) drives the system which then takes its direction
purely by chance. Like descending water breaking through the weakest of all barri-
ers in its wake, biological species are coupled in a global comparative process that
persistently washes away the least fit. In this process, unlikely but highly adapted
structures surface inadvertently. Optimal adaptation thus emerges naturally, from
the dynamics, simply through a selection *against* the extremely "bad". In fact, this
process prevents the inflexibility inevitable in a controlled breeding of the "good".

Various models relying on extremal processes have been proposed to explain
the phenomenon of self-organization [**37**]. In particular, the Bak-Sneppen model
of biological evolution is based on this principle [**4, 9**]. Assuming an unspecified

interdependency between species, it produces salient nontrivial features of pale-ontological data such as broadly distributed lifetimes of species, large extinction events, and punctuated equilibrium.

In the Bak-Sneppen model, species are located on the sites of a lattice, and have an associated "fitness" value between 0 and 1. At each time step, the one species with the smallest value (poorest degree of adaptation) is selected for a random update, having its fitness replaced by a new value drawn randomly from a flat distribution on the interval $[0, 1]$. But the change in fitness of one species impacts the fitness of interrelated species. Therefore, all of the species at neigh-boring lattice sites have their fitness replaced with new random numbers as well. After a sufficient number of steps, the system reaches a highly correlated state known as self-organized criticality (SOC) [**5**]. In that state, almost all species have reached a fitness above a certain threshold. These species, however, possess *punc-tuated equilibrium*: only one's weakened neighbor can undermine one's own fitness. This coevolutionary activity gives rise to chain reactions called "avalanches", large fluctuations that rearrange major parts of the system, potentially making any con-figuration accessible.

Although coevolution may not have optimization as its exclusive goal, it serves as a powerful paradigm. We have used it as motivation for a new approach to approximate hard optimization problems [**10**]. The heuristic we have introduced, called extremal optimization (EO), follows the spirit of the Bak-Sneppen model, updating those variables which have among the "worst" values in a solution and replacing them by random values without ever explicitly improving them.

2. Extremal Optimization

To introduce EO, let us consider a spin glass [**34**] as a specific example of a hard optimization problem. It consists of a d-dimensional hyper-cubic lattice of length L with periodic boundary conditions, with a spin variable $x_i \in \{-1, 1\}$ at each site i, $1 \leq i \leq n \; (= L^d)$. A spin is connected to each of its nearest neighbors j via a bond variable $J_{ij} \in \{-1, 1\}$, assigned at random. The configuration space Ω consist of all configurations $S = (x_1, \ldots, x_n) \in \Omega$ where $|\Omega| = 2^n$.

We wish to minimize the cost function, or Hamiltonian

$$(1) \qquad\qquad C(S) = H(\mathbf{x}) = -\frac{1}{2} \sum_{\langle i,j \rangle} J_{ij} x_i x_j,$$

where the sum extends over all nearest-neighbor pairs of spins. Due to frustra-tion [**34**], ground state configurations $S_{\min}$ are hard to find, and it has been shown that for $d > 2$ the problem is among the hardest optimization problems [**6**].

To find near-optimal solutions for a particular optimization problem, EO per-forms a neighborhood search on a single configuration $S \in \Omega$. As in the spin problem in Eq. (1), S consists of a large number n of variables x_i. We assume that each S possesses a neighborhood $N(S)$ that rearranges the state of merely a small number of the variables. This is a characteristic of a local search, in contrast to a genetic algorithm, say, where cross-overs may effect $O(n)$ variables on each update. The cost $C(S)$ is assumed to consist of the individual cost contributions, or "fit-nesses", λ_i for each variable x_i (analogous to the fitness values in the Bak-Sneppen model [**4, 10**]). The fitness of each variable assesses its contribution to the total cost:

$$C(S) = -\sum_{i=1}^{n} \lambda_i. \tag{2}$$

Typically, the fitness λ_i depends on the state of x_i in relation to connected variables.

For example, for the Hamiltonian in Eq. (1), we assign to each spin x_i the fitness

$$\lambda_i = x_i \left(\frac{1}{2} \sum_j J_{ij} x_j \right), \tag{3}$$

so that Eq. (2) is satisfied. Each spin's fitness thus corresponds to (the negative of) its local energy contribution to the overall energy of the system. In similarity to the Bak-Sneppen model, EO then proceeds through a neighborhood search of Ω by sequentially changing variables with "bad" fitness on each update, for instance, via single spin-flips. After each update, the fitnesses of the changed variable and of all its connected neighbors are reevaluated according to Eq. (3). The basic EO algorithm proceeds as follows:

> (1) Initialize configuration S at will; set $S_{\text{best}} := S$.
> (2) For the "current" configuration S,
>   (a) evaluate λ_i for each variable x_i,
>   (b) find j satisfying $\lambda_j \leq \lambda_i$ for all i, i.e., x_j has the "worst fitness",
>   (c) choose $S' \in N(S)$ such that x_j *must* change,
>   (d) accept $S := S'$ *unconditionally*,
>   (e) if $C(S) < C(S_{\text{best}})$ then set $S_{\text{best}} := S$.
> (3) Repeat at step (2) as long as desired.
> (4) Return S_{best} and $C(S_{\text{best}})$.

The algorithm operates on a single configuration S at each step. Each variable x_i in S has a fitness, of which the "worst" is identified. This *ranking* of the variables provides the only measure of quality on S, implying that all other variables are "better" in the current S. In the move to a neighboring configuration, typically only a small number of variables change state, so only a few connected variables need to be re-evaluated [step (2a)] and re-ranked [step (2b)]. Note that there is not a single parameter to adjust for the selection of better solutions. It is merely the memory encapsulated in the ranking that directs EO into the neighborhood of increasingly better solutions. Similar to the Bak-Sneppen model, those "better" variables possess punctuated equilibrium: their memory only get erased when they happen to be connected to one of the variables forced to change. On the other hand, in the choice of move to S', there is no consideration given to the outcome of such a move, and not even the worst variable x_j itself is guaranteed to improve its fitness. Accordingly, large fluctuations in the cost can accumulate in a sequence of updates. Merely the bias *against* extremely "bad" fitnesses produces improved solutions.

Tests have shown that this basic algorithm is very competitive for optimization problems where EO can choose randomly among many $S' \in N(S)$ that satisfy step (2c), as is the situation for graph bipartitioning [10, 7]. But in cases such as the single spin-flip neighborhood for the spin Hamiltonian, focusing on only the worst fitness [step (2b)] leads to a deterministic process, leaving no choice in

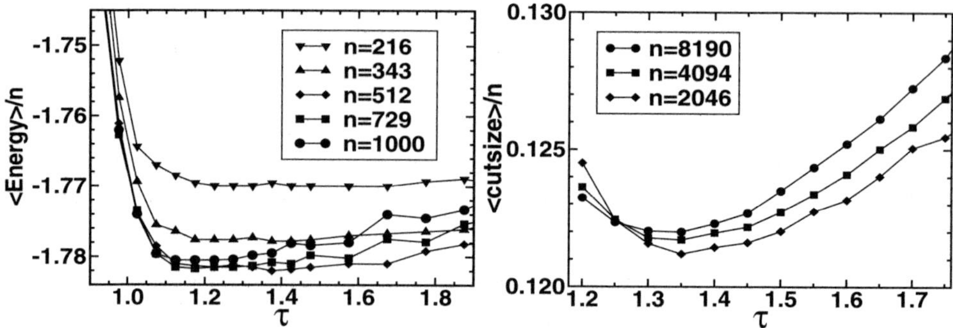

FIGURE 1. Plot of the average costs obtained by EO for a $\pm J$ spin glass (left) and for graph bipartitioning (right), both as a function of τ. For each size n, a number of instances were generated. For each instance, 10 different EO runs were performed at each τ. The results were averaged over runs and instances. Although both problems are quite distinct, in either case the best results are obtained at a value of τ that moves slowly toward $\tau \to 1^+$ for $n \to \infty$.

step (2c): If the "worst" spin x_j *has* to flip and any neighbor S' differs by only one flipped spin from S, it must be $S' = (S/x_j) \cup \{-x_j\}$. This deterministic process inevitably will get stuck near some poor local minimum. To avoid these "dead ends" and to improve results [**10**], we introduce a single parameter into the algorithm. To this end, we rank all x_i according to fitness λ_i, *i.e.*, we find a permutation Π of the variable labels i with

$$(4) \qquad \lambda_{\Pi(1)} \leq \lambda_{\Pi(2)} \leq \ldots \leq \lambda_{\Pi(n)}.$$

The worst variable x_j [step (2b)] is of rank 1, $j = \Pi(1)$, and the best variable is of rank n. Now, consider a scale-free probability distribution over the *ranks* k,

$$(5) \qquad P_k \propto k^{-\tau}, \qquad 1 \leq k \leq n,$$

for a fixed value of the parameter τ. At each update, select a rank k according to P_k. Then, modify step (2c) so that the variable x_j with $j = \Pi(k)$ changes its state.

For $\tau = 0$, this "τ-EO" algorithm is simply a random walk through Ω. Conversely, for $\tau \to \infty$, the process approaches a deterministic local search, only updating the lowest-ranked variable, and is bound to reach a dead end (see Fig. 1). However, for finite values of τ the choice of a *scale-free* distribution for P_k in Eq. (5) ensures that no rank gets excluded from further evolution, while still maintaining a bias against variables with bad fitness. In all problems studied so far, a value of

$$(6) \qquad \tau - 1 \sim 1/\ln n \qquad (n \to \infty)$$

seems to work best [**11, 12**]. We have studied a simple model problem for which the asymptotic behavior of τ-EO can be solved exactly [**8**]. The model reproduces Eq. (6) exactly in cases where the model develops a "jam" amongst its variables, which is quite a generic feature of frustrated systems.

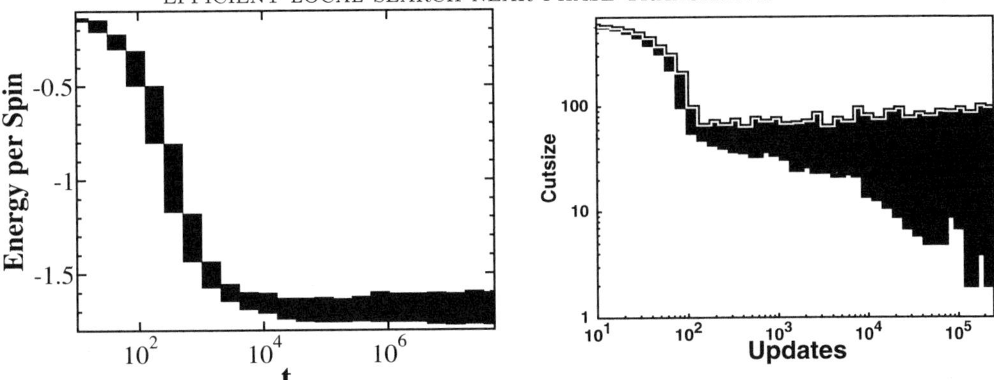

FIGURE 2. Plots of the range of states attained by EO ($\tau = 1.1$) during exponentially increasing time periods of a single run on a particular instance (of an $n = 1000$ spin glass (left) and of an $n = 500$ geometric graph bipartition (right). Note that the data is binned logarithmically in time, i. e. that every bin spans as much time as all bins to its left taken together. A transition at $t \sim n$ is clearly visible between an initial "mop-up" phase, and the ultimate "steady state" in which EO fluctuates widely through near-optimal configurations.

In Fig. 2 we show the range of states that are sampled during a typical run of EO, here for a spin-glass instance with $n = 1000$ and for the partitioning of a geometric graph with $n = 500$. Starting with a random initial condition, for the first $O(n)$ update steps EO establishes local order, leading to a rapid decrease in the energy. After that EO searches through a wide band of states with frequent returns to near-optimal configurations. (In fact, further investigation shows that EO's error decays proportional to a power of the run time [12].) This behavior is very similar to the evolution towards the critical state of the Bak-Sneppen model [37].

In a series of projects [13], we have demonstrated the success of this heuristic in obtaining near-optimal solutions for the graph bipartitioning problem [10, 12, 7], the 3-coloring of graphs, and the Edwards-Anderson spin-glass problem [11]. The algorithmic results show remarkable similarities on these diverse problems. As Fig. 1 suggests, in each of these problems the choice of the parameter τ exhibits universal features that may allow an *a priori* prediction for the best setting of τ, as suggested by Eq. (6) and Ref. [8]. While τ-EO proves to be successful in many applications, like the spin-glass problem, our experiments near phase transitions, as for partitioning [7] and coloring [11] below, indicate that EO's large fluctuations allow it to succeed where other methods fail.

Some of the shortcomings of the algorithm are already apparent:

- Many optimization problems may not allow a decomposition into individual variables with "fitnesses" that represent their contribution to the cost function. And even with distinct variables, a definition of "fitness" may not be unique.

- Accepting every move [step (2d)] may put EO at a disadvantage over other methods if acceptance almost always destroys a good local arrangement. EO for the Traveling Salesperson Problem is limited by this effect, but can be modified to obtain reasonable results [**10**].
- If variables are highly connected (average connectivity c), the re-evaluation of fitnesses λ_i [step (2a)] may become costly, since the ranking of fitnesses [step (2b)] produces a extra runtime factor of at least $c \ln n$ [**10**].

For many problems, these disadvantages do not apply or are surmountable, for instance, the ranking of variables often are only marginally effected by ambiguities in the fitness. As we have shown, problems that can be converted into the form of the Hamiltonian in Eq. (1) are readily adapted to the EO algorithm. This includes the important optimization class MAX-SNP [**39**], which fits naturally into the EO-framework. In a MAX-SNP problem we have boolean variables and a collection of bounded-arity boolean terms and seek an assignment satisfying as many terms as possible. Such problems have a natural choice of fitness functions, and typically have low variable connectivity. Indeed, some complete problems for the class have bounded connectivity in the worst case. Existing surveys [**2**] suggest many such optimization problems for potential study. In particular we propose to study MAX-K-SAT, K-COL, and MAXCUT (similar to a spin glass [**26**]), whose EO-implementations are discussed below.

3. Empirical Studies of Extremal Optimization

3.1. EO for the Graph Bipartitioning Problem. In the GBP, variables x_i are given by a set of n vertices, where n is even. "Edges" connect certain pairs of vertices to form an instance of a graph. The problem is to find a way of partitioning the vertices into two subsets, each constrained to be *exactly* of size $n/2$, such that a *minimal* number of edges cut across the partition. The cost function $C(S)$ (called "cutsize") counts the number of such edges. Instances are typically parameterized by the average connectivity c of its vertices.

To implement τ-EO for the GBP, we attribute to each vertex x_i a local cost $\lambda_i = -b_i/2$, where b_i is the number of "bad" edges crossing the partition (equally shared with the vertex on the other end of that edge). Note that Eq. (2) is satisfied. The simplest neighborhood $N(S)$ for the GBP is an "exchange" of one vertex from each subset. For EO we choose two numbers k_1, k_2 according to Eq. (5) and exchange x_{j_1} and x_{j_2} with $j_1 = \Pi(k_1)$ and $j_2 = \Pi(k_2)$ as in Eq. (4), subject to the restriction that x_{j_1} and x_{j_2} be in opposite subsets. For the sizes of graphs studied, a value of $\tau = 1.4 - 1.6$ worked best [see Fig. 1(b)].

Table 1 summarizes τ-EO's results on large-n graphs, using $\tau = 1.4$ and best-of-10 runs. On each graph, we used as many update steps t as appeared productive for EO to reliably obtain stable results. This varied with the particularities of each graph, from $t = 2n$ to $200n$, and the reported runtimes are of course influenced by this. It is worth noting, though, that EO's *average* performance has been varied. For instance, half of the *Brack2* runs returned cutsizes near 731, but the other half returned cutsizes of above 2000. This may be a product of an unusual structure in this particular graph.

Studies on the average rate of convergence toward better-cost configurations as a function of runtime t indicate power-law convergence, roughly like $C(S_{\text{best}})_t \sim C(S_{\text{min}}) + A\, t^{-0.4}$ [**12**], also found by Ref. [**17**]. Of course, it is not easy to assert for

TABLE 1. Best cutsizes (and allowed runtime) for a testbed of large graphs. GA results are the best reported [32] (at 300MHz). τ-EO results are from our runs (at 200MHz). Comparison data for three of the large graphs are due to results from heuristics in Ref. [23] (at 50MHz). METIS is a partitioning program based on hierarchical decomposition instead of local search [28], obtaining extremely fast deterministic results (at 200MHz).

Large Graph		GA	τ-EO	Ref. [23]	p-METIS
Hammond	$(n = 4720; c = 5.8)$	90(1s)	90(42s)	97(8s)	92(0s)
Barth5	$(n = 15606; c = 5.8)$	139(44s)	139(64s)	146(28s)	151(0.5s)
Brack2	$(n = 62632; c = 11.7)$	731(255s)	731(12s)	—	758(4s)
Ocean	$(n = 143437; c = 5.7)$	464(1200s)	464(200s)	499(38s)	478(6s)

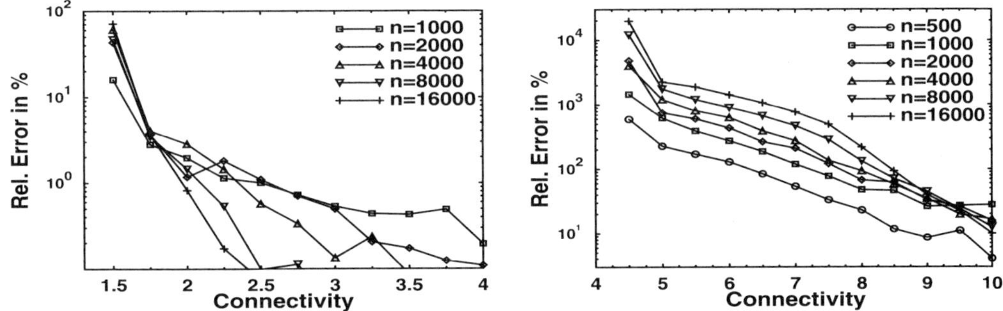

FIGURE 3. Plot of the error in the best result of SA relative to EO's on identical instances of random graphs (left) and geometric graphs (right) as function of the average connectivity c. The critical points for the GBP are at $c = 2\ln 2 = 1.386$ [33] for random and at $c \approx 4.5$ [7] for geometric graphs. SA's error near the critical point in both cases rises with n.

graphs of large n that those runs in fact converge closely to the optimum $C(S_{\min})$, but finite-size scaling analysis for random graphs seems to justify that expectation [12].

In an extensive numerical study on random and geometric graphs [7] we have shown that τ-EO outperforms simulated annealing (SA) [29, 25] significantly near phase transitions [15], where cutsizes first become non-zero. To this end, we have compared the averaged best results obtained for both methods for a large number of instances for increasing n at a fixed parameter setting ($\tau = 1.4$ for EO). Fig. 3 shows that the SA implementation produces increasingly worse results near the phase transition relative to EO. In turn, it was found that EO reproduces many features of the transition quite accurately [7].

3.2. EO for Graph Coloring (K-COL). In K-COL we are given K different colors to label the vertices of a graph in a way that minimizes the number of "monochromatic" edges, connecting vertices of identical color. To implement τ-EO, we again choose for each vertex $\lambda_i = -b_i/2$, where b_i is its number of monochromatic edges on i so that Eq. (2) is satisfied. The simplest neighborhood $N(S)$ for K-COL is a change of color for any one of the vertices in S.

The phase transition in 3-COL has been investigated in Refs. [**15, 1, 16**]. Here, we used EO to completely enumerate *all* optimal solutions $S_{\min}$ near the critical point for 3-COL of random graphs to determine the "backbone," *i. e.*, the fraction of *constrained variables* that are found in an identical state in *all* $S_{\min}$. Instances of random graphs typically possess a large number of equally optimal solutions $S_{\min}$. In Ref. [**36**] it was shown that at the phase transition of 3-SAT its backbone discontinuously jumps to a non-zero value. In 3-COL, unlike 3-SAT, the cost is invariant to global permutations of the vertex colors. Such symmetries often affect critical behavior [**19**], hence 3-COL poses a worthy challenge to the conjectured generality of the discontinuous phase transition in the backbone for NP-hard problems. (Due to the symmetry, we have to define the backbone as the fraction of all *pairs* of vertices locked into the same relative state, same or opposite, for all $S_{\min}$.)

To test the conjecture for the 3-COL, we generated a large number of random graphs and explored Ω for as many ground states as EO could find. To that end, we have fixed runtimes for EO at $O(n^3)$. With that implementation we indeed reproduced all ground states $S_{\min}$, as verified with an exact method (DSATUR) on a (necessarily small) number of test-instances of $n \leq 256$. For each instance, we measured the optimal cost and the backbone fraction of fixed pairs of vertices. The results in Fig. 4 allow us to estimate precisely the location of the transition and the scaling behavior of the cost function. The backbone fraction (Fig. 4, on the left) appears to be discontinuous at the transition, consistent with the conjecture of Ref. [**36**].

3.3. Spin Glasses. Of significant physical relevance are the low temperature properties of spin glasses [**34, 18**] we introduced in Sec. 2. With this implementation [**11**], we have reproduce previous results [**38, 21, 22**] for the ground states of spin glasses in $d = 3$ and 4. Since EO never freezes into a local minimum, it is well suited to enumerate the distribution of near-optimal states of a system, which is of great importance for the low-temperature dynamics in glassy materials [**18, 20, 31**]. (Of course, the frequency of states visited by EO itself cannot produce thermodynamic quantities directly!)

We have run the τ-EO algorithm with $\tau = 1.15$ on a large number of realizations of the J_{ij}, for $n = L^d$ with $L = 5, 6, 7, 8, 9, 10, 12$ in $d = 3$, and with $L = 3, 4, 5, 6, 7$ in $d = 4$. To reduce variances, we fixed $|\sum J_{ij}| \leq 1$. For each instance, we have run EO with 5 restarts from random initial conditions, retaining only the lowest energy state obtained, and then averaging over instances. Inspection of the convergence results for the genetic algorithms in Refs. [**38, 24**] suggest a runtime scaling at least as n^3–n^4 for consistent performance. Indeed, using $\sim n^4/100$ updates enables EO to reproduce its lowest energy states on about 80% to 95% of the restarts, for each n. Our results are listed in Table 2. A fit of our data with $e_d(n) = e_d(\infty) + A/n$ for $n \to \infty$ predicts $e_3(\infty) = 1.7865(3)$ for $d = 3$ and $e_4(\infty) = 2.093(1)$ for $d = 4$. Both values are consistent with the findings of Refs. [**38, 21, 22**], providing independent confirmation of those results, with far less parameter tuning.

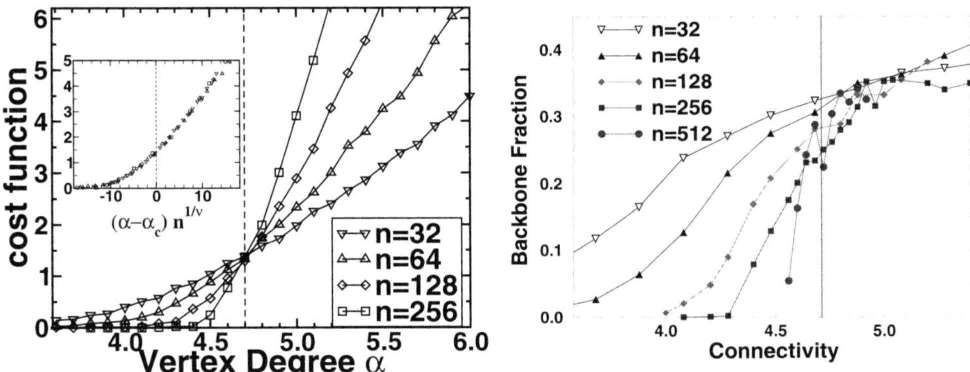

FIGURE 4. Plot of the average cost (left) and of the backbone fraction (right) as a function of the average connectivity c for random graph 3-coloring. The data collapse in the insert on the left predicts a critical point for random graphs at $c \approx 4.72$ (indicated by a vertical line) and $\nu = 1.53(5)$. At each value of c we generated 10000, 5000, 1300, and 650 instances for $n = 32$, 64, 128, and 256, respectively.

TABLE 2. EO approximations to the average ground-state energy per spin $e_d(n)$ of the $\pm J$ spin glass in $d = 3$ on the left, compared with genetic algorithm results from Refs. [38, 21], and in $d = 4$ on the right (see also Ref. [22]). For each size $n = L^d$ we have studied a large number I of instances.

L	I	$e_3(n)$	Ref. [38]	Ref. [21]	I	$e_4(n)$
3	40100	-1.6712(6)	-1.67171(9)	-1.6731(19)	10000	-2.0214(6)
4	40100	-1.7377(3)	-1.73749(8)	-1.7370(9)	4472	-2.0701(4)
5	28354	-1.7609(2)	-1.76090(12)	-1.7603(8)	2886	-2.0836(3)
6	12937	-1.7712(2)	-1.77130(12)	-1.7723(7)	283	-2.0886(6)
7	5936	-1.7764(3)	-1.77706(17)		32	-2.0909(12)
8	1380	-1.7796(5)	-1.77991(22)	-1.7802(5)		
9	837	-1.7822(5)				
10	777	-1.7832(5)	-1.78339(27)	-1.7840(4)		
12	30	-1.7857(16)	-1.78407(121)	-1.7851(4)		

To gauge EO's performance for larger n, we have run our implementation also on two $3d$ lattice instances, *toruspm*3-8-50 and *toruspm*3-15-50, with $n = 8^3$ and $n = 15^3$, considered in the 7th DIMACS challenge for semi-definite problems [26]. Bounds [27] on the ground-state cost established for the larger instance are $C_{\text{lower}} =$

-6138.02 (from semi-definite programming) and $C_{\text{upper}} = -5831$ (from branch-and-cut). EO found $C(S_{\text{best}}) = -6049$ (or $C/n = -1.7923$), a significant improvement on the upper bound and even lower than $e(\infty)$ from above. Furthermore, in that single run EO found 116 different states of that energy with Hamming-distances as far separated as 1500 mutually distinct spins! For the smaller instance the bounds given are -922 and -912, resp., while EO finds -916 (or $C/n = -1.7891$). Here we found 10^5 such states before we terminated the run.

3.4. Satisfiability (MAX-K-SAT). Satisfiability problems are archetypical for models of artificial intelligence, and MAX-K-SAT in particular is an example of a typical MAX-SNP problem [39]. Instances of MAX-K-SAT consist of a formula composed of M clauses. Each clause contains K literals, *i.e.*, x_i or $\neg x_i$, drawn randomly from a pool of n boolean variables x_i. A clause is verified if at least one of its K literals is true (logical "or"), and the entire formula is verified only if every clause is true (logical "and"). The optimization problem consists of minimizing the number of false (or MAXimizing true) clauses by some configuration of the variables. This problem has two parameters, K and M.

We intend to study MAX-K-SAT using an obvious EO-implementation for this problem: For each variable we set $\lambda_i = -1/K \times \{\# \text{ of false clauses containing } x_i\}$. Again, Eq. (2) holds. Typically, $K = O(1)$ and $M = O(n)$ so that each variable appears only in a few ($\approx M/n$) clauses, each connecting it to $\sim K$ other variables. The phase transition in 2-SAT and 3-SAT has been investigated in Refs. [30, 36] on small instances using exact methods, as well as with replica methods [35]. Reproducing the results of Refs. [30, 36] would provide a good test of EO's capabilities. Ideally, we intend to investigate the validity of the "backbone"-conjecture in Ref. [36] also for much larger instances than those accessible with exact enumeration methods.

Acknowledgements

I would like to thank Allon Percus and Michelangelo Grigni for their collaboration on many aspects of the work presented here. I would like to thank the University Research Committee at Emory and especially the organizers of the workshop for support, and the participants for many valuable discussions.

References

[1] See *Frontiers in problem solving: Phase transitions and complexity*, Special issue of Artificial Intelligence **81**:1–2 (1996).

[2] G. Ausiello et al., *Complexity and Approximation* (Springer, Berlin, 1999). See also http://www.nada.kth.se/~viggo/problemlist/compendium.html.

[3] P. Bak, *How Nature Works* (Springer, New York, 1996).

[4] P. Bak and K. Sneppen, *Punctuated Equilibrium and Criticality in a simple Model of Evolution*, Phys. Rev. Lett. **71**, 4083-4086 (1993).

[5] P. Bak, C. Tang, and K. Wiesenfeld, *Self-Organized Criticality*, Phys. Rev. Lett. **59**, 381 (1987).

[6] F. Barahona, *On the computational complexity of Ising spin glass models*, J. Phys. A: Math. Gen. **15**, 3241-3253 (1982).

[7] S. Boettcher, *Extremal Optimization and Graph Partitioning at the Percolation Threshold*, J. Math. Phys. A: Math. Gen. **32**, 5201-5211 (1999).

[8] S. Boettcher and M. Grigni, *Jamming Model for the Extremal Optimization Heuristic*, J. Math. Phys. A: Math. Gen. **35**, 1109-1123 (2002).

[9] S. Boettcher and M. Paczuski, *Exact results for spatiotemporal correlations in a self-organized critical model of punctuated equilibrium*, Phys. Rev. Lett. **76**, 348 (1996).

[10] S. Boettcher and A. G. Percus, *Nature's Way of Optimizing*, Artificial Intelligence **119**, 275-286 (2000).

[11] S. Boettcher and A. G. Percus, *Optimization with Extremal Dynamics*, Phys. Rev. Lett. **86**, 5211-5214 (2001).

[12] S. Boettcher and A. G. Percus, *Extremal Optimization for Graph Partitioning*, Phys. Rev. E **64**, 026114 (2001).

[13] S. Boettcher, A. G. Percus, and M. Grigni, *Optimizing through Co-Evolutionary Avalanches*, Lecture Notes in Computer Science **1917**, 447-456 (2000).

[14] M. Cieplak, A. Giacometti, A. Maritan, A. Rinaldo, I. Rodriguez-Iturbe, and J. R. Banavar, *Models of fractal river basins*, J. Stat. Phys. **91**, 1-15 (1998).

[15] P. Cheeseman, B. Kanefsky, and W. M. Taylor, *Where the really hard Problems are*, in Proc. of IJCAI-91, eds. J. Mylopoulos and R. Rediter (Morgan Kaufmann, San Mateo, CA, 1991), pp. 331–337.

[16] J. Culberson and I. P. Gent, *Frozen Development in Graph Coloring*, available at http://www.apes.cs.strath.ac.uk/apesreports.html.

[17] J. Dall, *Searching Complex State Spaces with Extremal Optimization and other Stochastic Techniques*, Thesis, Fysisk Institut, Syddansk Universitet Odense, 2000 (in danish).

[18] K. H. Fischer and J. A. Hertz, *Spin Glasses* (Cambridge University Press, Cambridge, 1991).

[19] N. Goldenfeld, *Lectures on Phase Transitions and the Renormalization Group* (Addison-Wesley, Reading, 1992).

[20] A. K. Hartmann, *Ground-state behavior of the three-dimensional +/- J random-bond Ising model*, Phys. Rev. B **59**, 3617-3623 (1999).

[21] A. K. Hartmann, *Evidence for existence of many pure ground states in 3d $\pm J$ Spin Glasses*, Europhys. Lett. **40**, 429 (1997).

[22] A. K. Hartmann, *Calculation of ground-state behavior of four-dimensional $\pm J$ Ising spin glasses*, Phys. Rev. E **60**, 5135-5138 (1999).

[23] B. A. Hendrickson and R. Leland, *A multilevel algorithm for partitioning graphs*, in: Proceedings of Supercomputing '95, San Diego, CA (1995).

[24] J. Houdayer and O. C. Martin, *Renormalization for discrete optimization*, Phys. Rev. Lett. **83**, 1030-1033 (1999).

[25] D. S. Johnson, C. R. Aragon, L. A. McGeoch, and C. Schevon, *Optimization by Simulated Annealing - an Experimental Evaluation. 1. Graph Partitioning*, Operations Research **37**, 865-892 (1989).

[26] *7th DIMACS Implementation Challenge on Semidefinite and related Optimization Problems*, eds. D. S. Johnson, G. Pataki, and F. Alizadeh (to appear, see http://dimacs.rutgers.edu/Challenges/Seventh/#PC).

[27] M. Jünger and F. Liers, private communication.

[28] G. Karypis and V. Kumar, *METIS, a Software Package for Partitioning Unstructured Graphs*, see http://www-users.cs.umn.edu/~karypis/metis/main.shtml, (METIS is copyrighted by the Regents of the University of Minnisota).

[29] S. Kirkpatrick, C. D. Gelatt, and M. P. Vecchi, *Optimization by simulated annealing*, Science **220**, 671-680 (1983).

[30] S. Kirkpatrick and B. Selman, *Critical Behavior in the Satisfiability of Random Boolean Expressions*, Science **264**, 1297-1301 (1994).

[31] T. Klotz and S. Kobe, *"Valley structures" in the phase space of a finite 3D Ising spin glass with +or-I interactions*, J. Phys. A: Math. Gen. **27**, L95-L100 (1994).

[32] P. Merz and B. Freisleben, *Memetic algorithms and the fitness landscape of the graph bipartitioning problem*, Lect. Notes Comput. Sc. **1498** 765-774 (1998).

[33] M. Mezard and G. Parisi, *Mean-Field Theory of Randomly Frustrated Systems with Finite Connectivity*, Europhys. Lett. **3**, 1067 (1987).

[34] M. Mezard, G. Parisi, and M. A. Virasoro, *Spin Glass Theory and Beyond* (World Scientific, Singapore, 1987).

[35] R. Monasson and R. Zecchina, *Entropy of the K-satisfiability problem*, Phys. Rev. Lett **76**, 3881-3885 (1996).

[36] R. Monasson, R. Zecchina, S. Kirkpatrick, B. Selman, and L. Troyansky, *Determining computational complexity from characteristic 'phase transitions,'* Nature **400**, 133-137 (1999), and Random Struct. Alg **15**, 414-435 (1999).

[37] M. Paczuski, S. Maslov, and P. Bak, *Avalanche dynamics in evolution, growth, and depinning models,* Phys. Rev. E **53**, 414-443 (1996).

[38] K. F. Pal, *The ground state energy of the Edwards-Anderson Ising spin glass with a hybrid genetic algorithm,* Physica A **223**, 283-292 (1996).

[39] C. H. Papadimitriou and M. Yannakakis, *Optimization, Approximation, and Complexity Classes,* Journal of Computer and System Sciences **43**, 425-440 (1991).

[40] E. Somfai, A. Czirok, and T. Vicsek, *Power-Law Distribution of Landslides in an Experiment on the Erosion of a Granular Pile,* J. Phys. A: Math. Gen. **27**, L757-L763 (1994).

PHYSICS DEPARTMENT, EMORY UNIVERSITY, ATLANTA, GEORGIA 30322, USA
E-mail address: sboettc@emory.edu

DIMACS Series in Discrete Mathematics
and Theoretical Computer Science
Volume **63**, 2004

On the Sampling Problem for H-Colorings
on the
Hypercubic Lattice

Christian Borgs, Jennifer T. Chayes, Martin Dyer, and Prasad Tetali

ABSTRACT. We consider the problem of random H-colorings of rectangular subsets of the hypercubic lattice $\mathbb{Z}^d$, with weight $\lambda_i \in (0, \infty)$ for the color i. First, we assume that H is non-trivial in the sense that it is neither the completely looped complete graph nor the complete bipartite graph. We consider quasi-local Markov chains on a periodic box of even side length L, that is, Markov chains that do not change more than a fraction $\rho < 1$ of the sites in the box in any single move. For any $\rho < 1$ and any finite, connected, non-trivial H, we show that there are weights $\{\lambda_i\}$ such that all quasi-local ergodic Markov chains have slow mixing in the sense that the mixing time is exponential in L^{d-1}. Under the same conditions, we prove phase coexistence in the sense that there are at least two extremal Gibbs states. We also prove that, for a large subclass of graphs H, one can choose weights $\{\lambda_i\}$ such that the corresponding Gibbs measure has exponentially fast spatial mixing.

1. Introduction

Let $H = (W, F)$ be a finite graph without multiple edges, but possibly with loops, and let $G = (V, E)$ be a simple, locally finite graph with countably many vertices. An H-coloring of G is a homomorphism from G to H, i.e., a mapping $\omega : V \to W : x \mapsto \omega(x)$ such that $\langle \omega(x), \omega(y) \rangle \in F$ for all edges $\langle x, y \rangle \in E$. As usual, we denote the set of all H-colorings of G by $\mathrm{Hom}(G, H)$.

When sampling H-colorings, it is often useful to assign different weights to the different colors in W. Given a set of weights $\{\lambda_i\}_{i \in W}$, $0 < \lambda_i < \infty$, and a finite graph G, we thus consider the distribution μ with weights

$$(1) \qquad \mu(\omega) = \frac{1}{Z} \prod_{x \in V} \lambda_{\omega(x)},$$

2000 *Mathematics Subject Classification.* 05C15, 60K35, 68W20, 82B26.

Key words and phrases. graph homomorphisms, Monte Carlo Markov chains, mixing time, conductance, phase coexistence, Pirogov-Sinai theory.

Research of P. Tetali supported in part by the NSF Grant No. DMS–9800351.

where Z is the normalization factor

$$(2) \qquad Z = \sum_{\omega \in \mathrm{Hom}(G,H)} \prod_{x \in V} \lambda_{\omega(x)}.$$

If G is infinite, we consider the set of Gibbs measures with weights $\{\lambda_i\}_{i \in W}$ on $\mathrm{Hom}(G, H)$. To make this precise, we first define a local observable as a function $f : \mathrm{Hom}(G, H) \to \mathbb{R} : \omega \mapsto f(\omega)$ which does not depend on ω_x for all but a finite number of vertices $x \in V$. We then equip $\mathrm{Hom}(G, H)$ with the minimal sigma algebra $\mathcal{F}$ which makes all local observables measurable. As usual, a probability measure μ on $(\mathrm{Hom}(G, H), \mathcal{F})$ is called a Gibbs measure with weights $\{\lambda_i\}_{i \in W}$ if, given any finite set $\Lambda \subset V$, and any function ω_{Λ^c} from $\Lambda^c = V \setminus \Lambda$ into W which can be extended to an H-coloring of G, the conditional measure $\mu(\cdot \mid \omega_{\Lambda^c})$ is given by

$$(3) \qquad \mu(\omega_\Lambda \mid \omega_{\Lambda^c}) = \begin{cases} \dfrac{1}{Z(\omega_{\Lambda^c})} \displaystyle\prod_{x \in \Lambda} \lambda_{\omega(x)} & \omega \in \mathrm{Hom}(G, H) \\[2ex] 0 & \omega \notin \mathrm{Hom}(G, H). \end{cases}$$

Here $Z(\omega_{\Lambda^c})$ is the normalization factor

$$(4) \qquad Z(\omega_{\Lambda^c}) = \sum_{\omega_\Lambda : \omega \in \mathrm{Hom}(G,H)} \prod_{x \in \Lambda} \lambda_{\omega(x)}.$$

As usual, we will often use the term Gibbs state for a Gibbs measure μ with weights $\{\lambda_i\}_{i \in W}$, and call such a measure extremal if it is not possible to write μ as a convex combination of two different Gibbs measures μ_A and μ_B with weights $\{\lambda_i\}_{i \in W}$.

In the past few years, there has been a good deal of work on the H-coloring problem. First, there are a few papers concerning the complexity of the H-coloring problem. An important result of Hell and Nešetřil [21] characterized the complexity of the decision problem, i.e., under what conditions does there exist an H-coloring: if H has no loops and is not bipartite, then they showed that the decision problem is NP-complete; otherwise it is trivially in P. More recently, Dyer and Greenhill [18] characterized the complexity of the counting problem, i.e., the size of $\mathrm{Hom}(G, H)$: If H has a component that is neither the completely looped complete graph, K_n^{loop}, nor the complete bipartite graph, $K_{n,m}$, then the counting problem is $\sharp$P-complete; otherwise, it is trivially in P. The $\sharp$P-completeness remains true even if G has bounded degrees. Henceforth, we will say that H *is trivial* if all its connected components are completely looped complete graphs or complete bipartite graphs.

There have been many results on almost uniform sampling (and approximate counting), using the Markov chain Monte Carlo (MCMC) method, of H-colorings or weighted H-colorings for specific H, including independent sets, the Widom-Rowlinson model, the beach model, and when H is a tree. Positive (fast mixing) results include those of Jerrum [22], Luby and Vigoda [24], Vigoda [31], Dyer and Greenhill [19] and Cooper, Dyer and Frieze [13]. Negative (slow-mixing) results include those of Thomas [30], Borgs, Chayes, Frieze, Kim, Tetali, Vigoda and Vu [4], Dyer, Frieze and Jerrum [17], and Cooper, Dyer and Frieze [13]. The third and the fourth of these slow mixing results are of the form: for uniform weights on H, there exists a graph G such that the mixing is slow. The first and second results are of a very different nature: if G is the hypercubic lattice, there exist weights on the graph H such that the mixing is slow; the specific H in [30] being an edge with loops on both the vertices, corresponding to the Ising model, and in [4] being an edge with a loop on a single vertex, capturing the independent set model. The slow-mixing results we derive in the present work are a significant generalization of the latter results to all non-trivial coloring graphs H.

The final category of previous results concerns Gibbs states for H-colorings. In 1968, Dobrushin showed that, provided that $\mathrm{Hom}(G, H) \neq \emptyset$, there exists at least one Gibbs measure for any set of activities $\{\lambda_i\}_{i \in W}$. It is then natural to ask for what sets of activities this measure is unique, and when there are non-unique measures. That there are activities for which the measure is unique was shown by van den Berg [2] for $G = \mathbb{Z}^d$ and H being the independent set model or Widom-Rowlinson model, by Burton and Steif [12] if $G = \mathbb{Z}^d$ and H the beach model, by Brightwell and Winkler [8] if G is a tree, and again by Brightwell and Winkler [9] if G is of bounded degree and H satisfies a certain condition called "dismantlability." Burton and Steif [11] also established the existence of weights for which there are non-unique measures if $G = \mathbb{Z}^d$ and H the beach model, thereby proving a phase transition in this case. Similarly, Brightwell and Winkler established the existence of weights for which the Gibbs measure is non-unique if G is a tree and if either H is what they called "fertile" [8] or if H is not dismantlable and satisfies a few additional constraints [9]; in the latter case, the model had infinitely many Gibbs measures. Some other results on non-existence or existence of phase transitions for specific H are described in Brightwell, Häggström and Winkler [7].

In this work, we will focus on G being the hypercubic lattice, and show that, for a large class of graphs H, there are weights for which the measure is unique and others for which it is non-unique, thus establishing the existence of a phase transition. While the precise statements of our results are described in the next section, we point out that our main objective here is a complete characterization of H-colorings not only in terms of unique and nonunique Gibbs measures, but fast and slow mixing of the associated Markov chains.

2. Statements of Results

In this paper we consider random H-coloring of the hypercubic lattice $\mathbb{Z}^d$ and the d-dimensional torus $(\mathbb{Z}/L\mathbb{Z})^d$, where L is assumed to be even. As usual, two vertices $x = (x_1, \ldots, x_d)$ and $y = (y_1, \ldots, y_d)$ in $\mathbb{Z}^d$ are joined by an edge whenever there is a j so that $|x_i - y_i| = \delta_{i,j}$ for all i; similarly, $x, y \in (\mathbb{Z}/L\mathbb{Z})^d$ are joined by an edge if there is a j so that $|x_i - y_i| = \delta_{i,j} \mod L$. We denote the corresponding graphs by $G = (V, E)$ and $G_L = (V_L, E_L)$, and the set of all H-colorings of G and G_L by Ω and Ω_L. On Ω_L, we consider the measure

$$(5) \qquad \mu_L(\boldsymbol{\omega}) = \frac{1}{Z_L} \prod_{x \in V_L} \lambda_{\omega(x)},$$

where

$$(6) \qquad Z_L = \sum_{\boldsymbol{\omega} \in \Omega_L} \prod_{x \in V_L} \lambda_{\omega(x)}$$

while on Ω, we consider the set of all Gibbs measures defined by (3). As usual, we say that a Gibbs measure μ has exponentially decaying correlations if there exists an $\epsilon > 0$ such that for all local observables f one has

$$(7) \qquad \sum_{x \in V} \left| E_\mu[f\, T_x f] - E_\mu[f]\, E_\mu[T_x f] \right| e^{\epsilon |x|} < \infty.$$

Here E_μ denotes expectations with respect to μ, $T_x f$ stands for the translate of f by x, i.e., $[T_x f](\boldsymbol{\omega}) := f(\boldsymbol{\omega}^{(x)})$, where $\boldsymbol{\omega}^{(x)}$ is the "shifted" coloring defined by $\omega^{(x)}(y) := \omega(y - x)$, and $|x|$ denotes the ℓ_1 norm on $\mathbb{Z}^d$.

Let $H = (W, F)$ be bipartite, with $W = W_{\text{even}} \cup W_{\text{odd}}$. We call the vertices in W_{even} even, and the vertices in W_{odd} odd. In a similar way, the sets V and V_L can be split into sets of even and odd vertices. Following the usual convention, we call a vertex x in V or V_L even if the sum of its coordinates $\sum_{i=1}^d x_i$ is even, and odd otherwise, and write V^{even} or V_L^{even} for the set of even, and V^{odd} or V_L^{even} for the set of odd vertices.

If H is bipartite, the measure μ_L can be naturally decomposed as

$$(8) \qquad \mu_L = \frac{1}{2}(\mu_L^+ + \mu_L^-),$$

where μ_L^+ lives on the space Ω_L^+ of H-colorings $\boldsymbol{\omega} \in \Omega_L$ that map even vertices into even vertices, and μ_L^- lives on $\Omega_L^- = \Omega_L \setminus \Omega_L^+$. More explicitly, μ_L^+ and μ_L^- are given by

$$(9) \qquad \mu_L^\pm(\boldsymbol{\omega}) = \begin{cases} \dfrac{1}{Z_L^\pm} \displaystyle\prod_{x \in V_L} \lambda_{\omega(x)} & \text{if } \boldsymbol{\omega} \in \Omega^\pm \\[2ex] 0 & \text{otherwise,} \end{cases}$$

where

$$(10) \qquad Z_L^\pm = \sum_{\boldsymbol{\omega} \in \Omega_L^\pm} \prod_{x \in V_L} \lambda_{\omega(x)}.$$

In a similar way, any Gibbs measure on Ω decomposes naturally into a convex combination of Gibbs measures on Ω^+ and Ω^-, where Ω^+ is the space of H-colorings $\boldsymbol{\omega} \in \Omega$ that map even vertices into even vertices, and $\Omega^- = \Omega \setminus \Omega^+$.

Given two H-colorings $\omega, \omega' \in \Omega_L$, let $D(\omega, \omega')$ be the number of vertices $x \in V_L$ such that $\omega(x) \neq \omega'(x)$. For a Markov chain $\mathcal{M}_L$ on Ω_L, let $D(\mathcal{M}_L)$ be the maximum of $D(\omega, \omega')$ over all ω and ω' for which the transition probability is non-zero. We say that $\mathcal{M}_L$ is local if $D(\mathcal{M}_L)$ is bounded uniformly in L, and we say that it is ρ-quasi-local if $D(\mathcal{M}_L) \leq \rho L^d$ for some $\rho < 1$ which is independent of L. The reader should note that this is a very weak notion of quasi-local, which might more accurately be called "almost global." As usual, the mixing time of an ergodic Markov chain $\mathcal{M}_L$ on Ω_L is defined as the time after which the variational distance from the stationary measure is small enough, say smaller than $1/2e$; see (29) in Section 5 for a precise definition. If H is bipartite, no local or quasi-local Markov chain connects the two components Ω_L^+ and Ω_L^- of Ω_L. We therefore consider the restrictions of $\mathcal{M}_L$ to Ω_L^+ and Ω_L^- separately.

Henceforth, we will consider only connected, non-trivial graphs H, i.e., connected graphs which are neither $H = K_n^{\text{loop}}$ nor $H = K_{n,m}$. Recall that trivial graphs H lead to counting problems in P.

Our first set of results establish the existence of weights that lead to fast spatial mixing (exponentially decaying correlations) and a unique Gibbs state for a large class of constraint graphs.

THEOREM 2.1. *Let H be a non-trivial connected graph, and let $d \geq 1$.*
i) If H contains at least one loop, there are weights $\{\lambda_i\}$ such that the limit

$$(11) \qquad \mu = \lim_{L \to \infty} \mu_L$$

exists and describes a translation-invariant, extremal Gibbs measure with exponentially decaying correlations.

ii) If H is bipartite, the analogue of statement (i) holds separately for the limits of each of the measures $\mu_L^{\pm}$,

$$(12) \qquad \mu^{\pm} = \lim_{L \to \infty} \mu_L^{\pm},$$

except that the translation invariance is replaced by periodicity with period 2.

REMARK 2.1. *Assume in addition that H is dismantlable in the sense of [9]. Combining the methods of [9] with those presented here, we can show that the weights $\{\lambda_i\}_{i \in W}$ can be chosen in such a way that the Gibbs state is unique and is given by (11).*

REMARK 2.2. *If W contains a vertex i such that $\langle i, j \rangle \in F$ for all $j \in W$, the single-site Dobrushin criterion is satisfied provided λ_i is chosen large enough. This immediately implies uniqueness of the Gibbs state and fast mixing for the standard single-site heat bath algorithm.*

REMARK 2.3. *Theorem 2.1 does not cover all connected, non-trivial graphs H. E.g., it does not apply to loopless graphs with at least one odd cycle, such as K_3. Indeed, if $H = K_3$, it is believed that for d large enough and any weights $\{\lambda_i\}_{i \in W}$, the limiting measure (11) is a convex combination of at least two extremal Gibbs states.*

Our next two results establish the existence of weights that lead to non-unique Gibbs states and slow mixing for all non-trivial connected H.

THEOREM 2.2. *Let H be a non-trivial connected graph, and let $d \geq 2$. If H is not bipartite, then there are weights $\{\lambda_i\}$ such that the limit (11) exists and is a convex combination*

$$(13) \qquad \mu = \frac{1}{2}(\mu_A + \mu_B),$$

where μ_A and μ_B are extremal Gibbs states with exponentially decaying correlations. If H is bipartite, the analogous statement holds separately for the measures $\mu^{\pm}$, so that there are at least four extremal Gibbs states.

THEOREM 2.3. *Let H be a non-trivial connected graph, let $\rho < 1$ and let $d \geq 2$. If H is not bipartite, then there are weights $\{\lambda_i\}$ such that for all L sufficiently large, the mixing time τ_L of any ρ-quasi-local ergodic Markov chain with stationary distribution μ_L is exponentially large, i.e.,*

$$(14) \qquad \tau_L \geq e^{K_1 L^{d-1}/(\log L)^2},$$

where K_1 is a constant that depends on d, ρ and the weights $\{\lambda_i\}$. If H is bipartite, the analogous statement holds for any ρ-quasi-local ergodic Markov chains on Ω_L^+ and Ω_L^-.

These results are proved using expansion methods. Theorem 2.1 is a so-called "high-temperature" or "disordered phase" result, and the proof is relatively straightforward. Namely, we find weights $\{\lambda_i\}$ so that the H-coloring problem maps into what is called a dilute polymer model, and then use standard Mayer expansions [29, 10, 16, 25] for the dilute polymer model to prove Theorem 2.1. This is sketched in Section 3. The details may be found in [5].

Theorems 2.2 and 2.3 are the so-called "low-temperature" or "ordered phase" results, the proofs of which are much more involved than those of the high-temperature results. First, we find a suitable classification of all non-trivial coloring graphs H. Within each class, we find weights $\{\lambda_i\}$ so that the H-coloring problem maps into what is called a dilute contour model. Here the contours separate different ordered phases. We then use rather involved expansion methods, namely Pirogov-Sinai theory [27, 28] in the form developed by Borgs and Imbrie [6], to control this contour model and prove Theorem 2.2. This is sketched in Section 4. Finally, in Section 5, we sketch how these Pirogov-Sinai methods can be combined with the conductance bounds of [4] to prove Theorem 2.3. Again, the details will be given in the full version of this paper; we also hope that the full version will further illustrates the usefulness of the statistical physics techniques (such as Mayer expansions and Pirogov-Sinai theory) in the context of Markov chain Monte-Carlo algorithms.

3. Polymer Expansions and Fast Spatial Mixing

In this section, we sketch the proof of Theorem 2.1. To this end, we map the partition function of our model to an abstract polymer system with sufficiently small weights. As usual, an abstract polymer system is a triple $\mathbf{\Gamma} = (\Gamma, \leftrightarrow, z(\cdot))$, where Γ is a finite set, $\leftrightarrow$ is a symmetric, reflexive relation on Γ, and $z(\cdot)$ is a complex-valued function on Γ. The elements of Γ are called polymers. Two polymers $\gamma, \gamma' \in \Gamma$ are said to be incompatible if $\gamma \leftrightarrow \gamma'$, and compatible otherwise. Finally, $z(\gamma)$ is called the weight or activity of the polymer γ. The partition function of the polymer system $\mathbf{\Gamma}$ is defined as

$$(15) \qquad \mathcal{Z}(\mathbf{\Gamma}) = \sum_{\tilde{\Gamma} \subset \Gamma} \phi(\tilde{\Gamma}) \prod_{\gamma \in \tilde{\Gamma}} z(\gamma),$$

where $\phi(\tilde{\Gamma}) = 0$ whenever there is a pair of polymers $\gamma, \gamma' \in \tilde{\Gamma}$ such that $\gamma \leftrightarrow \gamma'$, and $\phi(\tilde{\Gamma}) = 1$ otherwise. In other words $\phi(\tilde{\Gamma}) = 1$ whenever the polymers in $\tilde{\Gamma}$ are pairwise compatible, and $\phi(\tilde{\Gamma}) = 0$ otherwise.

In order to prove Theorem 2.1, we will choose weights λ_i such that the partition functions (6) and (10) can be written in terms of a polymer system with small weights. Together with a similar representation for the measures (5) and (9), the general theory of Mayer expansions for abstract polymer systems [29, 10, 16, 25] then gives Theorem 2.1. For the case where H has at least one loop, this will be sketched in Subsection 3.1, and for the case where H is bipartite this will be sketched in Subsection 3.2.

3.1. At least one loop present. Let H be a graph with at least one loop, and without loss of generality let us assume that the vertices in W are labelled in such a way that the loop $\ell_1 = \langle 1, 1 \rangle \in F$. We then set

$$(16) \qquad \lambda_i = \begin{cases} \lambda & \text{if } i = 1 \\ 1 & \text{otherwise.} \end{cases}$$

If λ is large, the configuration $\boldsymbol{\omega}$ with largest weight is the one where every site has color $\omega(x) = 1$. For a general configuration $\boldsymbol{\omega}$, we define a vertex $x \in V_L$ to be excited whenever $\omega(x) \neq 1$, and call the connected components of the set of excited vertices the polymers corresponding to $\boldsymbol{\omega}$. A configuration $\boldsymbol{\omega}$ with polymers $\gamma_1, \ldots, \gamma_n$ then has weight $\lambda^{|V_L|} \prod_{i=1}^{n} \lambda^{-|\gamma_i|}$, where $|V_L|$ and $|\gamma_i|$, $i = 1, \ldots, n$, denotes the number of vertices in V_L and γ_i, respectively. This motivates the following definition of a polymer system $\mathbf{\Gamma}_1$: The set of polymers, $\mathbf{\Gamma}_1$, is defined as the set of connected subsets $\gamma \subset V_L$. Two polymers γ and γ' are called incompatible if $\gamma \cup \gamma'$ is a connected subset of V_L. The weight $z(\gamma)$ of a polymer γ is defined as $\lambda^{-|\gamma|}$ times the number of H-colorings ω with $\omega^{-1}(\{1\}) = V_L \setminus \gamma$.

LEMMA 3.1. *Let $\mathbf{\Gamma}_1$ be the polymer system defined above, and let Z_L be the partition function defined in* (6). *Then*

$$(17) \qquad Z_L = \lambda^{|V_L|} \mathcal{Z}(\mathbf{\Gamma}_1).$$

PROOF. Given a subset $W \subset V_L$, let $N_{V_L}(W)$ be the number of H-colorings ω such that W is the set of excited sites corresponding to ω, i.e., $N_{V_L}(W) = |\{\omega \colon \omega^{-1}(\{1\}) = V_L \setminus W\}|$. By definition, the polymers corresponding to an H-coloring ω form a set of pairwise compatible polymers $\tilde{\Gamma} \subset \mathbf{\Gamma}_1$. Since the weight of an H-coloring with polymers $\gamma_1, \ldots, \gamma_n$ is $\lambda^{|V_L|} \prod_{i=1}^{n} \lambda^{-|\gamma_i|}$, the left hand side of (17) is therefore equal to

$$(18) \qquad \lambda^{|V_L|} \sum_{n \geq 0} \sum_{\{\gamma_1, \ldots, \gamma_n\}} N_{V_L}(\gamma_1 \cup \cdots \cup \gamma_n) \prod_{i=1}^{n} \lambda^{-|\gamma_i|},$$

where the second sum goes over sets of pairwise compatible polymers. Since the the weight $z(\gamma)$ of a polymer γ can be rewritten as $\lambda^{-|\gamma|} N_{V_L}(\gamma)$, the proof of the lemma reduces to establishing the fact that $N_{V_L}(W) = \prod_{i=1}^{n} N_{V_L}(W_i)$ whenever $W_1, \ldots, W_n$ are the connected components of W. The proof of this fact is elementary and is left to the reader. $\qquad\square$

Given Lemma 3.1, the partition function (6) can be analyzed using the general theory of Mayer expansions for abstract polymer systems. In order to apply this theory, one has to show that the weights $z(\gamma)$ are small enough. A sufficient condition is that

$$(19) \qquad \sum_{\gamma \in \Gamma \colon \gamma \leftrightarrow \gamma'} z(\gamma) e^{|\gamma|} \leq |\gamma'|$$

for all $\gamma' \in \Gamma$, see [**10, 16, 25**]. Since the number of connected sets $\gamma \subset V_L$ of size s that have distance 1 or less from a given point $x \in V_L$ can be bounded by $(2de)^s$, while $N_{V_L}(\gamma)$ is obviously at most $(|W|-1)^{|\gamma|}$, the condition (19) is satisfied whenever

$$(20) \qquad \lambda > 4de^2(|W| - 1).$$

For $\lambda > 4de^2(|W| - 1)$, the partition function Z_L can therefore be analyzed with the help of the general theory of Mayer expansion for abstract polymer systems. Together with a similar representation for the measure $\mu_L(\cdot)$, one obtains a proof of Theorem 2.1 (i); see [**5**] for details.

3.2. Bipartite. Assume without loss of generality that the vertices in W are labelled in such a way that F contains the edge $\langle 1, 2 \rangle$. We then set

$$(21) \qquad \lambda_i = \begin{cases} \lambda & \text{if } i = 1 \text{ or } 2 \\ 1 & \text{otherwise.} \end{cases}$$

If λ is large, the configurations with maximal weight are the two checkerboard configurations ω_A and ω_B: in ω_A, every even site in V_L has color 1 and every odd site has color 2, and vice versa in ω_B. Restricting ourselves to the configurations in Ω_L^+ rules out either ω_A or ω_B, so that we are left with only one configuration of maximal weight. For a general configuration $\omega \in \Omega_L^+$, we define the set of excited vertices as the set of vertices x with color $\omega(x) \in W \setminus \{1, 2\}$. The connected components of the set of excited vertices are again called the polymers corresponding to ω.

As in the argument in the last subsection, this leads to a polymer representation of the partition function which can be analyzed using convergent Mayer expansions if λ is sufficiently large. The only difference here is the fact that the weights $z(\gamma)$ are no longer translation invariant. Again, see [5] for details.

4. Contour Representations and Phase Coexistence

In this section we derive a contour representation for random H-colorings which will allow us to sketch the proofs of Theorems 2.2 and 2.3. To this end, we first note that all connected, non-trivial graphs H fall into one of the following four classes:

    1) all loops present

    2) not all loops, but at least one loop present

    3) no loops present, but at least one odd cycle

    4) bipartite (and thus loopless)

Our contour model will be defined differently in each case, but in all cases contours will be pairs $\gamma = (\operatorname{supp}\gamma, \omega_\gamma)$, where $\operatorname{supp}\gamma$ is a $*$-connected subset of W, and ω_γ is an H-coloring of $\operatorname{supp}\gamma \cup \partial\operatorname{supp}\gamma$. Here $\Lambda \subset V_L$ is called $*$-connected if for every pair of vertices $x, y \in \Lambda$ there is a path of vertices $x_1, \ldots, x_k \in \Lambda$ with $x_1 = x$ and $x_k = y$ such that for all $i = 1, \ldots, k-1$, x_i and x_{i+1} have Euclidean distance $\sqrt{2}$ or less. As usual, the boundary $\partial\Lambda$ of a set $\Lambda \subset V_L$ is the set of all vertices in $V_L \setminus \Lambda$ that are connected to Λ via an edge in E_L. Throughout this section, two contours γ and γ' will be called compatible if the Euclidean distance between $\operatorname{supp}\gamma$ and $\operatorname{supp}\gamma'$ is strictly larger than $\sqrt{2}$, and a set $\{\gamma_1, \ldots, \gamma_n\}$ of contours will be called compatible if γ_i and γ_j are compatible for all pairs $i \neq j$. Finally the size $|\gamma|$ of a contour $\gamma = (\operatorname{supp}\gamma, \omega_\gamma)$ will be defined as the number of vertices in $\operatorname{supp}\gamma$.

4.1. All loops present. Since by assumption H is not the completely looped complete graph, there must be vertices i and j in W such that $\langle i, j \rangle$ is not an edge in H; without loss of generality we assume that the vertices in W are labelled in such a way that $\langle 1, 2 \rangle \notin F$. We then set

$$
(22) \qquad \lambda_i = \begin{cases} \lambda_A & \text{if } i = 1 \\ \lambda_B & \text{if } i = 2 \\ 1 & \text{otherwise.} \end{cases}
$$

If λ_A and λ_B are large, the dominant configurations $\omega \in \Omega_L$ will be the constant configurations $\omega_A \equiv 1$ and $\omega_B \equiv 2$, with weights $\lambda_A^{|V_L|}$ and $\lambda_B^{|V_L|}$, respectively.

For a general configuration ω, we define the "ground state regions" $V_A(\omega)$ and $V_B(\omega)$ as $V_A(\omega) = \omega^{-1}(\{1\})$ and $V_B(\omega) = \omega^{-1}(\{2\})$, and the "set of excited vertices" $V^*(\omega)$ as $V^*(\omega) = V_L \setminus (V_A(\omega) \cup V_B(\omega))$. The contours corresponding to the H-coloring ω are then defined as the pairs $\gamma_1 = (\operatorname{supp}\gamma_1, \omega_{\gamma_1})$, $\ldots$, $\gamma_n = (\operatorname{supp}\gamma_n, \omega_{\gamma_n})$, where $\operatorname{supp}\gamma_1, \ldots, \operatorname{supp}\gamma_n$ are the $*$-connected components of $V^*(\omega)$, and $\omega_{\gamma_1}, \ldots, \omega_{\gamma_n}$ are the restrictions of ω to $\operatorname{supp}\gamma_1 \cup \partial\operatorname{supp}\gamma_1$, $\ldots$, supp $\gamma_n \cup \partial$supp γ_n, respectively. The weight of an H-coloring $\boldsymbol{\omega}$ can then be written as a product of suitable weights for the contours and ground state regions corresponding to $\boldsymbol{\omega}$,

$$(23) \qquad \prod_{x \in V_L} \lambda_{\omega(x)} = e^{-e_A |V_A(\boldsymbol{\omega})|} e^{-e_B |V_B(\boldsymbol{\omega})|} \prod_{i=1}^{n} \rho(\gamma_i),$$

with $e_A = -\log \lambda_A$, $e_B = -\log \lambda_B$ and $\rho(\gamma) = \prod_{x \in \text{supp} \, \gamma} \lambda_{\omega(x)} = 1$ for all contours γ (recall that $\lambda_i = 1$ for $i \neq 1, 2$). As is usual in cluster expansions, we will need a Peierls' bound of the form

$$(24) \qquad \rho(\gamma) \leq e^{-\tau |\gamma|} e^{-e_0 |\gamma|},$$

where $e_0 = \min\{e_A, e_B\}$ is the ground state energy and $\tau > 0$ is a "suppression factor" used to control the expansion. Here, since $\rho(\gamma) = 1$ and $e_0 = -\max\{\log \lambda_A, \log \lambda_B\}$, the contour weights trivially have an exponential suppression with respect to the corresponding ground state weights provided λ_A and λ_B are large, and (24) holds with $\tau = \max\{\log \lambda_A, \log \lambda_B\}$.

A general configuration thus consists of regions in one of the two ground states, separated by regions which have much smaller weight. If the vertices 1 and 2 are "isomorphic" vertices (i.e., if there is an automorphism which transposes 1 and 2) in H, we set $\lambda_A = \lambda_B = \lambda$. For large λ, a Peierls' argument then implies the existence of at least two translation invariant Gibbs states μ_A and μ_B related to each other by the symmetry $1 \leftrightarrow 2$. Here μ_A consists of small fluctuations around the ground state $\boldsymbol{\omega}_A$, and μ_B consists of small fluctuations around the ground state $\boldsymbol{\omega}_B$, in close analogy to the two low temperature states of the Ising model which are small perturbation of the ground states where all spins are up and down, respectively.

If 1 and 2 are not isomorphic, the fluctuations about the ground states $\boldsymbol{\omega}_A$ and $\boldsymbol{\omega}_B$ will in general favor one of the two, and only one of them will contribute to the limiting state (11) for $\lambda_A = \lambda_B$. To correct for this, we will set $\lambda_A = \lambda e^h$ and $\lambda_B = \lambda e^{-h}$. For λ large and a suitable choice of h (which will in general depend on λ) the difference in the weight for the ground state $\boldsymbol{\omega}_A$ and $\boldsymbol{\omega}_B$ will exactly compensate for the difference induced by the fluctuations, so that we again get two different, translation invariant Gibbs states μ_A and μ_B which are small perturbations of the ground states $\boldsymbol{\omega}_A$ and $\boldsymbol{\omega}_B$. The precise argument is rather complicated; it uses the version of Pirogov-Sinai theory [**27**] developed in [**6**] and is given in [**5**]. These methods also imply exponential decay of correlations and extremality for μ_A and μ_B, whether or not the vertex 1 and 2 are related by symmetry. To prove that the Gibbs state obtained via the limit (11) is a convex combination of μ_A and μ_B with equal weight for μ_A and μ_B (see (13)), we use the methods of Section 5 of [**6**]; see again [**5**] for details.

4.2. Not all, but at least one loop present. Consider two vertices i and j in W such that the loop $\ell_i \notin F$ and the loop $\ell_j \in F$. Since H is connected, there must be a pair of vertices $\tilde{i}$ and $\tilde{j}$ in W such that $\ell_{\tilde{i}} \notin F$, $\ell_{\tilde{j}} \in F$ and $\langle \tilde{i}, \tilde{j} \rangle \in F$. Without loss of generality we assume that H has been labelled in such a way that $\tilde{i} = 1$ and $\tilde{j} = 2$, so that $\ell_1 \notin F$, $\ell_2 \in F$ and $\langle 1, 2 \rangle \in F$. We then set

$$(25) \qquad \lambda_i = \begin{cases} \lambda & \text{if } i = 1 \\ 1 & \text{otherwise.} \end{cases}$$

If λ is large, one would like to set the color of all vertices to 1. But since $\ell_1 \notin F$, this is not a configuration in $\Omega_L = \mathrm{Hom}(G_L, H)$. The best one can do is to put color 1 on every second vertex, i.e., either to color all even or all odd vertices with the color 1.

In general, there are many configurations $\omega \in \Omega_L$ with $\omega(x) = 1$ for all $x \in V_L^{\mathrm{even}}$ or $\omega(x) = 1$ for all $x \in V_L^{\mathrm{odd}}$, namely $|\mathcal{N}(1)|^{|V_L|/2}$, where $\mathcal{N}(1)$ is the neighborhood of 1 in H, $\mathcal{N}(1) = \{i \in W : \langle 1, i \rangle \in F\}$. In the language of statistical physics, this means that these states have ground state entropy $S_0 = s_0|V_L|$, with $s_0 = \frac{1}{2}\log|\mathcal{N}(1)|$. Note that in contrast to the situation in the last subsection, a "ground state" of the system is not a single configuration, but rather a measure supported on many configurations. Here the two ground state measures, ν_A and ν_B, are defined by

$$(26) \qquad \nu_A(\omega) = \prod_{x \in V_L^{\mathrm{even}}} \mathbb{I}(\omega(x) = 1) \prod_{y \in V_L^{\mathrm{odd}}} \mathbb{I}(\omega(y) \in \mathcal{N}(1))$$

and a similar equation for ν_B with the roles of V_L^{even} and V_L^{odd} switched. Here $\mathbb{I}(\mathcal{E})$ stands for the indicator function of the event $\mathcal{E}$.

To define the contours corresponding to a configuration $\omega \in \Omega_L$, we consider again three regions V_A, V_B and V^*. The set V_A consists of all even vertices x with color $\omega(x) = 1$ and all odd vertices which have only neighbors of color 1, the set V_B consists of all odd vertices x with color $\omega(x) = 1$ and all even vertices which have only neighbors of color 1, and the set V^* consists of the remaining vertices in V_L. For the ground state configurations described above, the set V^* is empty, and either V_A or V_B consists of all vertices in V_L. The contours corresponding to a general configuration $\omega \in \Omega_L$ are again defined as the pairs $\gamma_1 = (\mathrm{supp}\,\gamma_1, \omega_{\gamma_1}), \ldots,$ $\gamma_n = (\mathrm{supp}\,\gamma_n, \omega_{\gamma_n})$, where $\mathrm{supp}\,\gamma_1, \ldots, \mathrm{supp}\,\gamma_n$ are the $*$-connected components of V^*, and $\omega_{\gamma_1}, \ldots, \omega_{\gamma_n}$ are the restrictions of ω to $\mathrm{supp}\,\gamma_1 \cup \partial\mathrm{supp}\,\gamma_1, \ldots,$ $\mathrm{supp}\,\gamma_n \cup \partial\mathrm{supp}\,\gamma_n$, respectively.

The weight of a set of contours $\Gamma = \{\gamma_1, \ldots, \gamma_n\}$ can again be written as a product of suitable weights for the contours and ground state regions V_A and V_B; however, the explicit formula is slightly more complicated due to the existence of the extra entropy factors $|\mathcal{N}(1)|$ for the sites in V_A and V_B whose color is not fixed. Nevertheless, it is possible to rewrite the weight of an H-coloring ω with contours $\gamma_1, \ldots, \gamma_n$ and ground state regions V_A and V_B in the form (23), with $e_A = e_B = -\log(\lambda^{1/2}|\mathcal{N}(1)|^{1/2})$ and weights $\rho(\gamma)$ for the contours which obey a bound of the form (24) with $\tau = \Theta\left(\frac{1}{d}\log\lambda\right)$; see [5] for details.

Again, a general configuration consists of regions in one of the two ground states, separated by regions which have much smaller weight. For large λ, a modified Peierls' argument like the one used by Dobrushin in his proof of a phase transition for the independent set model [15] then implies the existence of at least two Gibbs states μ_A and μ_B related to each other by the symmetry of shifting each contour by one lattice unit. Here μ_A consists of small fluctuations around the ground state ν_A, and μ_B consists of small fluctuations around the ground state ν_B. While the emerging picture is very similar to that of the independent set model on the level of contours, it is slightly more complicated on the level of configurations, since the ground states we perturb around are now product measures, not delta functions on a single configuration. Nevertheless, the techniques developed for the independent set model [15] can be generalized to this case. Combined with the methods of

Section 5 of [**6**], this leads to the proof of Theorem 2.2 in the case where H has at least one loop, but does not have all loops. Again see [**5**] for details.

4.3. No loop present, at least one odd cycle. This case is completely analogous to the case with at least one, but not all loops present. Indeed, let i be a vertex contained in an odd cycle, without loss of generality the vertex labelled by 1. If we set

$$
(27) \qquad \lambda_i = \begin{cases} \lambda & \text{if } i = 1 \\ 1 & \text{otherwise,} \end{cases}
$$

there are again two ground state measures for large λ, namely the measure ν_A defined in (26) and its analog ν_B. While the precise contour weights will be different from those for the case where at least one, but not all loops are present, the general features of the resulting contour model will be very similar. In particular, we can again use a modified Peierls' argument to infer the existence of at least two Gibbs states μ_A and μ_B related to each other by the symmetry of shifting each contour by one lattice unit, with μ_A consisting of small fluctuations around the ground state ν_A, and μ_B consisting of small fluctuations around the ground state ν_B.

4.4. Bipartite. We recall that the vertex set of a bipartite graph H naturally splits into two subsets W_{odd} and W_{even}, such all edges in F are of the form $\langle i, j \rangle$, $i \in W_{\text{odd}}, j \in W_{\text{even}}$. Since H is not the complete bipartite graph by the assumption that it is non-trivial, there exist two vertices $\widetilde{i} \in W_{\text{odd}}, \widetilde{j} \in W_{\text{even}}$ such that $\langle \widetilde{i}, \widetilde{j} \rangle \notin F$. Without loss of generality, assume $\widetilde{i} = 1$ and $\widetilde{j} = 2$. Obviously, neither ℓ_1 nor ℓ_2 are in F by the assumption that H is bipartite. We set

$$
(28) \qquad \lambda_i = \begin{cases} \lambda_A & \text{if } i = 1 \\ \lambda_B & \text{if } i = 2 \\ 1 & \text{otherwise.} \end{cases}
$$

Restricting ourselves to colorings in Ω_L^+, we obtain two ground states: the ground state ν_A, where every odd vertex has color 1, while the colors of the even vertices are chosen independently at random from $\mathcal{N}(1)$, and the ground state ν_B, where every even vertex has color 2, while the colors of the odd vertices are chosen independently at random from $\mathcal{N}(2)$. The weights of these ground states are $\lambda_A^{|V_L|/2} |\mathcal{N}(1)|^{|V_L|/2}$ and $\lambda_B^{|V_L|/2} |\mathcal{N}(2)|^{|V_L|/2}$, respectively.

For a general configuration $\omega \in \Omega_L^+$, the ground state region V_A now consists of all vertices x with color $\omega(x) = 1$ and all vertices y which have only neighbors of color 1, and the ground state region V_B consists of all vertices x with color $\omega(x) = 2$ and all vertices which have only neighbors of color 2. Setting again $V^* = V_L \setminus (V_A \cup V_B)$ we define contours as before, obtaining again a representation of the form (23), with $e_A = -\log\left(\lambda_A^{1/2} |\mathcal{N}(1)|^{1/2}\right)$, $e_B = -\log(\lambda_B^{1/2} |\mathcal{N}(2)|^{1/2})$, and contour weights $\rho(\gamma)$ which obey a bound of the form (24) provided λ_A and λ_B are large enough. If 1 and 2 are isomorphic vertices in H, then a Peierls' argument proves that typical configurations are either small perturbations of ground state ν_A, or small perturbations of ground state ν_B if $\lambda_A = \lambda_B$. If there is no symmetry, one has to adjust the ratio of λ_A and λ_B to correct for entropic preference of one of the two states. The proof again uses Pirogov-Sinai theory, and is carried out in [**5**]. Of course each of the above Gibbs states has a ghost sister supported in Ω^-

where the even and odd sublattices of V_L exchange their roles, leading to a total of at least four extremal Gibbs states.

5. Mixing Time and Conductance Bounds

In this section, we sketch the key ideas of the proof of Theorem 2.3. The proof uses the notion of conductance, first introduced to the field of MCMC by Jerrum and Sinclair in [23]. We start with a few general definitions.

Let $\mathcal{M}$ be an ergodic Markov chain on a finite state space Ω, with transition probabilities $P(\omega, \widetilde{\omega})$, $\omega, \widetilde{\omega} \in \Omega$. Let π denote the stationary distribution of $\mathcal{M}$. For $\omega_0 \in \Omega$, we denote by $P_{t,\omega_0}(\omega)$ the probability that the system is in the state ω at time t given that ω_0 is the initial state. The mixing time of the Markov chain $\mathcal{M}$ is defined as

$$(29) \qquad \tau = \min \left\{ t : \max_{\omega \in \Omega} d(P_{t,\omega}, \pi) \leq \frac{1}{2e} \right\},$$

where $d(P_{t,\omega}, \pi)$ is the *variational distance* between $P_{t,\omega}$ and π,

$$(30) \qquad d(P_{t,\omega}, \pi) = \max_{S \subseteq \Omega} |P_{t,\omega}(S) - \pi(S))| .$$

The conductance of a set of states $\emptyset \neq S \subset \Omega$ is

$$(31) \qquad \Phi_S = \sum_{\omega \in S} \sum_{\widetilde{\omega} \in \Omega \setminus S} \frac{\pi(\omega) P(\omega, \widetilde{\omega})}{\pi(S) \pi(\Omega \setminus S)},$$

and the conductance of the chain itself is simply $\Phi_{\mathcal{M}} = \min_{S \neq \emptyset} \Phi_S$.

We prove our lower bounds on mixing time by showing that $\Phi_{\mathcal{M}}$ is small and then using the well-known bound [1] (see [14] or Claim 2 of the journal version of [17] for the non-reversible case):

$$(32) \qquad \tau^{-1} = O(\Phi_{\mathcal{M}}).$$

Here the finite state space is the space of all H-colorings Ω_L (or the spaces $\Omega_L^{\pm}$ if H is bipartite). Our goal is to decompose Ω_L as a disjoint union of three sets $\Omega_{L,A}, \Omega_{L,B}$ and Ω_L^* such that $\pi(\Omega_L^*) = O(e^{-KL^{d-1}/(\log L)^2})$, while both $\pi(\Omega_{L,A})$ and $\pi(\Omega_{L,B})$ are $\Theta(1)$, and such that *for any* ρ-quasi-local Markov chain, the transition probability $P(\omega, \widetilde{\omega})$ is zero whenever $\omega \in \Omega_{L,A}$ and $\widetilde{\omega} \in \Omega_{L,B}$. Taking $S = \Omega_{L,A} \cup \Omega_L^*$, we then have

$$(33) \qquad \Phi_{\mathcal{M}} \leq \Phi_S = \sum_{\omega \in \Omega_L^*} \sum_{\widetilde{\omega} \in \Omega_{L,B}} \frac{\pi(\omega) P(\omega, \widetilde{\omega})}{\pi(\Omega_{L,A} \cup \Omega_L^*) \pi(\Omega_{L,B})} \leq \frac{\pi(\Omega_L^*)}{\pi(\Omega_{L,A}) \pi(\Omega_{L,B})}.$$

The right hand side of the above equation is of order $O(e^{-KL^{d-1}/(\log L)^2})$, which combined with (32) gives Theorem 2.3.

More precisely, we define $\Omega_{L,A}, \Omega_{L,B}$ and Ω_L^* in such a way that $|V_A(\omega)| > (1 - \epsilon)|V_L|$ for all $\omega \in \Omega_{L,A}$ and $|V_B(\omega)| > (1 - \epsilon)|V_L|$ for all $\omega \in \Omega_{L,B}$ and then combine the methods developed in [4] with Pirogov-Sinai theory to show that for λ sufficiently large $\mu_L(\Omega_L^*) \leq e^{-KL^{d-1}/(\log L)^2}$, $\mu_L(\Omega_{L,A}) = \Theta(1)$ and $\mu_L(\Omega_{L,B}) = \Theta(1)$; see [5] for details. If we choose ϵ as $\frac{1}{2}(1 - \rho)$, then we have that for all ρ-quasi-local Markov chains the transition probability $P(\omega, \widetilde{\omega})$ is zero whenever $\omega \in \Omega_{L,A}$ and $\widetilde{\omega} \in \Omega_{L,B}$, and (33) implies the statement of Theorem 2.3.

In the rest of this section, we give the proof of a weakened form of (33) in a particularly simple case. As already noted in [4], it is significantly easier to prove a bound of the form (33) in which the exponential factor of $L^{d-1}/(\log L)^2$ is replaced by $L/(\log L)^2$. Here we sketch a proof of this weakened inequality for a subclass of the graphs H considered in Subsection 4.1.

Suppose that H has all loops present, so that W contains two vertices (w.l.o.g., we assume they are labelled 1 and 2) such that $\langle 1, 2 \rangle \notin F$; see Section 4.1. Suppose further that 1 and 2 are isomorphic vertices in H, so that the statements of Theorem 2.2 hold whenever the weights $\{\lambda_i\}_{i \in W}$ are chosen according to (22) with $\lambda_A = \lambda_B = \lambda$ sufficiently large. In particular this latter assumption allows us to use a Peierls' argument rather than the considerably more involved methods of Pirogov-Sinai theory.

In a first step, we introduce the space $\widetilde{\Omega}_L^*$ as the set of H-colorings ω such that at least one of the contours corresponding to ω has size L or larger. In order to bound $\pi(\widetilde{\Omega}_L^*)$, we use the fact that the probability that a given contour γ is among the contours corresponding to a configuration ω can be bounded by $\rho(\gamma)$. (Since vertices 1 and 2 are isomorphic, so that the states A and B are related by a symmetry, this can be proved via a standard Peierls' argument; see [26], [20].) The weight of the space $\widetilde{\Omega}_L^*$ can therefore be bounded by

$$(34) \qquad \mu_L(\widetilde{\Omega}_L^*) \leq \sum_{\gamma:|\gamma|\geq L} \tilde{\rho}(\gamma),$$

where the sum goes over all contours of size $|\gamma| \geq L$ and $\tilde{\rho}(\gamma) := \rho(\gamma)e^{e_0|\gamma|}$ measures the suppression with respect to the ground state energy (recall that $e_0 = e_A = e_B$ due to our symmetry assumptions). To estimate this sum, we note that the number of vertices $y \in V_L$ that have distance at most $\sqrt{2}$ from a given vertex $x \in V_L$ is equal to $2d+2d(2d-2)$, implying that the number of $*$-connected sets of size s in V_L can be bounded by $|V_L|(2d(2d-1)e)^s$. As a consequence, the number of contours γ that have size $|\gamma| = s$ can be bounded by $L^d C(d,|W|)^s$ for some constant $C(d,|W|)$ depending on d and the size of W. Together with the bound (24) and the fact that $\tau = \log \lambda$, we therefore get that whenever λ is sufficiently large, the weight of $\widetilde{\Omega}_L^*$ is bounded by

$$(35) \qquad \mu_L(\widetilde{\Omega}_L^*) \leq L^d \sum_{s \geq L} \left(C(d,|W|)\lambda^{-1} \right)^s = O\left(L^d \left(C(d,|W|)\lambda^{-1} \right)^L \right).$$

Consider now a contour γ corresponding to a configuration $\omega \in \Omega_L \setminus \widetilde{\Omega}_L^*$, i.e., a contour γ with $|\gamma| < L$. Since the diameter of the torus G_L is L, such a contour can be embedded in $\mathbb{Z}^d$. Defining $V(\gamma)$ as the union of $\text{supp}\,\gamma$ with all finite components of $\mathbb{Z}^d \setminus \text{supp}\,\gamma$, we then introduce the exterior of γ as $V_L \setminus V(\gamma)$. This in turn allows us to define the exterior $\text{Ext}(\omega)$ of a configuration $\omega \in \Omega_L \setminus \widetilde{\Omega}_L^*$ as the intersection of the exteriors of all contours corresponding to ω. Using our definition of contours as pairs $(\gamma, \text{supp}\,\gamma)$, where $\text{supp}\,\gamma$ is a $*$-connected component of $V_L^*(\omega)$, it is not hard to verify that the exterior $\text{Ext}(\omega)$ is a connected set, and that on $\text{Ext}(\omega)$, ω is constant and either equal to 1 or equal to 2. This allows us to decompose $\omega \in \Omega_L \setminus \widetilde{\Omega}_L^*$ into two sets: one, denoted by $\widetilde{\Omega}_{L,A}$, for which $\text{Ext}(\omega) \subset V_A(\omega)$, and one, denoted by $\widetilde{\Omega}_{L,B}$, for which $\text{Ext}(\omega) \subset V_B(\omega)$.

At this point, the proof is a straightforward application of the methods developed in [**4**]: we define

$$\Omega_{L,A} = \{\boldsymbol{\omega} \in \widetilde{\Omega}_{L,A} : \mathrm{Ext}\,(\boldsymbol{\omega}) > (1 - \epsilon)|V_L|\}, \tag{36}$$

and similarly for $\Omega_{L,B}$. Since $\mathrm{Ext}\,(\boldsymbol{\omega}) \subset V_A(\boldsymbol{\omega})$ if $\boldsymbol{\omega} \in \Omega_{L,A} \subset \widetilde{\Omega}_{L,A}$, we obviously have $|V_A(\boldsymbol{\omega})| > (1 - \epsilon)|V_L|$ if $\boldsymbol{\omega} \in \Omega_{L,A}$, and similarly for $\boldsymbol{\omega} \in \Omega_{L,B}$. Defining $\Omega_L^* = \Omega_L \setminus (\Omega_{L,A} \cup \Omega_{L,B})$, we thus are left with a proof of the inequality $\mu(\Omega_L^*) \leq O(e^{-KL/(\log L)^2})$, for some absolute constant $K > 0$. To this end, we consider a configuration $\boldsymbol{\omega} \in \widetilde{\Omega}_{L,A} \setminus \Omega_{L,A}$, and the set of contours Γ corresponding to $\boldsymbol{\omega}$. Using the isoperimetric inequality of Bollobás and Leader [**3**], we then have

$$\epsilon L^d \leq |V_L \setminus \mathrm{Ext}\,(\boldsymbol{\omega})| \leq \sum_{\gamma \in \Gamma} |V(\gamma)| \leq \Theta(1) \sum_{\gamma \in \Gamma} |\gamma|^{d/(d-1)} \leq \Theta(1) \left(\sum_{\gamma \in \Gamma} |\gamma| \right)^{d/(d-1)}, \tag{37}$$

and hence $\sum_{\gamma \in \Gamma} |\gamma| \geq \Theta(L^{d-1})$. Proceeding with a multi-scale Peierls' argument as in the proof of Lemma 8 in [**4**], one gets the bound

$$\mu_L(\widetilde{\Omega}_{L,A} \setminus \Omega_{L,A}) \leq O(e^{-KL^{d-1}/(\log L)^2}), \tag{38}$$

and similarly for $\mu_L(\widetilde{\Omega}_{L,B} \setminus \Omega_{L,B})$. Combining (35) with (38) and the fact that $\mu_L(\Omega_L^*) \leq \mu(\widetilde{\Omega}_L^*) + \mu_L(\widetilde{\Omega}_{L,A} \setminus \Omega_{L,A}) + \mu_L(\widetilde{\Omega}_{L,B} \setminus \Omega_{L,B})$, we thus get the desired bound $\mu_L(\Omega_L^*) \leq O(e^{-KL/(\log L)^2})$.

References

[1] D. Aldous, J. Fill: *Reversible Markov Chains and Random Walks on Graphs, in preparation.* http://stat-www.berkeley.edu/pub/users/aldous/book.html.

[2] J. van den Berg: *A uniqueness condition for Gibbs measures with appliction to the 2-dimensional Ising antiferromagnet.* Commun. Math. Phys. **152** (1993) 161–163.

[3] B. Bollobás and I. Leader: *Edge-isoperimetric inequalities in the grid,* Combinatorica **11** (1991) 299-314.

[4] C. Borgs, J. T. Chayes, A. Frieze, J. H. Kim, P. Tetali, E. Vigoda, V. Vu: *Torpid mixing of some MCMC algorithms in statistical physics.* Proc. 40th IEEE Symp. on Found. of Comp. Sc. (1999) 218–229.

[5] C. Borgs, J. T. Chayes, M. Dyer, P. Tetali, Full version of this paper, in preparation.

[6] C. Borgs, J. Imbrie: *A Unified approach to phase diagrams in field theory and statistical mechanics.* Commun. Math. Phys. **123** (1989) 305–328.

[7] G.R. Brightwell, O. Häggström, P. Winkler: *Nonmonotonic behavior in hard-core and Widom-Rowlinson models.* J. Stat. Phys. **94** (1999) 415–435.

[8] G.R. Brightwell, P. Winkler: *Graph homomorphisms and phase transitions.* J. Combin. Theory **B77** (1999) 221–262.

[9] G.R. Brightwell, P. Winkler: *Gibbs measures and dismantlable graphs.* J. Combin. Theory **B78** (2000) 141–166.

[10] D. Brydges: *A short course on cluster expansions.* Proc. of the 1984 Les Houches Summer School. North Holland, Amsterdam, 1986, 129 – 1893.

[11] R. Burton, J. Steif: *Nonuniqueness of measures of maximal entropy for subshifts of finite type.* Ergodic Theory and Dynamical Systems **14** (1994) 213–235.

[12] R. Burton, J. Steif: *New results on measures of maximal entropy.* Israel J. of Math. **89** (1995) 275–300.

[13] C. Cooper, M. Dyer, A. Frieze: *On Markov chains for randomly H-colouring a graph.* preprint (2000).

[14] F. Chen, L. Lovász, I. Pak: Unpublished appendix for *Lifting Markov chains to speed up mixing.* Proc. 31st ACM Symp. on Theory of Comp. (1999) 275–281.

[15] R.L. Dobrushin: *The problem of uniqueness of a Gibbsian random field and the problem of phase transitions.* (Russian) Funkcional. Anal. i Priložen. **2** (1968) 44–57.

[16] R.L. Dobrushin: *Estimates on semi-invariants for the Ising model at low temperatures.* In: Topics in Statistical and Theoretical Physics, Amer. Math. Soc. Trans., Ser. 2 **177** (1996) 59–81.

[17] M.E. Dyer, A. Frieze, M. Jerrum: *On counting independent sets in sparse graphs.* Proc. 40th IEEE Symp. on Found. of Comp. Sc. (1999) 210–217.; journal version (submitted).

[18] M.E. Dyer, C.S. Greenhill: *The complexity of counting graph homomorphisms.* In Proc. 11th ACM/SIAM Symp. on Disc. Alg. (2000) 246–255.

[19] M.E. Dyer, C.S. Greenhill: *On Markov chains for independent sets.* J. Algorithms **35** (2000) 17–49.

[20] R. Griffiths: *Peierls' proof of spontaneous magnetization in a two-dimensional Ising ferromagnet.* Phys. Rev.**136A** (1964) 437.

[21] P. Hell, J. Nešetřil: *On the complexity of H-colouring.* J. Combin. Theory B **48** (1990) 92–110.

[22] M. Jerrum: *A very simple algorithm for estimating the number of k-colorings of low-degree graphs.* Rand. Struc. Alg. **7** (1995) 157–165.

[23] M. Jerrum, A. Sinclair: *Approximate counting, uniform generation and rapidly mixing Markov chains.* Information and Computation **82** (1989) 93–133.

[24] M. Luby, E. Vigoda: *Approximate counting up to four.* Proc. 29th ACM Symp. on Theory of Comp. (1997) 682–687.

[25] S. Miracle-Solé: *On the convergence of cluster expansions.* Physica **A279** (2000) 244–249.

[26] R. Peierls: *Ising's model of ferromagnetism.* Proc. Cambridge Phil. Soc. **32** (1936) 477.

[27] S.A. Pirogov, Ya.G. Sinai: *Phase diagrams of classical lattice systems.* Theor. Math. Phys. **25** (1975) 1185–1192.

[28] S.A. Pirogov, Ya.G. Sinai: *Phase diagrams of classical lattice systems. Continuation.* Theor. Math. Phys. **26** (1976) 39–49.

[29] E. Seiler: *Gauge Theories as a Problem of Constructive Quantum Field Theory and Statistical Mechanics.* Lecture Notes in Physics **159**. Springer, New York, 1982.

[30] L. Thomas: *Bound on the mass gap for finite volume stochastic Ising models at low temper-ature*, Commun. Math. Phys. **126** (1989) 1–11.

[31] E. Vigoda: *Improved bounds for sampling colorings.* Proc. 40th IEEE Symp. on Found. of Comp. Sc. (1999) 51–59.

MICROSOFT RESEARCH, 1 MICROSOFT WAY, REDMOND, WA 98052
E-mail address: `borgs@microsoft.com`

MICROSOFT RESEARCH, 1 MICROSOFT WAY, REDMOND, WA 98052
E-mail address: `jchayes@microsoft.com`

SCHOOL OF COMPUTING, UNIVERSITY OF LEEDS
E-mail address: `dyer@comp.leeds.ac.uk.`

SCHOOL OF MATHEMATICS, GEORGIA TECH, ATLANTA, GA 30332
E-mail address: `tetali@math.gatech.edu`

DIMACS Series in Discrete Mathematics
and Theoretical Computer Science
Volume **63**, 2004

Graph Homomorphisms
and Long Range Action

Graham R. Brightwell and Peter Winkler

ABSTRACT. We show that if a graph H is k-colorable, then $(k-1)$-branching
walks on H exhibit long range action, in the sense that the position of a token
at time 0 constrains the configuration of its descendents arbitrarily far into
the future.

This long range action property is one of several investigated herein; all
are similar in some respects to chromatic number but based on viewing H as
the range, instead of the domain, of a graph homomorphism.

The properties are based on combinatorial forms of probabilistic concepts
from statistical physics, although we argue that they are natural even in a
purely graph-theoretic setting. They behave well in many respects, but quite
a few fundamental questions remain open.

1. Introduction

Suppose some token takes a walk on a connected graph H, stepping from node
to adjacent node at each tick of a clock. If we know its position at time 0, can we
deduce anything about its position at time t for large t?

Certainly we can if H is bipartite, and conversely if $\chi(H) > 2$ then for large t
($t \geq 2|H| - 2$ will do) the token could be anywhere. Thus in the realm of connected
graphs we could take the property to be an alternate definition of 2-colorability,
and seek an extension analogous to k-colorability.

Suppose, for instance, that $H = K_3$ and imagine that the token takes a *2-
branching* walk on H; at each step the token divides into two (labeled) tokens both
of which step at the next tick, so that one token at time 0 yields 2^t at time t,
many of which may occupy the same node. If it happens that the two "children"
always take different steps, then it is clear that the positions of the 2^t descendents
will uniquely determine the starting node. We take this property of the constraint
graph K_3 as analogous to being 3-colorable.

A d-branching walk on H is nothing more than a graph homomorphism from
the complete d-branching tree T^d to H. (We define T^d to be the regular Cayley tree
of degree $d+1$, although it is often convenient to assume, as here, that the root r

2000 *Mathematics Subject Classification.* 05C15 (82C20).

Key words and phrases. graph homomorphism, long range action, chromatic number, circular
chromatic number.

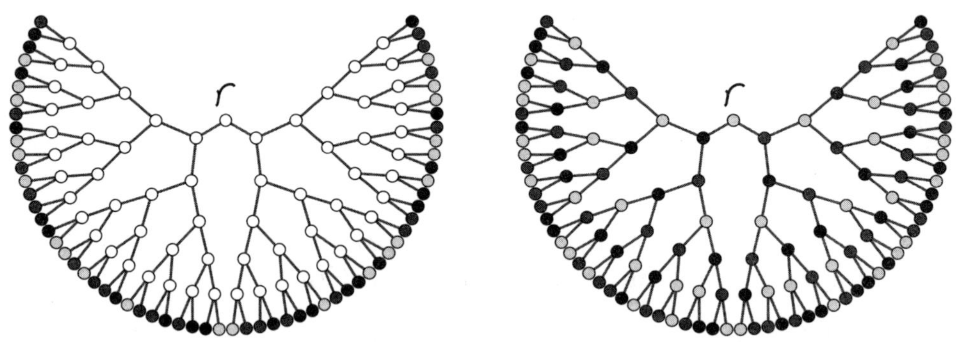

FIGURE 1. Reconstructing a 2-branching walk on K_3

has degree only d.) Fig. 1 illustrates the reconstruction of a particular 2-branching walk on K_3, viewed as a 3-coloring of T^2, from positions at time $t = 6$.

We denote the set of homomorphisms from a graph G to a graph H by $\mathrm{Hom}(G, H)$; later we will endow $\mathrm{Hom}(G, H)$ with its own graph structure. To avoid confusion we will call vertices of H "nodes" (usually denoted by a, b or c) and vertices of G "sites" (usually denoted by x, y or z). In this context G, which is often infinite but always countable and locally finite, will be called the "board" and the elements of $\mathrm{Hom}(G, H)$ *labelings* of G.

To simplify notation we will often confuse a graph with its set of vertices. The "constraint graph" H will always be finite and connected, and (unlike G) may have some loops; a loop at a node $a \in H$ allows a homomorphism in $\mathrm{Hom}(G, H)$ to affix the label a to adjacent sites of G. Some of what follows will be uninteresting for looped constraint graphs, however, for the reason that a looped node causes the chromatic number to become infinite. We do require that H contain at least one edge, thus it cannot consist of a single unlooped node.

We take the *distance* $d(x, y)$ between two sites of G to be the length of a shortest path between them, and for sets X and Y of sites, $d(X, Y) = \min\{d(x, y) \mid x \in X, \ y \in Y\}$.

We will say that $\mathrm{Hom}(G, H)$ exhibits *action at distance k* if there are sets X and Y of sites of G with $d(X, Y) \geq k$, and maps φ and ψ in $\mathrm{Hom}(G, H)$, such that no $\theta \in \mathrm{Hom}(G, H)$ agrees with φ on X and with ψ on Y. If $\mathrm{Hom}(G, H)$ exhibits action at all distances, then we say it exhibits *long range action*. This is precisely the *negation* of the property "strongly irreducible" of [**1**].

If $\mathcal{G}$ is a *class* of graphs, then $\langle \mathcal{G}, H \rangle$ is said to exhibit long range action if $\mathrm{Hom}(G, H)$ exhibits long range action for some $G \in \mathcal{G}$.

When G is a Cayley tree, we shall show that we can always take X to consist just of the root. A map φ in $\mathrm{Hom}(T^d, H)$ is said to be *cold* if there is a node a of H such that for any k, no $\psi \in \mathrm{Hom}(T^d, H)$ agrees with φ on the sites at distance k from the root r but has $\psi(r) = a$. $\mathrm{Hom}(T^d, H)$ itself is said to be cold if it contains a cold map, i.e. if some label can be forbidden at the root by values arbitrarily far away. We will see later that $\mathrm{Hom}(T^d, H)$ is cold if and only if it exhibits long range action.

It is easy to verify that if $\text{Hom}(T^d, H)$ is cold then so is $\text{Hom}(T^{d'}, H)$ for any $d' > d$. We say that H is d-*warm* if $\text{Hom}(T^{d-2}, H)$ is *not* cold, and define the warmth $w(H)$ of H to be the greatest d such that H is d-warm; equivalently, the least d such that $\text{Hom}(T^{d-1}, H)$ is cold. If H is d-warm for every d, we say $w(H) = \infty$. Every constraint graph H is 2-warm, and $w(H) = 2$ if and only if H is bipartite.

Although we hope to persuade the reader that this notion is natural and interesting within graph theory, some explanation of its origin may be in order here. A statistical system in physics is said to exhibit "long range order" if its state in one region of space gives non-vanishing information about its state in other regions far away. Thus, for example, a magnetized bar exhibits long range order because the spin of a particle at one end is non-trivially correlated with the spin of a particle at the other end.

In our system the space is an infinite graph G, often the discrete Cayley tree T^d; the states are nodes of H constrained by requiring adjacent sites to be in adjacent states. Our combinatorial notion of long range action is stronger than the notion from physics in requiring configurations to be forbidden rather than merely discouraged, but weaker in asking only for possibility, not probability, of long range effects. We choose to use the term "action" instead of "order" to distinguish the two notions. Long range action does arise in the consideration of "frozen" and "semi-frozen" Gibbs measures in [**3**] to which, along with [**2**] and [**4**], the reader is referred for a more complete explanation.

2. Warmth

We establish first that warmth and chromatic number coincide for complete graphs.

THEOREM 2.1. *For any integer $d \geq 2$, the complete graph K_d has warmth d.*

PROOF. As we have already seen in the $d = 3$ case, $\text{Hom}(T^{d-1}, K_d)$ is cold; a cold map is obtained by assigning all possible labels to the children of each site. To complete the proof we need to show that $\text{Hom}(T^{d-2}, K_d)$ is *not* cold; but this is obvious because any labeling of a site and its grandchildren can be extended to its children. $\qquad\square$

It will be useful to note the very nice behavior of warmth with respect to products and retracts. The product at issue is the "categorical" graph product, in which $(a, a') \sim (b, b')$ in $H \times H'$ if and only if $a \sim b$ in H and $a' \sim b'$ in H'. A *retraction* of H is a homomorphism ρ from H to some (necessarily induced) subgraph H^- of H, called a *retract*, such that the restriction $\rho \upharpoonright H^-$ is the identity.

THEOREM 2.2. *(i) For any H and H', $w(H \times H') = \min(w(H), w(H'))$; (ii) If H^- is a retract of H then $w(H^-) \geq w(H)$.*

PROOF. Both parts are just a matter of chasing down the definition. A map in $\text{Hom}(T^d, H \times H')$ is nothing more or less than the pointwise product of maps in $\text{Hom}(T^d, H)$ and $\text{Hom}(T^d, H')$ and is cold precisely if one of the two factor maps is cold. For the second statement, let ρ retract H to H^- and suppose φ is a cold map in $\text{Hom}(T^d, H^-)$ which forbids the label $a \in H^-$ at the root r. Then φ, regarded as a homomorphism to H, also forbids a at r within $\text{Hom}(T^d, H)$, because if $\theta \in \mathrm{Hom}(T^d, H)$ agrees with φ on the sites at distance k from r and $\theta(x) = a$ then $\rho \circ \theta$ contradicts the coldness of φ. $\qquad\square$

We remark that the chromatic number analog of statement (ii), namely "$\chi(H^-) \geq \chi(H)$", holds with equality. However, for the analog of statement (i), "$\chi(H \times H') \leq \min(\chi(H), \chi(H'))$" is easy but equality is a notorious conjecture of Hedetniemi [**7**].

It follows from Theorem 2.2 that $w(H) \leq \chi(H)$ whenever H contains a clique of size $\chi(H)$, since a $\chi(H)$-coloring of H can then be regarded as a retraction. In fact, we will see later that the clique condition can be dropped.

If A is a subset of (the nodes of) a constraint graph H, we let $N(A) := \{b \in H \mid b \sim a$ for some $a \in A\}$. A collection $\{A_1, \ldots, A_s\}$ of subsets of H is said to *produce* a subset A of H if $\bigcap N(A_i) = A$. The idea is that, if what we know about the labels of s neighbors of a site x is that neighbor x_i has a label from the set A_i, then what we can deduce is exactly that x has a label from A. A subset of H is deemed *non-trivial* if it is neither the empty set nor H itself. A family $\mathcal{A}$ of non-trivial subsets of H is *d-stable* if every set in $\mathcal{A}$ can be produced from some collection of at most d sets in $\mathcal{A}$.

THEOREM 2.3. *Given a constraint graph H and a natural number $d \geq 1$, the following are equivalent:*

(i) $\mathrm{Hom}(T^d, H)$ *exhibits long range action;*

(ii) *there is a cold H-labeling of T^d (i.e., H is not $(d+2)$-warm);*

(iii) *there is a d-stable family of subsets of H;*

(iv) *there is no ordering $A_1, \ldots, A_N$ of the non-trivial subsets of H such that each d-tuple $(A_{i_1}, \ldots, A_{i_d})$ of sets produces either $\emptyset$, H, or a set A_j with $j > \min\{i_k\}$.*

PROOF. $(ii) \implies (i)$ is obvious, taking $X = \{r\}$ and Y the sites at distance k from the root. We are using rather degenerately the fact that H is connected since otherwise the forbidden label a might be an isolated node of H, preventing us from constructing φ.

$(iii) \implies (ii)$. Suppose there is a d-stable family $\mathcal{A}$. Then we define an H-labeling of T^d as follows. Associate the root r with a pair (A_r, a_r) where $A_r \in \mathcal{A}$ and $a_r \in A_r$. Now work out from the root. If y is associated with a pair (A_y, a_y), and $y_1, \ldots, y_d$ are the children of y, let $\{A_{y_1}, \ldots, A_{y_d}\}$ be a collection of sets from $\mathcal{A}$ producing A_y, and let a_{y_i} be a node in A_{y_i} adjacent to a_y. When all the sites of T^d have been treated in this manner, the labeling ψ where $\psi(y) = a_y$ for each y is a member of $\mathrm{Hom}(T^d, H)$. Furthermore, for any $k \in \mathbb{N}$ and any H-labeling θ such that $\theta \upharpoonright (T^d \backslash N_k(x)) = \psi \upharpoonright (T^d \backslash N_k(x))$, where $N_k(x)$ is the set of sites at distance at most k from x, we see (by working in towards the root) that $\theta(y) \in A_y$ for every $y \in N_k(x)$. In particular, $\theta(x)$ can only take values in $A_x \neq H$, so ψ is a cold H-labeling of T^d.

$(iv) \implies (iii)$. Suppose there is no d-stable family and let $\mathcal{A}_0$ be the family of all non-trivial sets; this is not a d-stable family, so there is some set $A_1 \in \mathcal{A}_0$ that cannot be produced by d sets in $\mathcal{A}_0$. Now let $\mathcal{A}_1 = \mathcal{A}_0 \backslash \{A_1\}$, and continue, thus generating an ordering which contradicts (iv).

$(ii) \implies (iv)$. Suppose $A_1, \ldots, A_N$ is an ordering forbidden by (iv), where $N = 2^{|H|} - 2$. Let ψ be any H-labeling of T^d. For any site x at distance $\ell < N$ from r, let S_x be the set of sites y at distance N from r such that x is on the r-y

path, and let C_y be the set of nodes b such that there is an extension of $\psi \restriction S_x$ in which x gets label b. Note that C_x is never empty, since it contains $\psi(x)$. Also note that C_x is exactly the set produced by $\{C_{x_1}, \ldots, C_{x_d}\}$, where $x_1, \ldots, x_d$ are the children of x. Therefore, by induction, either $C_x = H$ or $C_x = A_j$ for some $j \geq N - \ell + 1$. In particular, $C_r = H$, so that all labels are possible for x, and ψ is not cold.

$(i) \implies (ii)$. Suppose from now on that $\mathrm{Hom}(T^d, H)$ exhibits long range action.

Fix a $k \in \mathbb{N}$. Suppose that, for all *finite* sets X, whenever Y is a set with $d(X, Y) \geq k$, and $\varphi, \psi \in \mathrm{Hom}(T^d, H)$, then there is some $\theta \in \mathrm{Hom}(T^d, H)$ extending both $\varphi \restriction X$ and $\psi \restriction Y$. We claim that (T^d, H) fails to exhibit action at distance k, which will be a contradiction. Indeed, let $Z = \{z_1, z_2, \ldots, \}$ be an infinite set of sites, let Y be a set with $d(Z, Y) \geq k$, and let φ, ψ be homomorphisms. Set $Z_n = \{z_1, \ldots, z_n\}$, for all n. Then there are homomorphisms $\theta_1, \theta_2, \ldots$ such that each θ_n agrees with φ on Z_n and with ψ on Y. Now there is some subsequence of (θ_n) that tends to a limit, and this limit is a homomorphism that agrees with φ on all of Z and with ψ on Y, which is the required contradiction.

It follows that, for each $k \in \mathbb{N}$, there is a finite set X, a set Y with $d(X, Y) \geq 2k$, and homomorphisms φ, ψ such that $\varphi \restriction X$ and $\psi \restriction Y$ cannot be simultaneously extended. We can also take Y to be finite (for instance we can assume it consists of the sites at distance exactly $2k$ from X). Now let X be minimal with this property, and then take Y minimal with the property. Take any shortest path from X to Y, and let x be the site on this path at distance k from X. The site x is thus at distance at least k from $X \cup Y$, and separates $X \cup Y$. By minimality of X and Y, for each branch B from x there is some homomorphism θ_B of $\mathrm{Hom}(T^d, H)$ such that θ_B agrees with φ on $X \cap B$ and with ψ on $Y \cap B$. Since we cannot glue these homomorphisms together to make a homomorphism θ extending all of $\varphi \restriction X$ and $\psi \restriction Y$, there must be some branch B and some label $a \in H$ such that there is no homomorphism θ with $\theta(x) = a$, agreeing with θ_B on $(X \cup Y) \cap B$.

Therefore, for all k, there is a singleton set x (which we may take to be the root of T^d in each case), a set Y_k with $d(x, Y_k) \geq k$ (which we may take to be $T^d \backslash N_k(x)$), a label a_k, and a homomorphism ψ_k, such that there is no θ with $\theta \restriction Y_k = \psi_k \restriction Y_k$ and $\theta(x) = a_k$. By taking a subsequence, we may assume that the label a_k is always equal to a. Now there is some subsequence of the ψ_k that tends to a limit ψ. It is now clear that ψ is a cold H-labeling of T^d. $\qquad \square$

3. Examples

We have seen that $w(K_d) = d$; note that the singletons comprise a $(d-1)$-stable family. In fact in any d-colorable graph the color classes are candidates for being a $(d-1)$-stable family; if the graph is *uniquely* d-colorable then they indeed will form such a family, since every node is then adjacent to nodes of all other colors. We could extend this argument to show that containing a uniquely d-colorable graph is enough to prevent $(d+1)$-warmth, but, again, we will prove a stronger result later.

Having girth at least 5, even for looped constraint graphs, already prevents 4-warmth. For, let $a_1 a_2 \cdots a_s a_1$ be a shortest cycle in H, with $s \geq 5$, and let $\mathcal{A} = \{\{a_1\}, \ldots, \{a_s\}\}$; then $\{a_i\}$ is produced by $\{\{a_{i-1}\}, \{a_{i+1}\}\}$. Indeed, this construction works whenever there is a long cycle in H not sharing two consecutive edges with a 4-cycle (or 3-cycle, when loops are present). We can use this idea

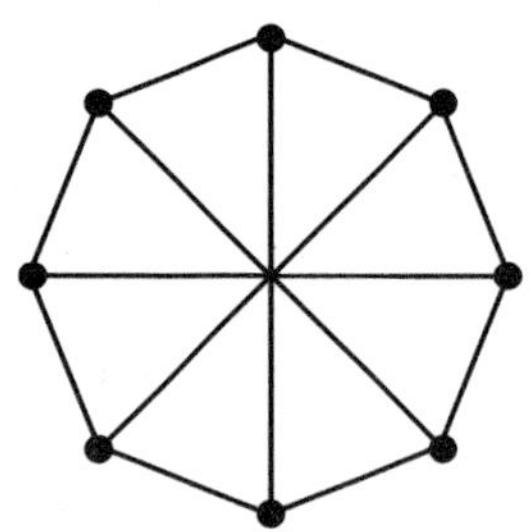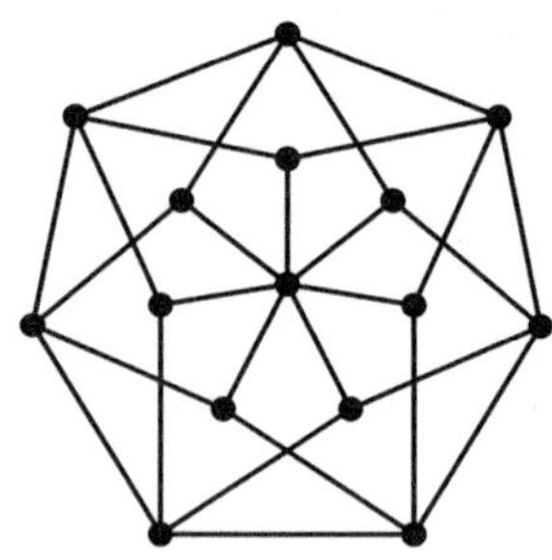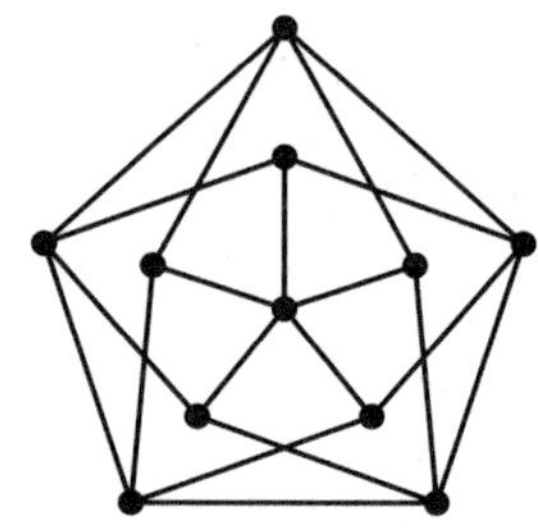

FIGURE 2. H_8, G_7 and G_5

to produce examples H that are not 4-warm but have arbitrarily large cliques, for instance. This in turn shows that we can have $w(H^-) > w(H)$ for H^- a retract of H (cf. Theorem 2.2).

For another example that is not 4-warm, consider the graph H_8 defined by taking an 8-cycle $a_1 a_2 \cdots a_8 a_1$, and adding the four "diagonal" edges $\{a_i a_{i+4}\}$ (see left-hand side of Fig. 2.) This graph has chromatic number 3, and is not 4-warm; the family of singletons is 2-stable, since each $\{a_i\}$ is produced by $\{\{a_{i-1}\}, \{a_{i+1}\}\}$. However, this family has no obvious connection to a 3-coloring, or to a shortest odd cycle. There is another quite different 2-stable family, namely $\{\{a_1, a_4\}, \{a_2, a_5\},$ $\ldots, \{a_8, a_3\}\}$; note that $\{\{a_8, a_3\}, \{a_2, a_5\}\}$ produces $\{a_1, a_4\}$.

For the next two examples we use the Grötzsch construction, defined as follows: if H has vertices $a_1, \ldots, a_n$ then $\mathrm{Gr}(H)$ has vertices $a_1, \ldots, a_n, b_1, \ldots, b_n$ and c; if $a_i \sim a_j$ in H then $a_i \sim a_j$ and $b_i \sim a_j$ in $\mathrm{Gr}(H)$, and c is adjacent to every b_i. One of the nice properties of the construction is that always $\chi(\mathrm{Gr}(H)) = 1 + \chi(H)$.

In the graph $G_7 := \mathrm{Gr}(C_7)$ (center, Fig. 2) the family

$$\mathcal{A} = \{\{a_1\}, \{a_2\}, \ldots, \{a_7\}, \{b_7, b_2\}, \{b_1, b_3\}, \ldots \{b_6, b_1\}, \{c\}\}$$

is 2-stable: for instance $\{a_1\}$ is produced by $\{\{a_7\}, \{b_2, b_4\}\}$, $\{b_7, b_2\}$ is produced by $\{\{a_1\}, \{c\}\}$, and $\{c\}$ is produced by $\{\{b_1, b_3\}, \{b_2, b_4\}\}$. Therefore G_7 is not 4-warm. However, there is no 2-stable family consisting of *disjoint* sets—our later proof will show that disjoint sets are always obtainable if H is 3-colorable.

The standard Grötzsch graph $G_5 := \mathrm{Gr}(C_5)$ (right-hand side of Fig. 2) also fails to be 4-warm. One nice 2-stable family is

$$\mathcal{A} = \{\{a_1\}, \ldots, \{a_5\}, \{a_1, b_1\}, \ldots, \{a_5, b_5\}, \{b_1, b_3\}, \{b_2, b_4\}, \ldots, \{b_5, b_2\}, \{c\}\} \ .$$

Notice that $\mathcal{A}$ contains some pairs of sets related by inclusion, and simply taking the inclusion-minimal elements of $\mathcal{A}$ does *not* yield a 2-stable family. Note however that, for every set A in $\mathcal{A}$, there is an element $u \in A$ such that A is minimal in $\mathcal{A}$ subject to containing u—we say that A is *semi-minimal* in $\mathcal{A}$. In general, if $\mathcal{A}$ is any d-stable family in a graph H, the family of semi-minimal sets in $\mathcal{A}$ is again d-stable.

We expect that there are constraint graphs where inclusion relations in a d-stable family are unavoidable. However, for G_5, a careful pruning of the family $\mathcal{A}$ leaves the smaller 2-stable family

$$\{\{a_2\}, \{a_3\}, \{a_4\}, \{a_5\}, \{a_1, b_1\}, \{b_1, b_3\}, \{b_1, b_4\}, \{b_2, b_5\}, \{c\}\} \ ,$$

where no pair of sets is related by inclusion.

An example of a graph that *is* 4-warm is the 5-wheel W_5 (right-hand side of Fig. 3). To see this, let $a_1, \ldots, a_5$ be the nodes of the 5-cycle in W_5, each attached to the center c. Order the subsets of W_5 by size, beginning with the smaller sets, and listing those that contain c after those that do not. It is straightforward to check that each pair of sets produces a set later in this order than the earlier of the two sets.

4. Circular Chromatic Number

In the next section, we shall prove that $w(H) \leq \chi(H)$ for all unlooped constraint graphs H. To motivate our approach, we begin by considering the case of chromatic number 3, when we will be able to prove a little more. In a sense, it is not $\chi(H) \leq 3$ that forces 3-warmth but $\chi(H) < 4$.

For integers $1 \leq q \leq p$, a (p,q)-*coloring* of a loopless graph H is a function $\gamma : H \to \{0, \ldots, p-1\}$ such that, whenever $a \sim b$, $q \leq |\gamma(a) - \gamma(b)| \leq p-q$. We think of the nodes of H as being mapped to a circle of circumference p, adjacent nodes being mapped to points at least q apart. The *circular chromatic number* $\chi_c(H)$ of H is the infimum of the set $\{p/q : \text{ there is a } (p,q)\text{-coloring of } H\}$. This parameter is also known as the *star-chromatic number*.

The following facts were established by Vince [16].

- For all H, $\chi(H) = \lceil \chi_c(H) \rceil$.
- The infimum in the definition of $\chi_c(H)$ is always attained; furthermore if $\chi_c(H) = p/q$ with p and q positive and relatively prime, then there is a (p,q)-coloring of H.
- If γ is a (p,q)-coloring of a graph H with $\chi_c(H) = p/q$, there is a cycle $a_0 a_1 \cdots a_{s-1} a_0$ in H such that $\gamma(a_{i+1}) \equiv \gamma(a_i) + q \pmod{p}$. We call such a cycle *tight*.

Zhu [17] has written a useful survey of work relating to the circular chromatic number.

THEOREM 4.1. *If H is an unlooped constraint graph, with circular chromatic number less than 4, then there is a 2-stable family of disjoint subsets of H, and in particular H is not 4-warm.*

PROOF. Take a (p,q)-coloring γ of H, where $\chi_c(H) = p/q < 4$ and p and q are relatively prime. Let $a_0 a_1 \cdots a_{s-1} a_0$ be a tight cycle C, with $\gamma(a_{i+1}) \equiv \gamma(a_i) + q \pmod{p}$. Without loss of generality, $\gamma(a_0) = 0$, so that $\gamma(a_i) \equiv iq \pmod{p}$, and therefore s is a multiple of p.

For $i = 0, \ldots, p-1$, let C_i be the set of nodes of C that are assigned color iq, so $C_i = \{a_i, a_{i+p}, a_{i+2p}, \ldots, a_{i+s-p}\}$. Now define the sets $A_i \supseteq C_i$ recursively by putting node a in A_i whenever there are nodes a^- and a^+ adjacent to a with $a^- \in A_{i-1}$ and $a^+ \in A_{i+1}$.

We claim that $\gamma(j) = iq$ for every $a \in A_i$. We establish this recursively; it is true for $a \in C_i$, and if a is adjacent to nodes $a^- \in A_{i-1}$ and $a^+ \in A_{i+1}$ for which the claim is true, then $\gamma(j) \notin ((i-2)q, iq)$ and $\gamma(j) \notin (iq, (i+2)q)$; since $p < 4q$, iq is the only color available for a.

This proves that the A_i are disjoint, and in particular non-trivial. By construction, $\{A_0, \ldots, A_{p-1}\}$ is a 2-stable family, with $\{A_{i-1}, A_{i+1}\}$ producing A_i for each i.

This completes the proof.   □

5. Warmth and Chromatic Number

We are now in a position to connect these two parameters, one of which treats H as the range of a homomorphism, the other as the domain.

THEOREM 5.1. *For every unlooped H, the warmth of H is at most its chromatic number.*

PROOF. Let $\chi(H) = d+1$ with the object of constructing a cold labeling of T^d by H. We begin by finding a map Ψ from H to the unit vectors of $\mathbb{R}^d$ with the following property:

$$\text{If } a \sim b \text{ in } H \text{ then } \Psi(a) \cdot \Psi(b) < 0 \ .$$

To define a suitable Ψ we fix a regular simplex in $\mathbb{R}^d$ which is centered at the origin and sized so that its vertices are unit vectors, then color H properly with $d+1$ colors and map each color class to a different vertex of the simplex. This gives $\Psi(a) \cdot \Psi(b) = -1/d$ whenever $a \sim b$.

Now that we know such a map exists, we fix a Ψ which maximizes

$$\lambda = \min\{-\Psi(a) \cdot \Psi(b) \,\big|\, a \sim b\}$$

and then minimizes the number of nodes a of H for which, for some $b \sim a$, $\Psi(a) \cdot \Psi(b) = -\lambda$. We call these nodes "tight". Note that, in the case $d = 2$, a map with the property above exists if and only if $\chi_c(H) < 4$; in this case, the tight nodes are those lying on some tight cycle. Thus our approach here generalizes that in Theorem 4.1.

We now digress slightly to prove a geometric lemma. We say that a set A of vectors *forces* $\vec{v}$ if $\vec{v}$ is a unit vector which satisfies $\vec{v} \cdot \vec{u} \leq -\lambda$ for every $\vec{u} \in A$, but no unit vector $\vec{w}$ satisfies $\vec{w} \cdot \vec{u} < -\lambda$ for every $\vec{u} \in A$. We say that A *fixes* the unit vector $\vec{v}$ if (a) $\vec{v} \cdot \vec{u} = -\lambda$ for every $\vec{u} \in A$, and (b) if $\vec{w} \cdot \vec{u} \leq -\lambda$ for every $\vec{u} \in A$ then $\vec{w} = \vec{v}$.

LEMMA 5.2. *Fix λ with $0 < \lambda < 1$, and let A be a finite set of unit vectors in $\mathbb{R}^d$ which forces a certain unit vector $\vec{v}$. Then there is a subset C of A with $|C| \leq d$ which fixes $\vec{v}$.*

PROOF. We may assume without loss of generality that $\vec{v} = (v_1, \ldots, v_d) = (1, 0, \ldots, 0)$; let B consist of those members of A which lie on the $(d-2)$-sphere

$$S := \{\vec{w} \,\big|\, |\vec{w}| = 1, \ w_1 = -\lambda\} \ .$$

We claim that every closed hemisphere of S contains some $\vec{u} \in B$.

If not, we may assume the hemisphere $S^- := \{\vec{w} \in S \,\big|\, w_2 \leq 0\}$ is missed, so that $u_2 > 0$ for $\vec{u} \in B$. For $\varepsilon > 0$ let

$$\vec{v}(\varepsilon) := (\sqrt{1 - \varepsilon^2}, -\varepsilon, 0, 0, \ldots, 0)$$

so that

$$\vec{u} \cdot \vec{v}(\varepsilon) = -\lambda\sqrt{1 - \varepsilon^2} - \varepsilon u_2 < -\lambda$$

for $\vec{u} \in B$ and sufficiently small ε. If we also take ε small enough so that $\vec{u} \cdot \vec{v}(\varepsilon) < -\lambda$ for those $u \in A \backslash B$, we have a contradiction to A forcing $\vec{v}$.

It follows by the separating hyperplane theorem that the point $\vec{z} := (-\lambda, 0, 0, \ldots, 0)$ lies in the convex hull of B. Moreover, since the dimension of the hyperplane defined by $w_1 = -\lambda$ is $d-1$, Carathéodory's Theorem (see e.g. [10]) tells us that there is a subset $C \subseteq B$ of size at most d such that $\vec{z}$ already lies in the convex hull of C.

If some vector $\vec{w}$ satisfies $\vec{w} \cdot \vec{u} \leq -\lambda$ for each $\vec{u} \in C$ then it also satisfies $\vec{w} \cdot \vec{z} \leq -\lambda$, thus $w_1 \geq 1$, and cannot be a unit vector unless $\vec{w} = \vec{v}$. Hence C fixes $\vec{v}$. $\qquad\square$

We are now ready to finish the proof of Theorem 5.1. We define a labeling of T^d by H as follows.

Choose any tight node $a \in H$ to label the root r. The images of the neighbors of a must force $\Psi(a)$, since otherwise $\Psi(a)$ could be adjusted so that a would no longer be tight. By Lemma 5.2 there is a set X of at most d tight neighbors of a whose images under Ψ fix $\Psi(a)$. Use all the elements of X to label the children of r and proceed in like fashion to label the rest of T^d.

To see that this labeling is cold, imagine that all sites at distance less than k are unlabeled, and then relabeled in some consistent fashion. We claim that at every site the old and new labels have the same image under Ψ. This can be seen by induction working in from distance $k-1$.

It follows that $\mathrm{Hom}(T_d, H)$ is cold. $\qquad\square$

The methods used in the above proof call to mind the *vector chromatic number* of a graph H, defined by Karger, Motwani and Sudan [9] as the minimum k such that there exists a labeling Ψ of $V(H)$ by unit vectors in $\mathbb{R}^{|V(H)|}$ in which adjacent nodes a and b satisfy $\langle \Psi(a), \Psi(b) \rangle \leq -\frac{1}{k-1}$. Karger, Motwani and Sudan show that the vector chromatic number of H can be approximated arbitrarily closely in randomized polynomial time. They also show that the vector chromatic number of H is at most $\vartheta(\overline{H})$, where ϑ is the Lovász theta-function (see for instance Grötschel, Lovász and Schrijver [6]). For us, the key issue is not the extremal value of the inner product, but the minimum *dimension* in which the inner product of adjacent nodes can be made negative, and we know of no connection between warmth and the Lovász theta-function.

Our particular version of "vector labeling" has occurred before in a very different context connected with the Ramsey number $R(3, 3, \ldots, 3)$; see for instance the survey article by Nešetřil and Rosenfeld [12]. In the language of that paper, we are interested in the minimum d such that the graph H is α-embeddable in $\mathbb{R}^d$ for some $\alpha > \sqrt{2}$.

We now introduce two new graph parameters, "heat" and "mobility", but warmth will remain in the picture.

6. Heat

If in condition (i) of Theorem 2.3 the d-regular tree T^{d-1} is replaced by a general graph of maximum degree d, we obtain a strengthening of the notion of warmth as follows.

Let $\mathcal{G}_d$ be the class of all (locally finite) graphs of maximum degree at most d. A constraint graph H is said to be d-hot if the pair $\langle \mathcal{G}_{d-1}, H \rangle$ does not exhibit long range action, i.e., if $\mathrm{Hom}(G, H)$ does not exhibit long range action for any board of maximum degree at most $d - 1$. The *heat* $h(H)$ of H is the greatest d such that

H is d-hot, or, equivalently, the least d such that $\mathrm{Hom}(G, H)$ exhibits long range action for some board of maximum degree d.

THEOREM 6.1. *The following are equivalent for any $d \geq 2$ and any constraint graph H:*

    (i) *for all $k > 0$ there is a graph $G \in \mathcal{G}_{d-1}$ such that $\mathrm{Hom}(G, H)$ exhibits action at distance k;*

    (ii) *H is not d-hot (i.e., there is a single graph $G \in \mathcal{G}_{d-1}$ such that $\mathrm{Hom}(G, H)$ exhibits action at all distances k);*

    (iii) *there is a graph $G \in \mathcal{G}_{d-1}$ such that, for all $k > 0$, there are finite witnesses X, $Y \subset G$ to action at distance k.*

PROOF. Clearly we have $(iii) \implies (ii) \implies (i)$.

$(i) \implies (ii)$. Suppose that (i) holds and choose, for each $k > 0$, a $G_k \in \mathcal{G}_{d-1}$ together with subsets X_k and Y_k and maps φ_k and ψ_k in $\mathrm{Hom}(G_k, H)$ with no common extension of $\varphi_k \upharpoonright X_k$ and $\psi_k \upharpoonright Y_k$. If H is d-hot there is a distance k' which is enough to eliminate action on the disjoint union $G := \bigcup_{k=1}^{\infty} G_k$, but the sets $X_{k'}$ and $Y_{k'}$, and extensions to G of the maps $\varphi_{k'}$ and $\psi_{k'}$, testify otherwise.

$(ii) \implies (iii)$. Take a board $G \in \mathcal{G}_{d-1}$ such that $\mathrm{Hom}(G, H)$ exhibits long range action. So, for any distance k, there are sets X and Y with $d(X, Y) \geq k$, and labelings φ and ψ with no common extension of $\varphi \upharpoonright X$ and $\psi \upharpoonright Y$. Suppose however that no finite subset $X' \subset X$ can replace X. Then, as in the proof of Theorem 2.3, we get a contradiction by taking a nested sequence (X_i) of finite subsets of X whose union is X, and letting θ be any pointwise limit of the labelings θ_i obtained as common extensions of $\varphi \upharpoonright X_i$ and $\psi \upharpoonright Y$. Once X is finite we can limit Y to the (finite) set of sites at distance exactly k from X. $\qquad\Box$

Thus we have a range of conditions for H equivalent to having heat less than d. However, the situation is not quite as good as for warmth; we don't know whether we can strengthen condition (iii) further to find a single finite set X in a graph $G \in \mathcal{G}_{d-1}$ which can be used for each k. It is conceivable that there is a constraint graph H of heat less than d for which the sets $X = X_k$ and $Y = Y_k$ in (iii) necessarily grow with k, whatever board $G \in \mathcal{G}_{d-1}$ is chosen, but we know of no examples of this phenomenon.

Clearly the heat $h(H)$ is always at most the warmth $w(H)$ of H. Since $\mathrm{Hom}(G, H)$ may be empty when G is not a tree, there is a *tendency* for T^{d-1} to be the easiest board in $\mathcal{G}_d$ on which to exhibit long range action, in which case heat and warmth will be equal. For instance, $\mathrm{Hom}(T^3, K_4)$ exhibits long range action, whereas $\mathrm{Hom}(\mathbb{Z}^2, K_4)$ (4-coloring the plane grid) does not. Indeed it is easily checked that $h(K_d) = w(K_d) = d$ for complete graphs K_d. Also, as for warmth, every constraint graph H is 2-hot, and H is 3-hot unless it is bipartite.

However, the 5-wheel W_5, which we served earlier as an example of a 4-warm graph, is not 4-hot. We start with a copy of T^2 having a root of degree 2, and form $G \in \mathcal{G}_3$ by replacing each site by a triangle, each vertex of which becomes incident to one of the edges incident to the original site. Let z be the lone site in G of degree 2 and suppose that $\varphi \in \mathrm{Hom}(G, W_5)$ is chosen so that in every triangle the site nearest z is labeled by the center node c of W_5 (see Fig. 3). Since in any labeling one site from each triangle must map to c, any θ consistent with φ outside some neighborhood of z must also label z by c; we have long range action for $\mathrm{Hom}(G, W_5)$.

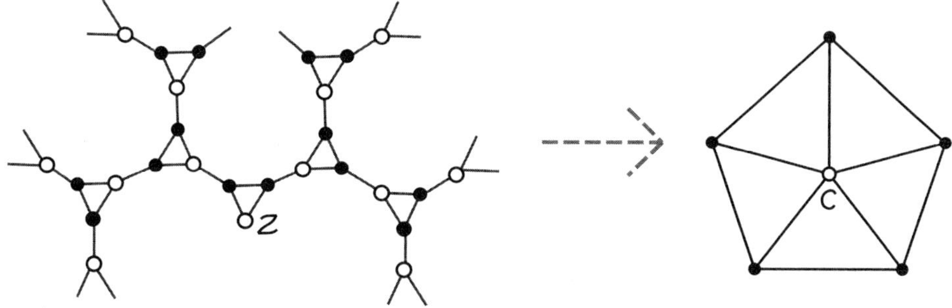

FIGURE 3. W_5 is not 4-hot

The other examples we considered earlier, namely H_8, $\mathrm{Gr}(C_7)$ and $\mathrm{Gr}(C_5)$, all have warmth 3 and are not bipartite, and therefore have heat 3.

7. Mobility

Our third new parameter is somewhat different from heat and warmth but again motivated by considerations from statistical physics. There is also a connection to the theory of computing. It is often useful to obtain a *random sample* from $\mathrm{Hom}(G,H)$ when the board G is large but finite. This can potentially be done by "heat bath", or "single-site Glauber dynamics", in which labels are randomly changed one site at a time in accordance with the constraints imposed by H. Let us define a graph structure on $\mathrm{Hom}(G,H)$ by making two maps adjacent when they differ on one site; then we see that the heat bath can only work if $\mathrm{Hom}(G,H)$ is connected.

We will say that H is *d-mobile* if $\mathrm{Hom}(G,H)$ is connected for any finite board $G \in \mathcal{G}_{d-2}$. The *mobility* $m(H)$ of H is the greatest d for which H is d-mobile, or equivalently the least d for which $\mathrm{Hom}(G,H)$ is disconnected for some finite $G \in \mathcal{G}_{d-1}$. If H is d-mobile for every d, i.e. $\mathrm{Hom}(G,H)$ is connected for *every* finite G, then we say $m(H) = \infty$. Every constraint graph is trivially 2-mobile (graphs in $\mathcal{G}_0$ being collections of isolated points) and as with heat and warmth, $m(H) = 2$ if and only if H is bipartite.

Mobility also matches the other parameters on complete graphs.

THEOREM 7.1. *For any integer $d \geq 2$, $m(K_d) = d$.*

PROOF. That K_d is not $(d+1)$-mobile is clear from taking $G = K_d$ itself, since $\mathrm{Hom}(K_d, K_d)$ consists of $d!$ isolated points.

To see that K_d is d-mobile, we repeat an argument from [8] where the objective was to use a heat bath to estimate the number of d-colorings of G. For the heat bath to work in polynomial time, connectivity is not enough; "rapid mixing" of the Markov chain is also required, but has been proven only when the maximum degree of G exceeds d by a constant factor (currently 11/6 [15]).

Let φ and ψ be any two labelings by K_d (i.e. proper d-colorings) of a graph G of maximum degree at most $d-2$. We change φ sequentially to obtain ψ, as follows. Our first goal is to ensure that $\varphi(y) = 1$ whenever $\psi(y) = 1$; to do this, we look at all those z such that $\varphi(z) = 1$—these form an independent set of sites, and for all of them there is some alternative label that can be used, and we do so. Now we can use label 1 on all the sites we want; we shall not relabel these sites again. We now repeat with label 2, and so on. $\qquad\square$

Among our earlier examples are some where mobility and warmth differ. For the 5-wheel W_5, we have $m(W_5) = 3 < 4 = w(W_5)$; to verify that W_5 is not 4-mobile, take the board to be K_3 and note (as we did when showing that $h(W_5) < 4$) that every labeling uses the center node of W_5 exactly once.

The Grötzsch graph $G_5 = \mathrm{Gr}(C_5)$ is an example of a graph whose mobility exceeds its warmth and heat. Recall that $w(G_5) = h(G_5) = 3$; we now demonstrate that G_5 is 4-mobile.

We need to show how to get from any G_5-labeling of a cycle C_n to any other. Note that there are no G_5-labelings of C_3, so we may take $n \geq 4$. Set $A = \{a_1, \ldots, a_5\}$ and $B = \{b_1, \ldots, b_5\}$, so $V(G_5) = A \cup B \cup \{c\}$.

Our first step is change any G_5-labeling to eliminate all uses of label c. If label c is used at any point in the cycle, the label to its left is some $b_i \in B$, and the label to the left of that is in $A \cup \{c\}$: in either case we have the option of changing the label b_i to at least one other b_j. There are at least three nodes of A adjacent to either b_i or b_j, and we similarly get a set of three possible nodes of A from the right side. Thus one element of A is possible from both sides, and the label c can be changed to this element by first changing its neighboring labels if necessary. Proceeding in this way, we can indeed eliminate all the uses of label c. Then of course we can replace each use of label b_i by the corresponding a_i.

We now have a homomorphism from C_n to $G_5 \restriction A$—a copy of C_5. It is easy to see that the graph $\mathrm{Hom}(C_n, C_5)$ in general falls into several connected components, with each component identified by the *winding number*, the number of times the sequence of labels winds around C_5; if n is a multiple of 5, there are also 10 isolated vertices of $\mathrm{Hom}(C_n, C_5)$, namely those homomorphisms with winding number $\pm n/5$. The winding number always has the same parity as n.

To complete the argument, it is enough to show that, working in $\mathrm{Hom}(C_n, G_5)$, we can reverse a sequence of labels such as $a_1 a_2 a_3 a_4 a_5 a_1$ that winds around G_5, hence changing the winding number by 2. To do this, we step through the following labelings in turn:

$$a_1 a_2 a_3 a_4 a_5 a_1, \ a_1 a_2 b_3 a_4 b_5 a_1, \ a_1 a_2 b_3 c b_5 a_1, \ a_1 a_2 b_1 c b_2 a_1,$$
$$a_1 a_5 b_1 c b_2 a_1, \ a_1 a_5 b_4 c b_2 a_1, \ a_1 a_5 b_4 a_3 b_2 a_1, \ a_1 a_5 a_4 a_3 a_2 a_1.$$

Our other example $\mathrm{Gr}(C_7)$ is not 4-mobile: take C_5 as a board. Figure 3 summarizes the parameter values for the various examples we have been considering.

In Section 4 we saw that if H has circular chromatic number less than 4, then H is not 4-warm; in fact it is not 4-mobile either. To see this, take a (p,q)-coloring γ of H, where $\chi_c(H) = p/q < 4$, and p and q are relatively prime, and take a tight cycle $a_0 a_1 \cdots a_{s-1} a_0$ in H, so that $\gamma(a_{i+1}) = \gamma(a_i) + q \pmod{p}$.

As we saw in the proof of Theorem 4.1, there are disjoint sets $\{A_0, \ldots, A_{p-1}\}$ such that A_i contains $a_i, a_{i+p}, a_{i+2p}, \ldots$, and the family $\{A_{i-1}, A_{i+1}$ produces A_i, for each i.

Now let the board G be an s-cycle $x_0 x_1 \cdots x_{s-1} x_0$, and consider the H-labeling defined by $\varphi(x_i) = a_i$. We see that, for any ψ in the component of $\mathrm{Hom}(G, H)$

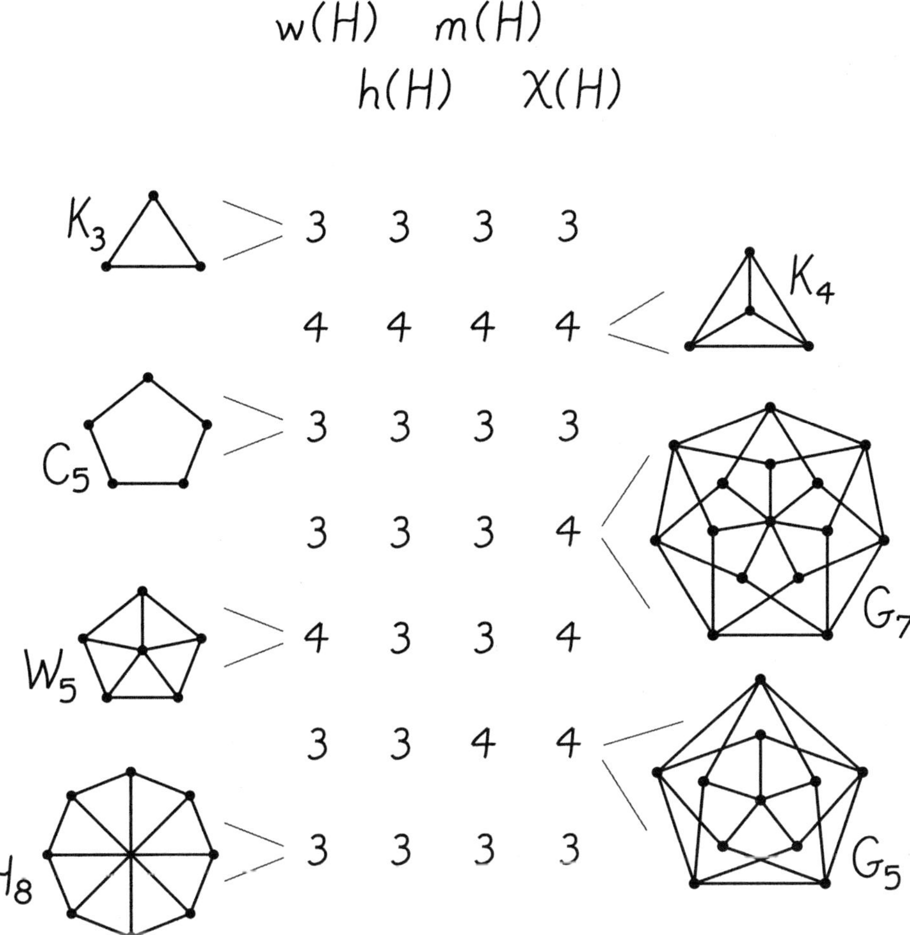

FIGURE 4. Warmth, heat, mobility and chromatic number

containing φ, $\psi(x_i) \in A_j$ for all i, where $j \equiv i \pmod{p}$. In particular, the H-labeling θ given by $\theta(x_i) = a_{i+1}$ is not in the same component of $\mathrm{Hom}(G, H)$ as φ.

The next two results connect mobility with heat and warmth.

THEOREM 7.2. *Any $(2d-3)$-hot constraint graph H is d-mobile.*

PROOF. Let k be an integer such that, whenever $G \in \mathcal{G}_{2d-4}$, X and Y are subsets of G with $d(X, Y) \geq k$, and φ and ψ are H-labelings of G, there is an H-labeling θ simultaneously extending $\varphi \restriction X$ and $\psi \restriction Y$. The fact that H is $(2d-3)$-hot guarantees that such an integer k exists, by Theorem 6.1.

Let G be any finite board in $\mathcal{G}_{d-2}$, let $x_1, x_2, \ldots, x_n$ be the sites of G, and form the board G' as follows. The set of sites of G' consists of $k+1$ copies $X_0, \ldots, X_k$ of the set of sites of G, with the mth copy of site x_i labeled $x_{i,m}$ $(0 \leq m \leq k)$. For each edge $x_i x_j$ of G, with $i < j$, and each m, there are edges $x_{i,m} x_{j,m}$ and $x_{j,m} x_{i,m+1}$ of G'. The graph G' thus has maximum degree at most $2d-4$, and also $d(X_0, X_k) \geq k$.

Now let φ and ψ be two H-labelings of G. These labelings lift to G' in the obvious way: set $\varphi'(x_{i,m}) = \varphi(x_i)$ for all i, m, and similarly for ψ. Now consider an H-labeling θ of G' simultaneously extending $\varphi' \restriction X_0$ and $\psi' \restriction X_k$.

The homomorphism θ tells us how to get from φ to ψ: we start with φ as our labeling, and look at the sites of G' in the order $x_{1,1}, x_{2,1}, \ldots, x_{n,1}, x_{1,2}, \ldots, x_{n,2}, \ldots, x_{1,k}, \ldots, x_{n,k}$. At each site, interpret $\theta(x_{i,m}) = a$ as an instruction to (re)label site x_i of G with label a. The fact that θ is a homomorphism of G' with $\theta \restriction X_0 = \varphi' \restriction X_0$ tells us exactly that this procedure is legitimate; the fact that $\theta \restriction Y_k = \psi' \restriction Y_k$ tells us that the final labeling of G is ψ. $\square$

THEOREM 7.3. *Any $(2d-2)$-mobile constraint graph H is d-warm.*

PROOF. We shall suppose that H is not d-warm, take a $(d-2)$-stable family $\mathcal{A}$ of subsets of H, and construct a finite graph G in $\mathcal{G}_{2d-4}$, together with two H-labelings of G that are not connected in $\mathrm{Hom}(G, H)$.

For each pair (A, a), with $A \in \mathcal{A}$ and $a \in A$, we choose $d-2$ pairs (A_1, a_1), $\ldots, (A_{d-2}, a_{d-2})$ such that all the A_i are in $\mathcal{A}$ and $\{A_1, \ldots, A_{d-2}\}$ produces A, and each a_i is a neighbor of a in A_i. The fact that $\mathcal{A}$ is a $(d-2)$-stable family ensures that we can do this.

Next we form a digraph D whose vertex set consists of all 4-tuples $(A, a; B, b)$ where $A, B \in \mathcal{A}$, $a \in A$, $b \in B$. We direct arcs from $(A, a; B, b)$ to each $(A_i, a_i; B_i, b_i)$, $i = 1, \ldots, d-2$. Thus every vertex of D has outdegree $d-2$.

Let N be the total number of vertices in D, and let $\mathbf{M}$ be the $N \times N$ incidence matrix of D, so the entry $m_{\alpha\beta}$ is equal to 1 if there is an arc from α to β, and 0 otherwise.

Note that $(1, \ldots, 1)$ is an eigenvector of $\mathbf{M}$, with largest eigenvalue $d-2$. Therefore there is a positive rational vector $\mathbf{r}$ such that $\mathbf{M}^{\mathrm{t}} \mathbf{r} = (d-2)\mathbf{r}$. By multiplying up we can take $\mathbf{r}$ to be an integer vector. Now take r_α copies of each vertex α of D. Our intention is to form a digraph D' on this blown-up vertex set by directing one arc from each copy of α to some copy of each β with $\alpha\beta$ an arc of D. If we do this, the total number of arcs arriving at the r_β copies of β is $\sum_{\alpha \to \beta} r_\alpha = (\mathbf{M}^{\mathrm{t}} \mathbf{r})_\beta = (d-2)r_\beta$, so we can distribute the incoming arcs so that every vertex of D' has indegree $d-2$. (We can also ensure at this stage that our digraph D' has no loops.)

Now we form our graph G by forgetting the orientation of all the arcs of D'. Thus the maximum degree of G is at most $2d-4$.

There are two H-labelings φ and ψ of G given by projections: in φ, each copy of the vertex $(A, a; B, b)$ of D is given label a; in ψ, each copy of $(A, a; B, b)$ gets label b. We claim that, in the component of $\mathrm{Hom}(G, H)$ containing φ, each copy of $(A, a; B, b)$ gets a label from A. Indeed, if θ satisfies this and θ' is an adjacent labeling, differing only on a copy γ of $(A, a; B, b)$, then γ is adjacent to some copy of each $(A_i, a_i; B_i, b_i)$, which are each given a label from A_i by θ—since the A_i produce A, $\theta'(\gamma) \in A$.

Therefore, considering any vertex $(A, a; B, b)$ in which $b \notin A$, we see that ψ is not in the component of φ. (It may be that G is not connected, in which case we take a component containing such a vertex.) This completes the proof. $\square$

It is convenient to collect here the various inequalities we have been able to prove between our parameters.

$$\begin{array}{rcll} h(H) & \leq & w(H) & \text{(trivial)} \\ w(H) & \leq & \chi(H) & \text{for unlooped } H \text{ (Theorem 5.1)} \\ h(H) & \leq & 2m(H) - 2 & \text{(Theorem 7.2)} \\ m(H) & \leq & 2w(H) - 1 & \text{(Theorem 7.3)} \end{array}$$

Some other inequalities can be deduced from these, notably that $m(H) \leq 2\chi(H) - 1$ for unlooped H. Furthermore, if $\chi(H) \leq 3$, then $\chi_c(H) < 4$ and so, as we noted earlier, $m(H) \leq 3$. Indeed, it seems very likely to us that, as has also been suggested by Lovász, $m(H) \leq \chi(H)$ in general. This would be of particular interest as it is a statement referring only to finite boards; another way of expressing it is as follows:

CONJECTURE 7.4. *For every unlooped H with chromatic number d, there is some finite graph G of maximum degree $d-1$ such that* $\mathrm{Hom}(G, H)$ *is disconnected.*

Alternatively, no graph of chromatic number d exhibits greater mobility than K_d. We have shown that the conjecture holds for $d = 2$ and $d = 3$; Lovász [11] has proved the $d = 4$ case. Theorem 7.3 says that for every unlooped H with chromatic number d, there is some finite graph G of maximum degree $2d-2$ such that $\mathrm{Hom}(G, H)$ is disconnected.

8. Loops and Dismantlability

We have seen that d-warmth, d-heat, and d-mobility all match for $d=2$ or 3; in fact a theorem from [3] shows that they match at the other end of the scale as well, that is, at $d = \infty$. We conclude with a short description and proof of this result.

The constraint graph consisting of a single looped node has infinite warmth, heat and mobility, but not every looped constraint graph is so lucky. For example, if H is a path on three nodes with a loop at each end but not in the middle, then $w(H) = 3$; to see that this H is not 4-warm, label T^2 in such a way that every node has children with two different labels.

The difference here is that this last H is not *dismantlable*. The notion of dismantlability goes back twenty years to the study of pursuit games on graphs (see e.g. [13, 14]) and reappeared in [3], where numerous equivalent conditions are given, among them the ones which interest us here. We may define the notion recursively by saying that the graph with one node and a loop is dismantlable, and if H has two distinct nodes a and b with $N(a) \subseteq N(b)$ then $H\backslash\{a\}$ is dismantlable. Note that an unlooped graph cannot be dismantlable. Some dismantlable and non-dismantlable graphs are illustrated in Fig. 5.

In any H, if $N(a) \subseteq N(b)$, the map which sends a to b and every other node of H to itself is a retraction of H onto $H\backslash\{a\}$, which we call a *fold* and denote by ρ_{ab}.

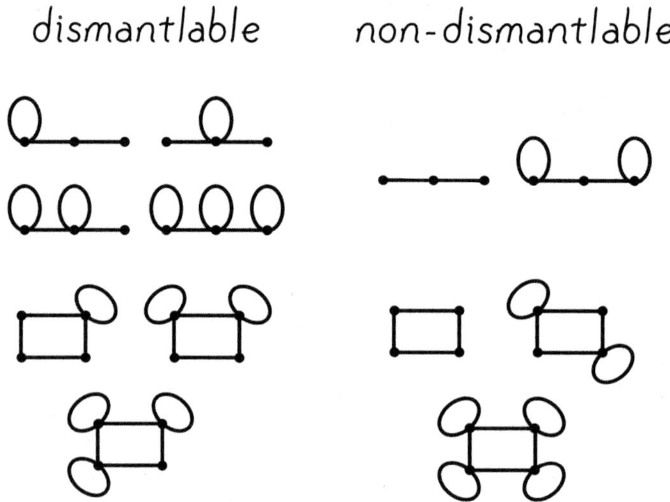

FIGURE 5. Some examples of dismantlable and non-dismantlable graphs

THEOREM 8.1. *If* $\rho_{ab} : H \to H^- = H\backslash\{a\}$ *is a fold then* $w(H) = w(H^-)$, $h(H) = h(H^-)$ *and* $m(H) = m(H^-)$.

PROOF. The first two statements follow if we show that for any board G, long range action for $\mathrm{Hom}(G, H^-)$ is equivalent to long range action for $\mathrm{Hom}(G, H)$. The forward direction is easy because if X, $Y \subset G$ with $d(X, Y) \geq k$ and φ^-, $\psi^- \in \mathrm{Hom}(G, H^-)$, then we can find $\theta \in \mathrm{Hom}(G, H)$ which agrees with $\varphi^- \restriction X$ and $\rho^- \restriction Y$, and $\rho_{ab} \circ \theta$ works similarly in $\mathrm{Hom}(G, H^-)$.

The reverse implication is not much harder. Choose a distance k over which there is no action in $\mathrm{Hom}(G, H)$, and let $d(X, Y) \geq k+2$. We need to expand X and Y slightly to $X' := X \cup N(X)$ and $Y' := Y \cup N(Y)$. Then $d(X', Y') \geq k$ and for any φ and ψ in $\mathrm{Hom}(G, H)$ we can find θ^- which matches $\rho_{ab} \circ \varphi$ on X' and $\rho_{ab} \circ \psi$ on Y'. Now we claim that the map θ which agrees with φ on X, ψ on Y and θ^- everywhere else is a legal H-labeling. Indeed, the only possible problem is for an edge from, say, $x \in X$ to $z \in N(X)\backslash X$ where $\varphi(x) = a$ and $\theta^-(x) = b$. If $\varphi(z) = a$ also then a is looped, thus $a \in N(a) \subseteq N(b)$ and it is legal for θ to label x with a and z with $\rho_{ab}(\varphi(z)) = b$. Otherwise $\theta(z) = \theta^-(z) = \varphi(z) \sim \varphi(x) = a$ and we still have a homomorphism.

For mobility it suffices to show that $\mathrm{Hom}(G, H)$ is connected if and only if $\mathrm{Hom}(G, H^-)$ is; the forward implication is easy because if $\theta_1, \ldots, \theta_k$ is a path in $\mathrm{Hom}(G, H)$ then so is $\rho_{ab} \circ \theta_1, \ldots, \rho_{ab} \circ \theta_k$ after redundant points have been discarded. For the reverse, we first convert φ and ψ to members of $\mathrm{Hom}(G, H^-)$ by changing a-labels to b-labels one by one. Then a path in $\mathrm{Hom}(G, H^-)$ completes the connection. $\square$

Now the definition of dismantlability, and induction on the number of nodes of H, gives the following result.

COROLLARY 8.2. *The following are equivalent for any constraint graph H:*

(i) H is dismantlable;

(ii) $w(H) = \infty$;
(iii) $h(H) = \infty$;
(iv) $m(H) = \infty$.

In the previous section, we raised the issue of whether $m(H) \leq \chi(H)$ for general constraint graphs H. A modest step in this direction is to prove that $m(H) \leq \Delta(H) + 1$, where $\Delta(H)$ is the maximum degree of H. This inequality, unlike that involving the chromatic number, makes sense even in the case when H has loops: we interpret a loop as adding one to the degree of the incident node. However, we have just seen that the inequality fails dramatically if H is dismantlable. This turns out to be the only obstacle.

THEOREM 8.3. *The mobility $m(H)$ of a non-dismantlable graph H is at most* $\Delta(H) + 1$.

PROOF. If x folds to y ($N(x) \subseteq N(y)$), then it is easy to see that $m(H) = m(H - x)$, while $\Delta(H) \geq \Delta(H - x)$. Therefore it is enough to prove the result for *stiff* graphs H, i.e., those admitting no folds.

Given such a graph H, our task is to construct a graph G of maximum degree at most $\Delta(H)$ such that $\mathrm{Hom}(G, H)$ is disconnected.

In [5], Cooper, Dyer and Frieze showed that, if H is not dismantlable but has at least one loop, then indeed $m(H) \leq \Delta(H) + 1$. Their construction of G was to take two node-disjoint copies of H, and replace each loop (in both copies) by an edge joining the two copies of the corresponding node. The natural H-coloring of G is seen to be an isolated point, since H is stiff, and it is not the whole of $\mathrm{Hom}(G, H)$ since it is possible to map all of G to a single looped vertex of H.

If H has no loops, simply taking $G = H$ does not suffice since the identity homomorphism may well be the only element of $\mathrm{Hom}(G, H)$. To get around this problem, we use a method from [3].

We start with the weak square $G' = H \times H$, noting as usual that the two projections are in $\mathrm{Hom}(G', H)$, but are isolated points in this graph. Indeed, it is enough to note that, for every node $(u, v) \in G'$, every neighbor w of u in H and every neighbor z of v, (u, v) is adjacent to *some* node (w, x) ($x \in N(v)$) and some node (x, z) ($x \in N(u)$).

Of course, the degree of G' is too large. To resolve this, we split each node of G' into pieces. Take any node $(u, v) \in H \times H$, set $s = d_H(u)$ and $t = d_H(v)$, and suppose without loss of generality that $s \leq t \leq \Delta(H)$. Denote the neighbors of u as $w_1, \ldots, w_s$, and the neighbors of v as $z_1, \ldots, z_t$. Now we replace the node (u, v) with s nodes, each of degree t. The edges of G' arriving at (u, v) are partitioned among the replacement nodes so that each node has at least one neighbor of the form (w_j, z), for each $j = 1, \ldots, s$, and at least one neighbor of the form (w, z_k), for each $k = 1, \ldots, t$.

Call the new graph, with at most $|V(H)|^2 \Delta(H)$ nodes, G. The two 'projections' are still distinct isolated points of $\mathrm{Hom}(G, H)$, and we are done. $\square$

9. Problems

Many basic questions about warmth, heat and mobility remain open; we list here some of our favorites.

(1) There are many missing bounds. Is there, for example, a function f such that $f(d)$-warm implies d-hot, or $f(d)$-warm implies d-mobile? Or such

that $f(d)$-mobile implies d-hot, or $f(d)$-mobile implies d-colorable? (As far as we know, $f(d) = d+1$ could work for all of these.) Are the bounds in Theorems 7.2 and 7.3 best possible?

(2) We know that girth at least 5 forces warmth at most 3, and it is similarly easy to show that it also forces mobility at most 3. There are examples of triangle-free graphs of warmth 4, but are there triangle-free graphs of arbitrarily large warmth? Heat? Mobility?

(3) Suppose we say that H is *strongly* d-mobile if for any board G, finite or infinite, in $\mathcal{G}_{d-2}$, whenever φ and ψ in $\mathrm{Hom}(G, H)$ differ on only a finite number of sites, there is a finite path in $\mathrm{Hom}(G, H)$ from φ to ψ. This is what we need for the one-site condition to be equivalent to the Gibbs condition for H-labelings of boards in $\mathcal{G}_{d-2}$ (see [2]). Does mobile imply strongly mobile? If not, is there at least a function f such that $f(d)$-hot implies strongly d-mobile?

(4) There are many computational issues concerning warmth, heat and mobility; we guess that the question of whether a graph is d-anything is NP-hard for any fixed $d > 3$. We do not know if the family of 4-warm graphs is even in $\mathrm{NP} \cup \mathrm{co\text{-}NP}$; it would be nice, for example, to have a decent bound on how far action can extend on $\mathrm{Hom}(T^{d-2}, H)$ when H is an n-node, d-warm graph; or, when it is not d-warm, on the size of a minimum $(d-2)$-stable family. Worse, it is not evident that there is *any* finite algorithm for determining whether a graph is 4-hot!

Acknowledgment

The authors benefited greatly from stimulation provided by the Workshop on Colourings and Homomorphisms, in Vancouver BC, August 2000, sponsored by the Pacific Institute for the Mathematical Sciences; and the DIMACS-DIMATIA Workshop on Graphs, Morphisms and Statistical Physics, in Piscataway, NJ, March 2001. We would also like to thank the referee for his or her many useful corrections.

References

[1] R. Burton and J. Steif, Nonuniqueness of measures of maximal entropy for subshifts of finite type, *Ergodic Theory and Dynamical Systems* **14** Part 2 (1994), 213–235.

[2] G.R. Brightwell and P. Winkler, Graph homomorphisms and phase transitions, *J. Comb. Theory (Series B)* **77** (1999), 221–262.

[3] G.R. Brightwell and P. Winkler, Gibbs measures and dismantlable graphs, *J. Comb. Theory (Series B)* **78** (2000), 141–166.

[4] G.R. Brightwell and P. Winkler, Random colorings of a Cayley tree, to appear in *Contemporary Combinatorics* (B. Bollobás ed.), Bolyai Society Mathematical Studies series.

[5] C. Cooper, M. Dyer and A. Frieze, On Markov chains for randomly H-colouring a graph, *Journal of Algorithms* **39** (2001), 117–134.

[6] M. Grötschel, L. Lovász and A. Schrijver, *Geometric Algorithms and Combinatorial Optimization*, Springer-Verlag, New York, 1987.

[7] S.T. Hedetniemi, Homomorphisms and graph automata, U. of Michigan Technical Report 03105-44-T (1966).

[8] M. Jerrum, A very simple algorithm for estimating the number of k-colourings of a low-degree graph, *Random Structures & Algorithms* **7** (1995), 157–165.

[9] D. Karger, R. Motwani and M. Sudan, Approximate graph coloring by semidefinite programming, *J. ACM* **45** (1998), 246–265.

[10] P.J. Kelly and M.L. Weiss, *Geometry and Convexity*, Wiley, New York, 1979.

[11] L. Lovász, private communication, 2001.

[12] J. Nešetřil and M. Rosenfeld, I. Schur, C.E. Shannon and Ramsey numbers, a short story, *Discrete Math.* **229** (2001), 185–195.

[13] R. Nowakowski and P. Winkler, Vertex-to-vertex pursuit in a graph, *Discrete Math.* **43** (1983), 235–239.

[14] A. Quilliot, Homomorphismes, points fixes, rétractions et jeux de pousuite dans les graphes, les ensembles ordonnés et les espaces métriques, Thése d'Etat, Université de Paris VI, Paris, France (1983).

[15] E. Vigoda, Improved bounds for sampling colorings, *40th Annual Symposium on Foundations of Computer Science*, IEEE Computer Society (1999), 51–59.

[16] A. Vince, Star chromatic number, *J. Graph Theory* **12** (1988), 551–559.

[17] X. Zhu, Circular chromatic number: a survey, *Discrete Math.* **229** (2001), 371–410.

DEPARTMENT OF MATHEMATICS, LONDON SCHOOL OF ECONOMICS, HOUGHTON ST., LONDON WC2A 2AE ENGLAND

E-mail address: `g.r.brightwell@lse.ac.uk`

BELL LABS 2C-365, LUCENT TECHNOLOGIES, 700 MOUNTAIN AVE., MURRAY HILL, NJ 07974 U.S.A.

E-mail address: `pw@lucent.com`

DIMACS Series in Discrete Mathematics
and Theoretical Computer Science
Volume **63**, 2004

Random Walks and Graph Homomorphisms

Amir Daneshgar and Hossein Hajiabolhassan

ABSTRACT. In this report (whose basic approach is based on [**12**]) we introduce
a general idea which gives rise to some necessary conditions for the existence
of graph homomorphisms (directed and undirected), which is mainly based
on available comparison techniques for Markov chains. We focus on the finite
strongly–connected case to propose the main ideas, however, there are also a
variety of conceivable extensions to weaker conditions or the infinite case.

In what follows we specially focus on a combination of Diaconis and Saloff–
Coste comparison technique for Markov chains and a generalization of Haemers
interlacing theorem to show one possible approach which results in a general
no–homomorphism theorem; and after that we also consider some special cases
and corollaries which are more appropriate for concrete applications. Specially,
when the range is a Cayley graph, we mention a theorem with a corollary about
the existence of homomorphisms to odd cycles. This, in particular, provides a
proof of the fact that the Coxeter graph is a core.

As some applications, we introduce a necessary condition for the spanning
subgraph problem, which also provides a generalization of a theorem of Mohar
(1992) as a necessary condition for Hamiltonicity.

In the end, we add some concluding notes about the possible extensions and
different aspects of our approach, which show some connections to the theory
of expanders and Ramanujan graphs as well as some applications of algebraic
number theory and group theory to the subject.

1. Introduction

The *homomorphism problem* for graphs, as the question about existence of natural maps between the set of vertices which preserve *adjacency*, is among the most fundamental problems in graph theory, as it is in any other branch of mathematics that considers abstract structures. It is not a surprising fact that the problem is usually very hard to answer [**30**], and this along with the importance of the concept in theoretical graph theory and its profound consequences in applications, motivates a study of *no–homomorphism* theorems through the study of necessary conditions which follow from the existence of such a mapping.

2000 *Mathematics Subject Classification.* Primary 05C15; Secondary 05C50, 60G50.

Key words and phrases. Graph colouring, Graph homomorphism, Graph spectra, Markov chains.

This research is partially supported by the Institute for Studies in Theoretical Physics and Mathematics (IPM).

Although, these conditions, in the current literature of the subject, vary from simple criteria in terms of the chromatic number or the odd girth of the graphs to some interesting theorems such as Albertson–Collins–Bondy–Hell no–homomorphism lemma, it seems that these results do not cover a vast class of graphs and are not flexible enough when needed (see Section 2.1).

The main objective of this report is to introduce a general idea which gives rise to no–homomorphism theorems that can be used for directed, undirected, finite and infinite graphs, by means of considering Markov chains on the base–graphs given as the domain and the range.

In a general sense, consider a parameter ζ_G which is somehow defined for a specific Markov chain on a base–graph G (we usually focus on the natural random walk of the graph as a special important case), and, moreover, assume that this parameter can also be expressed in the space of functions on the vertex set of G. Then if there exists a homomorphism σ from G to H, one may obtain a comparison between parameters ζ_G and ζ_H using the pullback along σ, which gives rise to some necessary condition(s) for the existence of σ.

To be more specific, as a special case, assume the existence of an onto homomorphism σ from G to H, and consider the spectra of the corresponding Markov chains (which, under suitable conditions, contain some information about the rate of convergence of these chains to their stationary distributions when they exist). These, following the standard formulations, can be expressed in $L^2(\pi_G)$ and $L^2(\pi_H)$ by means of Courant–Fischer Min-Max principle, if stationary distributions π_G and π_H exist, respectively. The comparison technique based on this idea was first considered by P. Diaconis and L. Saloff-Coste [**14, 15, 37**].

In what follows, after going through some basic definitions and results from graph theory in Section 2.1, we consider the basic background for natural random walks of graphs in Section 2.2, where we try to develop a unified language for what follows afterwards.

In Section 3.1 we introduce our main no–homomorphism theorem (see Theorem 3) which is based on bounds for two important parameters η^v and θ^e. Section 3.2 is devoted to some methods for estimating these parameters and contains our second main theorem (see Theorem 5), which is based on a technique to cancel out the parameters by changing to some new homomorphism problem. We also consider some applications, where we apply our results to test whether G is a spanning subgraph of H (see Theorem 6). It is interesting to note that this, as a special case, contains a corollary that can be used to prove that a graph or digraph is not Hamiltonian which generalizes a theorem of Mohar [**35, 36**]. Also, we provide a simple proof of the fact that the Coxeter graph is a core.

Section 4 contains our concluding remarks which considers different aspects of the subject and methods introduced in this report along with some new directions to be considered in further investigations.

2. Basic Definitions And Concepts

2.1. Some Preliminaries From Graph Theory. Throughout this report, for the sake of simplicity, we only consider finite graphs, however, the approach is also applicable to the infinite case with some modifications and with different objectives. In this report $\mathbf{N}$ and $\mathbf{R}$ are the sets of *natural* numbers and *real* numbers, respectively. A *directed graph* G (or a *digraph* for short) on n *vertices* with ε *edges* is an structure which is determined by an ordered pair $(V(G), E(G))$ such that $V(G) = \{v_1, ..., v_n\}$ is the set of n vertices and $E(G)$ is a family of ε ordered pairs (u, v) for some elements u and v of $V(G)$ called (directed) edges. We usually use $e = uv$ as an abbreviation for $e = (u, v) \in E(G)$; and we may write $u \to v$ to show that the vertex u is connected to the vertex v through the edge uv. Also, note that throughout this report we threat multiple edges as edges with multiplicities. Moreover, for any two subsets $A \subset V(G)$ and $B \subset V(G)$, the symbol $\overrightarrow{E}(A, B)$ denotes the set of edges from A to B, i.e.

$$\overrightarrow{E}(A, B) = \{uv \in E(G) \mid u \in A \ \& \ v \in B \ \& \ u \to v\}.$$

The symbol $|X|$ is used to show the number of elements of a set X. A digraph G is *finite* if $|V(G)| < \infty$ and $|E(G)| < \infty$.

An edge vv is called a (directed) *loop*; and a digraph is called *symmetric* if either u and v are not connected, or there are unique edges in each direction, i.e. $u \to v$ and $v \to u$. Note that in this case one may replace each pair of directed edges with an undirected edge which shows connection in both directions; and we may write $u \leftrightarrow v$ to show this two–sided connection. For this reason we always identify the concept of a *simple graph* with the concept of a symmetric loopless digraph. Also, We use the word *graph* as an abbreviation for *simple graph*.

For what does not appear in what follows, we assume the standard literature for graphs and digraphs. A *homomorphism* $f : G \longrightarrow H$ from a digraph G to a digraph H is a map $f : V(G) \longrightarrow V(H)$ such that $u \to v$ implies $f(u) \to f(v)$. A *homomorphism* $f : G \longrightarrow H$ of graphs, as well as the concepts of an *isomorphism* and an *automorphism* are defined accordingly. Moreover, $\mathrm{Hom}(G, H)$, $\mathrm{Hom}^v(G, H)$ and $\mathrm{Hom}^e(G, H)$ denote the sets of ordinary, *onto* (vertices) and *onto–edges* homomorphisms from G to H, respectively.

A homomorphism $f : G \longrightarrow H$ of graphs is sometimes called an H–coloring of G. This is mainly due to the special case $H = K_k$ in which one can interpret the situation as a coloring of $V(G)$ with k colors such that $u \to v$ implies that u and v take different colors. For this reason, a K_k–coloring of G is usually called an k–*vertex-coloring* (or a k–coloring) of G. The concepts of a k–*chromatic* digraph (graph) and a k–*critical* digraph (graph) can be defined considering the interpretation through homomorphisms.

It has been shown that the H–coloring problem is NP–complete for any non-bipartite graph H [**30**]. Therefore, it is very interesting to look for *conditions* under which one can guarantee the existence (or nonexistence) of a homomorphism $f : G \longrightarrow H$.

It is quite easy to see that if there exists a homomorphism $f : G \longrightarrow H$ of graphs, then $\chi(G) \leq \chi(H)$. Another easy observation is that $og(G) \geq og(H)$ in which $og(G)$, the *odd girth* of G, is the length of a minimal odd cycle in G ($og(G) = 0$ if there is no odd cycle). These *monotone* parameters give rise to simple *no-homomorphism Theorems* which usually have trivial consequences in graph colorings; however, by imposing some additional constraints on H one can obtain more powerful results as the following theorem of Albertson and Collins shows. Note that in this theorem we have $\mu(G) = |V(G)|/\alpha(G)$.

THEOREM A. [**1**] *If H is a vertex–transitive graph and there exists a homomorphism $f : G \longrightarrow H$ then $\mu(G) \leq \mu(H)$.*

In [7] Bondy and Hell define $\nu(G, K)$ for two graphs G and K as the maximum number of vertices in a subgraph of G that admits a homomorphism to K; and using this they give the following generalization of the above theorem in which $\mu(G, K) = |V(G)|/\nu(G, K)$.

THEOREM B. [7] *Let G, H, K be graphs where H is vertex–transitive. If there exists a homomorphism $f : G \longrightarrow H$ then $\mu(G, K) \leq \mu(H, K)$.*

It is clear that the no–homomorphism lemma of Albertson and Collins is a corollary of this theorem for $K = K_1$. Also, it is easy to see that this theorem has a generalization for digraphs.

2.2. Random Walks On Graphs. In this section we go through some classical known results about the natural random walk on a digraph. Note that all we need in this report about finite Markov chains can be found in [2, 8, 37] where we always confine ourselves to the natural random walk on a digraph. Moreover, note that, with suitable modifications, essentially what follows is also true for any Markov chain on the corresponding base–digraph having a stationary distribution with nonzero entries, however, we focus on the case of the natural random walk of strongly connected digraphs (which are irreducible) to show the main ideas. The natural random walk on a digraph G can be defined through the kernel

$$G(u, v) = \begin{cases} \dfrac{1}{d_G^+(u)} & u \to v \\ 0 & u \not\to v, \end{cases}$$

where one can also think of $G(u, v)$ as the entry of a $|V(G)| \times |V(G)|$ matrix G which is indexed by the row index u and the column index v. In the sequel, we use the same notation G for the digraph and the matrix $G = (G(u, v))_{u, v \in V(G)}$, which is actually the transition matrix of the corresponding Markov chain. Note that the matrix G is not necessarily symmetric even if the corresponding graph is a symmetric digraph (i.e. undirected graph), however, for regular digraphs, and of course for regular (undirected) graphs as special cases, the corresponding matrix is symmetric.

On the other hand, it is a classical observation that this chain is *irreducible* (i.e. for any u, v there exists a power $k = k(u, v)$ such that $G^k(u, v) > 0$) if and only if G is a strongly connected digraph. As a matter of fact, the methods which are used in this report are very much related to evaluation or bounding the rate of convergence of this sequence (e.g. see [8, 14, 15, 16, 17, 19, 20, 37]).

DEFINITION 1. For a random walk G with stationary distribution π_G we define

- $Q_G(u, v) \overset{\text{def}}{=} G(u, v)\pi_G(u)$.
- $\pi_G^M \overset{\text{def}}{=} \max_{u \in V(G)} \pi_G(u)$ and $\pi_G^m \overset{\text{def}}{=} \min_{u \in V(G)} \pi_G(u)$.
- $Q_G^M \overset{\text{def}}{=} \max_{u, v \in V(G)} Q_G(u, v)$ and $Q_G^m \overset{\text{def}}{=} \min\{Q_G(u, v) \mid Q_G(u, v) \neq 0\}$.

Also, G is called *reversible* if

$$(1) \qquad\qquad \forall\, u, v \in V(G) \quad Q_G(u, v) = Q_G(v, u).$$

It is easy to see that if condition (1) is satisfied for some irreducible Markov chain and for some function π_G, then π_G should be the stationary distribution of the chain. $\Diamond$

Now we address ourselves to the relationship between reversibility and symmetry, where we try to introduce the basics of how one may define a suitable *symmetric* (reversible) chain on a base–digraph G, however, before we delve into the details of this, we prefer to add some remarks which may be illuminating in some sense and will distinguish between what follows and what is usually found in the standard literature about the definition of *Laplacian*.

Let G be a weakly–connected digraph (i.e. its symmetric closure is connected as a simple graph), and assume that we have assigned a *conductance* $0 \neq c(uv) \in \mathbf{R}$ to each edge $e = uv$ as a usual setting in the theory of electrical networks. Then it is quite easy and well–known (at least for simple graphs) that if one writes down the Kirchhoff equations and obtains the corresponding Poisson equation,

$$\nabla_G(J) = f,$$

then one obtains the Laplacian as the operator $\nabla_G = I_G - H_G$, with the kernel

$$H_G(v, u) \stackrel{\text{def}}{=} \begin{cases} \dfrac{c(vu)+c(uv)}{c(v)} & v \stackrel{e}{\sim} u \\ \\ 0 & \text{otherwise} \end{cases}$$

where

$$c(v) \stackrel{\text{def}}{=} \sum_{v \stackrel{e}{\sim} u} (c(vu) + c(uv)),$$

and $v \stackrel{e}{\sim} u$ stands for $e = uv$ or $e = vu$. Also, note that this is in coherence with the standard approach through the definition of a natural *boundary operator* (the sum of outputs minus the sum of inputs) in the interpretation of a graph as a chain complex or the geometrical approach in which one considers the adjoint of this boundary operator, the difference operator, as the discretized version of the usual differential.

However, note that in this approach H_G is based on the *local* properties of the base–graph and always represents a *reversible* chain. On the other hand, there is another classical setup in probability theory in terms of the Hilbert space $L^2(\pi_G)$ which, of course, strongly depends on the existence of a stationary distribution with *nonzero* entries π_G. This, which follows in the sequel, is what we have chosen to define a suitable selfadjoint operator as the additive symmetric part of the kernel of the chain, which gives rise to a new definition for the Laplacian as $S_G \stackrel{\text{def}}{=} I_G - \frac{1}{2}(G + G^*)$, where, for instance, G is the natural random walk of the digraph G (by abuse of notation). It is important to note that, although, $S_G = \nabla_G$ for any simple graph G, but these operators may be quite different when G is a *nonsymmetric* digraph, since S_G is essentially defined in terms of some *global* behavior of the chain and heavily depends on the stationary distribution π_G. Hence, in what follows we assume that we are dealing with an irreducible Markov chain on a strongly connected base–graph (which guaranties the existence and uniqueness of the stationary distribution) for the sake of simplicity, although, the whole method also works whenever a stationary distribution with nonzero entries exists.

Now, in order to follow our own notations, let G be an irreducible random walk on $V(G)$ with the stationary distribution π_G, and look at G as a linear operator on $L^2(\pi_G)$ of all real functions on $V(G)$ with the following inner product

$$< f, g >_{\pi_G} \stackrel{\text{def}}{=} \sum_{v \in V(G)} f(v)g(v)\pi_G(v).$$

54 AMIR DANESHGAR AND HOSSEIN HAJIABOLHASSAN

Consequently

$$\|f\|_{\pi_G} = \left(\sum_{v \in V(G)} |f(v)|^2 \pi_G(v) \right)^{\frac{1}{2}},$$

and it is easy to see that the adjoint of G is defined by the kernel

$$G^*(u,v) = \frac{Q_G(v,u)}{\pi_G(u)},$$

where it is also clear that G is a reversible Markov chain if and only if it is selfadjoint in $L^2(\pi_G)$ i.e. $G = G^*$. Let $S_G = I_G - \frac{1}{2}(G + G^*)$ be the additive symmetric part of G where I_G is the identity operator. Then, all eigenvalues of S_G are real and positive with only the least eigenvalue equal to zero (this easily follows from Perron–Frobenius theorem and the fact that S_G is selfadjoint). Hence, if $|V(G)| = n$, then one may order the eigenvalues of S_G as follows

$$0 = \lambda_0^G < \lambda_1^G \leq \lambda_2^G \leq \cdots \leq \lambda_{n-1}^G,$$

which is always assumed throughout this report.

It should be noted that in the classical theory of Markov chains the eigenvalues of S_G are used to estimate the rate of convergence of the chain; while, above all, λ_1^G which is called the *spectral gap* of the chain, is effectively used to obtain upper bounds for the rate of convergence of the chain to its stationary distribution. Note that if G is selfadjoint then λ_1^G is the second smallest eigenvalue of $S_G = I_G - G$. The following theorem is actually invented by P. Diaconis and L. Saloff-Coste to be used in their comparison technique [14, 15, 37] and is one of the major tools in our approach. Although the theorem can be stated for general Hilbert spaces, we just state it in terms of the language we follow in this report.

THEOREM 1. [37] *Let G and H be two strongly connected digraphs with $|V(G)| = n$ and $|V(H)| = m \leq n$ whose random walks have stationary distributions π_G and π_H. Also, let λ_k^G and λ_k^H be the corresponding eigenvalues of S_G and S_H, respectively, and let $T : L^2(\pi_H) \longrightarrow L^2(\pi_G)$ be a linear map. Then, if $\mathcal{E}$ is the Dirichlet form and for some constants $a > 0$, $A < \infty$ and for any $f \in L^2(\pi_H)$ we have*

$$a\|f\|_{\pi_H}^2 \leq \|T(f)\|_{\pi_G}^2 \quad \text{and} \quad \mathcal{E}^G(T(f), T(f)) \leq A\,\mathcal{E}^H(f,f),$$

then $\frac{a}{A}\lambda_k^G \leq \lambda_k^H$ for any $1 \leq k \leq m - 1$.

Of course, it is usually hard to compute the whole spectrum when $|V(G)|$ is large, and one should usually obtain bounds or estimates for the eigenvalues λ_k^G. This is a classical subject in the theory of Markov chains and functional analysis, where we will use some of the techniques in the rest of this report, however, in what follows we first apply some interlacing which has its roots in combinatorics.

To do this, we should need more data which is actually produced as a culmination of ideas from algebraic combinatorics and probability theory, in which our basic objective is to reduce the size of the state space of a random walk G with $|V(G)| = n$ to a new (quotient) random walk G^σ, assuming a graph homomorphism $\sigma \in \mathrm{Hom}^v(G, H)$ with $|V(H)| = m \leq n$, in such a way that $|V(G^\sigma)| = |V(H)| = m$ and we can also say something about the relationship between the corresponding spectra.

One may use a more restricted version of the technique used in Theorem 1 by choosing the linear map to be a unitary operator. This has already been done for symmetric matrices in the celebrated work of Haemers [**24, 25, 26, 27**]. We should note that G^σ can also be considered as what is sometimes called the induced chain in the theory of Markov chains [**2**].

Let $\sigma \in \mathrm{Hom}^{\mathrm{v}}(G, H)$ where G and H are strongly connected digraphs and $n = |V(G)| \geq |V(H)| = m$. Also, assume that the corresponding random walks have stationary distributions π_G and π_H, respectively. For any $x \in V(H)$ define,

$$U_x \stackrel{\mathrm{def}}{=} \sigma^{-1}(x) \quad \text{and} \quad \mathcal{U} \stackrel{\mathrm{def}}{=} \{U_x \mid x \in V(H)\},$$

and note that each U_x is an independent set of vertices. Then we define the probability distribution $\pi_{\mathcal{U}}$ on the set $\mathcal{U}$ as

$$\pi_{\mathcal{U}}(U_x) = \sum_{v \in U_x} \pi_G(v),$$

and it is easy to see that for the induced chain, G^σ, defined through the kernel

$$Q_{G^\sigma}(U_x, U_y) = \sum_{u \in U_x} \sum_{v \in U_y} Q_G(u, v),$$

we have,

THEOREM 2. [**12**] *Let $\sigma \in \mathrm{Hom}^{\mathrm{v}}(G, H)$ where G and H are strongly connected digraphs and $n = |V(G)| \geq |V(H)| = m$. Then,*

$$\forall\, 1 \leq k \leq m - 1 \qquad \lambda_k^G \leq \lambda_k^{G^\sigma} \leq \lambda_{n-m+k}^G.$$

A couple of remarks are at hand. Note that if there exists a partition $\mathcal{U}$ of vertices such that for the corresponding homomorphism $\sigma \in \mathrm{Hom}^{\mathrm{v}}(G, H)$ we have $GA = AG^\sigma$, where columns of A are the characteristic functions of the partition classes, then following the classical literature of graph divisors, one may call $\mathcal{U}$ a *randomized divisor* of G. It is clear that numbers m for which there exists a randomized divisor of size m are very important, since in this case we know that the spectrum of G^σ is a subset of the spectrum of G, which makes the comparison more effective when $|V(H)| = m$. Therefore, the study of randomized divisors and their relationship to the symmetries of G is a very important subject, although, we do not continue this line of thought in this report (also see Section 4).

3. No–homomorphism Theorems

Let $\sigma \in \mathrm{Hom}^{\mathrm{v}}(G, H)$ and assume that the corresponding random walks have stationary distributions π_G and π_H, respectively. What we are going to claim in this section is that there are some relationship between the spectra of S_G and S_H which is actually related to σ. Clearly, this will provide no–homomorphism theorems which can be used later in our special applications.

The main setup is to consider G^σ and then compare it with the natural random walk of H. This will give rise to some necessary conditions in terms of the corresponding spectra (specially the spectral gap). Using these ideas we first focus on general onto homomorphisms and then we consider some special cases under some regularity or symmetry conditions.

3.1. Spectral Conditions For Onto Homomorphisms. In this section we usually assume that $\sigma \in \mathrm{Hom}^{\mathrm{v}}(G, H)$ where G and H are strongly connected digraphs and $n = |V(G)| \geq |V(H)| = m$; and we assume that the corresponding random walks have stationary distributions π_G and π_H, respectively. Then, one may apply comparison methods to compare the spectrum of G^{σ} with that of H. The comparison methods we use are mainly developed by P. Diaconis and L. Saloff-Coste, however, we consider the whole spectrum to get more information [**14, 15, 19, 37**]. We begin by a couple of definitions.

DEFINITION 2. If $\sigma \in \mathrm{Hom}^{\mathrm{v}}(G, H)$ we define,

- $\mathcal{M}_{\sigma} \stackrel{\text{def}}{=} \min_{x,y \in V(H)} \{|\vec{E}(\sigma^{-1}(x), \sigma^{-1}(y))| \mid \vec{E}(\sigma^{-1}(x), \sigma^{-1}(y)) \neq \emptyset\}.$

- $\mathcal{M}^{\sigma} \stackrel{\text{def}}{=} \max_{x,y \in V(H)} |\vec{E}(\sigma^{-1}(x), \sigma^{-1}(y))|.$

- $\mathcal{S}_{\sigma} \stackrel{\text{def}}{=} \min_{x \in V(H)} |\sigma^{-1}(x)| \quad \text{and} \quad \mathcal{S}^{\sigma} \stackrel{\text{def}}{=} \max_{x \in V(H)} |\sigma^{-1}(x)|.$

- $\eta_{\sigma} \stackrel{\text{def}}{=} \dfrac{\mathcal{M}^{\sigma}}{\mathcal{S}_{\sigma}} \quad \text{and} \quad \theta^{\sigma} \stackrel{\text{def}}{=} \dfrac{\mathcal{M}_{\sigma}}{\mathcal{S}^{\sigma}}.$

- $\eta^{\mathrm{v}}(G, H) \stackrel{\text{def}}{=} \begin{cases} \inf\{\eta_{\sigma} \mid \sigma \in \mathrm{Hom}^{\mathrm{v}}(G, H)\} & \mathrm{Hom}^{\mathrm{v}}(G, H) \neq \emptyset \\ -\infty & \mathrm{Hom}^{\mathrm{v}}(G, H) = \emptyset \end{cases}$

- $\theta^{\mathrm{v}}(G, H) \stackrel{\text{def}}{=} \begin{cases} \sup\{\theta^{\sigma} \mid \sigma \in \mathrm{Hom}^{\mathrm{v}}(G, H)\} & \mathrm{Hom}^{\mathrm{v}}(G, H) \neq \emptyset \\ +\infty & \mathrm{Hom}^{\mathrm{v}}(G, H) = \emptyset \end{cases}$

Also, $\eta^{\mathrm{e}}(G, H)$ and $\theta^{\mathrm{e}}(G, H)$ are define in terms of $\mathrm{Hom}^{\mathrm{e}}(G, H)$, similarly. $\Diamond$

We are ready to state our main no–homomorphism theorem which is mainly based on Theorems 1 and 2.

THEOREM 3. [**12**] *Let G and H be strongly connected digraphs with $n = |V(G)| \geq |V(H)| = m$. Then*

a) *If $\mathrm{Hom}^{\mathrm{v}}(G, H) \neq \emptyset$, then for all $1 \leq k \leq m - 1$,*

$$\lambda_k^G \leq \eta^{\mathrm{v}}(G, H) \frac{Q_G^M \pi_H^M}{Q_H^m \pi_G^m} \lambda_k^H.$$

b) *If $\mathrm{Hom}^{\mathrm{e}}(G, H) \neq \emptyset$, then for all $1 \leq k \leq m - 1$,*

$$\lambda_{n-m+k}^G \geq \theta^{\mathrm{e}}(G, H) \frac{Q_G^m \pi_H^m}{Q_H^M \pi_G^M} \lambda_k^H.$$

SKETCH OF PROOF . First assume that $\sigma \in \mathrm{Hom}^{\mathrm{v}}(G, H)$ and consider the identity map from G^{σ} to H. Note that since σ is a graph homomorphism, for all $x \neq y$

$$Q_{G^{\sigma}}(U_x, U_y) \neq 0 \quad \Rightarrow \quad Q_H(x, y) \neq 0.$$

Also, we have

$$\mathcal{E}^{G^{\sigma}}(f, f) = \frac{1}{2} \sum_{x,y \in V(H)} |f(y) - f(x)|^2 Q_{G^{\sigma}}(U_x, U_y)$$

and

$$\mathcal{E}^{H}(f, f) = \frac{1}{2} \sum_{x,y \in V(H)} |f(y) - f(x)|^2 Q_H(x, y).$$

But, since

$$Q_{G^{\sigma}}(U_x, U_y) \leq \mathcal{M}^{\sigma} Q_G^M,$$

we may write

$$\mathcal{E}^{G^{\sigma}}(f,f) \leq \mathcal{M}^{\sigma}\, \frac{Q^{M}_{G}}{Q^{m}_{H}}\, \mathcal{E}^{H}(f,f).$$

On the other hand, similarly, we have

$$\|f\|^{2}_{\pi_{G^{\sigma}}} \geq \mathcal{S}_{\sigma}\, \frac{\pi^{m}_{G}}{\pi^{M}_{H}}\, \|f\|^{2}_{\pi_{H}}\,;$$

and consequently, by Theorem 1 for $1 \leq k < m$,

$$\lambda^{G^{\sigma}}_{k} \leq \eta_{\sigma}\, \frac{Q^{M}_{G}\,\pi^{M}_{H}}{Q^{m}_{H}\,\pi^{m}_{G}}\, \lambda^{H}_{k}.$$

Conversely, if $\sigma \in \mathrm{Hom}^{e}(G,H)$, also for all $x \neq y$ we have,

$$Q_{H}(x,y) \neq 0 \quad \Rightarrow \quad Q_{G^{\sigma}}(U_{x},U_{y}) \neq 0.$$

Hence, using the same method we get

$$\lambda^{G^{\sigma}}_{k} \geq \theta^{\sigma}\, \frac{Q^{m}_{G}\,\pi^{m}_{H}}{Q^{M}_{H}\,\pi^{M}_{G}}\, \lambda^{H}_{k}.$$

The rest of the proof is an application of Theorem 2. $\qquad\square$

It should be noted that the full power of Theorem 2 appears in the second part, since the first part may also be obtained by a direct comparison of G and H through the pullback along σ.

Also, considering the case of regular undirected graphs is instructive. Note that in this case for an s–regular graph, $s\lambda^{G}_{k}$ is the kth eigenvalue of the Laplacian. Therefore, from this point of view S_{G} can be thought of as a generalized (normalized) version of the combinatorial Laplacian of G.

On the other hand, one should note that, since it is not always easy to compute the whole spectrum, it is usual to focus on the spectral gap, λ_{1}, or some other parameters which are mainly related to the rate of convergence of the chain, like the log–Sobolev or the Cheeger constants. It is interesting to note that one may also use the techniques of the previous sections to obtains some partial results concerning these parameters too. We also note that the role of these parameters can be quite different from the spectral gap, which, in a way, justifies their consideration as a separate subject (for more on these subjects see [**16, 17, 20, 22, 37**]).

For instance, we may consider one special type of the Cheeger constant as follows.

DEFINITION 3. Let G be a strongly connected digraph whose natural random walk has the stationary distribution π_{G}. Then

- For any edge $e = uv$ we define

$$Q_{G}(uv) \stackrel{\text{def}}{=} \frac{1}{2}(Q_{G}(u,v) + Q_{G}(v,u)) \quad \text{and} \quad df(uv) \stackrel{\text{def}}{=} f(v) - f(u).$$

- $\iota_{G} \stackrel{\text{def}}{=} \min\left\{ \frac{1}{\pi_{G}(A)} \sum_{u \in A, v \in A^{c}} Q_{G}(uv) \;\middle|\; A \subset V(G) \;\&\; \pi_{G}(A) \leq \frac{1}{2} \right\}.$

$\Diamond$

Then one may prove the following by the same approach.

THEOREM 4. [**12**] *Let G and H be strongly connected digraphs with $n = |V(G)| \geq |V(H)| = m$. Then if $\mathrm{Hom}^{\mathrm{v}}(G, H) \neq \emptyset$ we have,*

$$\iota_G \leq \eta^{v}(G, H) \, \frac{Q_G^M \pi_H^M}{Q_H^m \pi_G^m} \, \iota_H.$$

3.2. Estimations And Some Finer Tools. In this section we focus on some basic methods that can be applied to obtain some information about the parameters $\eta^{v}(G, H)$ and $\theta^{e}(G, H)$ which are needed in Theorem 3. Also, we consider methods that can be used to get rid of some of these parameters under special conditions. Needless to say, methods considered in this section are just some examples of a variety of approaches that can be chosen to apply Theorem 3 (also see Section 4). Note that as first estimates, we have the following rough bounds,

$$\eta^{v}(G, H) \leq \beta_2(G) \quad \text{and} \quad \alpha(G)^{-1} \leq \theta^{e}(G, H),$$

where $\alpha(G)$ is the *independence number* of G and $\beta_2(G)$ is the largest integer ε, such that there exists a subgraph H of G with $|E(H)| = \varepsilon$ and $\mathrm{Hom}^{\mathrm{v}}(H, K_2) \neq \emptyset$. Of course, there are a lot of known results and methods to obtain estimates for $\alpha(G)$ and $\beta_2(G)$ (specially for undirected graphs), however, getting better bounds asks for some more information about G. As an important example we mention the following simple lemma, which also shows the vast range of the cases Theorem 3 can be applied.

LEMMA 1. [**12**] *Let G and H be two digraphs, then*

a) *If H is vertex–critical with $\chi(H) = \chi(G)$, then $\mathrm{Hom}(G, H) = \mathrm{Hom}^{\mathrm{v}}(G, H)$ and $\mathcal{S}_\sigma$ is greater than or equal to the minimum number of vertices which is needed to reduce the chromatic number of G.*

b) *If H is edge–critical with $\chi(H) = \chi(G)$, then $\mathrm{Hom}(G, H) = \mathrm{Hom}^{\mathrm{e}}(G, H)$ and $\mathcal{M}_\sigma$ is greater than or equal to the minimum number of edges which is needed to reduce the chromatic number of G.*

PROOF . Note that deletion of any vertex of H in case (a) or any edge of H in case (b) strictly decreases the chromatic number, and hence the lemma follows.   □

In the next example we consider the set $\mathrm{Hom}(P, C_5)$, of all homomorphisms from the Petersen graph P to the 5–cycle C_5. Although, it will easily follow from our next results (see Corollary 1 and comments preceding Example 3) that $\mathrm{Hom}(P, C_5) = \emptyset$, but we follow the details of the next example to show how one may estimate the necessary parameters.

EXAMPLE 1. Let us consider a homomorphism $\sigma \in \mathrm{Hom}(P, C_5)$. First, note that since P contains two disjoint copies of C_5, we can deduce that $\mathcal{S}^\sigma = 2$ (note that this also follows from the fact that P is vertex–transitive and that C_5 is a core). Also, since C_5 is critical, the homomorphism should be in $\mathrm{Hom}^{\mathrm{e}}(P, C_5)$, and by Lemma 1($b$) we have $\mathcal{M}_\sigma \geq 3$. Hence, applying Theorem 3(b) for $k = 3$ shows that we should have

$$1.5 \leq \frac{\mathcal{M}_\sigma}{\mathcal{S}^\sigma} \leq \theta^{e}(G, H) \leq 1.38,$$

which is a contradiction. Hence, $\mathrm{Hom}(P, C_5) = \mathrm{Hom}^{\mathrm{e}}(P, C_5) = \emptyset$. (Note that Theorem 3($a$) is not applicable in this case.)   ♡

As it is clear, one of the main problems in applying Theorem 3 is to obtain good estimates for parameters η^v and θ^e. The following theorem presents one of the techniques which may be used to simplify this by excluding $\mathcal{S}_\sigma$ and $\mathcal{S}^\sigma$ from computations (also see Section 4).

THEOREM 5. [**12**] *Let G and H be strongly connected digraphs with $n = |V(G)| \geq |V(H)| = m$, where $H = \mathrm{Cay}(V(H), S)$ is also a Cayley graph in which S is closed under conjugation. Then if $\sigma \in \mathrm{Hom}^e(G, H)$, for all $1 \leq k \leq m - 1$ we have,*

$$\lambda^{G \square H}_{m(n-1)+k} \geq \left(1 + \frac{m\mathcal{M}_\sigma}{n}\right) \frac{Q^m_{G \square H} \pi^m_H}{Q^M_H \pi^M_{G \square H}} \lambda^H_k,$$

where $G \square H$ is the cartesian product of the two graphs G and H.

SKETCH OF PROOF . Let $V(G) = \{v_1, \ldots, v_n\}$, $V(H) = \{x_1, \ldots, x_m\}$, and consider a homomorphism $\sigma \in \mathrm{Hom}^e(G, H)$. Then define a map $\tilde{\sigma} : V(G \square H) \longrightarrow V(H)$ as follows,

$$\tilde{\sigma}((v_i, x_j)) = \sigma(v_i)x_j \quad i = 1, \ldots, n \ \ j = 1, \ldots, m,$$

in which we have used the group operation at the right-hand side.
It can be verified that $\tilde{\sigma}$ is a homomorphism, $\mathcal{S}^{\tilde{\sigma}} = \mathcal{S}_{\tilde{\sigma}} = n$ and $\mathcal{M}_{\tilde{\sigma}} \geq n + m\mathcal{M}_\sigma$. The rest of proof follows from Theorem 3. $\square$

Note that the results in [**32**] show that the property of being a Cayley graph for H in Theorem 5 is somehow the best possible. Also, it is possible to sharpen this result when H is a (directed) cycle as follows, in which there is no need to have an estimate for $\mathcal{M}_\sigma$.

COROLLARY 1. [**12**] *Let G be a strongly connected digraph with $n = |V(G)|$ and $\varepsilon = |E(G)|$, and let $H \in \{C_m, \vec{C}_m\}$ with $m \leq n$. Then if $\sigma \in \mathrm{Hom}^e(G, H)$, for all $1 \leq k \leq m - 1$ we have,*

$$\lambda^{G \square H}_{m(n-1)+k} \geq \left(1 + \frac{\varepsilon}{n}\right) \frac{Q^m_{G \square H} \pi^m_H}{Q^M_H \pi^M_{G \square H}} \lambda^H_k.$$

It is instructive to note that, as we mentioned before, the whole setup works whenever a stationary distribution with nonzero entries exists. This in a variety of cases (such as what appears in Theorem 5) will provide nice generalizations or a large degree of freedom since one may even use or build *disconnected* graphs and apply the method. However, we do not follow this approach and its consequences in this report for space limitations.
To mention some applications, we consider an important case in which G is a spanning subgraph of H. Note that in this case considering σ as the identity homomorphism we have $\mathcal{S}_\sigma = \mathcal{M}^\sigma = 1$. Hence, we have the following as a corollary of Theorem 3, which is a generalization of a theorem of Mohar [**35, 36**].

THEOREM 6. [**12**] *Let G and H be strongly connected digraphs with $|V(G)| = |V(H)| = n$. If there exists $1 \leq k < n$ such that*

$$\lambda^G_k > \frac{Q^M_G \pi^M_H}{Q^m_H \pi^m_G} \lambda^H_k,$$

then G is not a spanning subgraph of H.

PROOF . On the contrary, assume that G is a spanning subgraph of H and consider the identity map σ from $V(G)$ to $V(H)$. This map is trivially an onto graph homomorphism i.e. $\sigma \in \mathrm{Hom}^{\mathrm{v}}(G, H)$. Also, it is clear that $\eta_\sigma = \mathcal{M}^\sigma = \mathcal{S}_\sigma = 1$. Hence, by Theorem 3($a$), for all $1 \le k < n$ we should have

$$\lambda_k^G \le \eta^v(G, H)\, \frac{Q_G^M \pi_H^M}{Q_H^m \pi_G^m}\, \lambda_k^H \le \frac{Q_G^M \pi_H^M}{Q_H^m \pi_G^m}\, \lambda_k^H,$$

which is a contradiction. $\qquad\square$

EXAMPLE 2. To show that the Petersen graph is not Hamiltonian, we should show that $\mathrm{Hom}^{\mathrm{v}}(C_{10}, P) = \emptyset$, which follows from Theorem 6 for λ_5^P. $\qquad\heartsuit$

Also, one may consider the case of vertex–transitive cores. Let G be a vertex–transitive simple undirected graph on n vertices, and let

$$\mathcal{O}(n) \stackrel{\mathrm{def}}{=} \{m \in \mathbf{N} \mid m \text{ is odd and divides } n\}.$$

First, we note that if G is 3–regular, and there is no homomorphism from G to the cycle C_m for all $m \in \mathcal{O}(n)$, then G is a core. This can be effectively verified using Theorems 3 and 5.

EXAMPLE 3. By applying Corollary 1 to $\mathrm{Hom}(G, C_7)$, where G is the Coxeter graph and $k = 6$, we get

$$1.843 \ge (1 + \frac{3}{2}) \times \frac{2}{5} \times 1.9 = 1.9,$$

which shows that $\mathrm{Hom}(G, C_7) = \emptyset$. This proves that the Coxeter graph is a core. It is also interesting to note that the only other simple proof of this, that the authors are aware of, uses the fact that any connected 2–arc transitive non–bipartite graph is a core [23, 28]. $\qquad\heartsuit$

4. Concluding Remarks

In this section we add some notes and we point out some important aspects which can be considered in forthcoming research.

The concept of a randomized divisor of a graph G, as a homomorphic image whose spectrum is a subset of the spectrum of G is an important concept and can be studied separately. Note that these homomorphic images may also have loops.

Considering loops and multiple–edges suggests the study of *perturbations*, which we believe is an important and interesting subject. It should be noted that not only there are cases in which perturbation by one edge show unpredictable behaviour in the spectrum, but also this may completely change the type of the Markov chain, e.g. from *transient* to *recurrent* (also see [11]).

Also, it is evident from our approach that any kind of information about the spectrum of the natural random walk of a graph has important consequences on what we may deduce about its homomorphic images or the graphs that have homomorphisms to it. This, in a special case, may draw ones attention to *expanders* and *Ramanujan graphs* as an important class of random–like graphs with a relatively large spectral gap. Even it is quite interesting to consider families of s–regular graphs for which the spectral gap (in the classical sense) is greater than $s - 2\sqrt{s-1} + \epsilon$ for some $\epsilon > 0$ (note that by Alon–Boppana theorem there are only a finite number of such graphs for a fixed s (see [3, 9, 33, 34] and references therein)).

On the other hand, as we know that our necessary conditions depend on the whole spectrum, we may deduce that there are still some graphs which are interesting, however, they may not have a large spectral gap. This observation suggests the study of graphs whose spectrum is *almost uniformly* distributed in $[-1, 1]$. Needless to say, one needs a good formulation of this concept to be able to find some good examples and results.

Considering another aspects of the subject which can be approached from both probabilistic and graph theoretical point of view, one may refer to the Kneser graphs and their relation to some probabilistic phenomena as Bernoulli–Laplace diffusion process whose spectrum are studied using Fourier analysis and group representations. This along with some other important cases such as distance regular graphs and association schemes provide a very interesting arena of research which has a variety of aspects in combinatorics, group theory and probability [**2, 5, 6, 13, 19**]. Moreover, it is interesting to note that the subject of *electrical networks* and *potential theory* on graphs introduces another important part of the theory which have many fruitful aspects, although we have not considered them in this report. On one hand, the *commuting time* is one more parameter which is computable both using *effective conductance* and the corresponding Dirichlet forms, which naturally provides some information on some of our parameters as $\mathcal{S}_\sigma$ in the reversible case. On the other hand, this approach through potential theory is the main tool in the case of *infinite* graphs. We just mention that one could also consider the whole subject from this point of view and note some special kind of homomorphisms which are more related to the network through Rayleigh's principle (e.g. see Theorem 2.17 and Corollary 4.76 of [**38**] and also see [**39**]).

Considering some graph theoretical aspects of our investigation, there are still very interesting approaches which are natural sequels to this report. First, note that the study of parameters η^v and θ^e are of great importance, and consequently, finding effective methods to estimate these parameters or applying techniques (as in Theorem 5) which may help to simplify things in special cases are among the main problems. In particular, we believe it is very interesting to prove that the Coxeter graph is not Hamiltonian by this approach.

Also, it is interesting to note that when X is a union of conjugacy classes in the Cayley graph $\text{Cay}(\mathsf{G}, X)$, then it is possible to compute the spectrum of the corresponding random walk from the character table of G [**18, 33**], which introduces a vast class of graphs that can be analysed by the methods related to the subject of this report.

Acknowledgment

Both authors wish to thank the anonymous referees and M. Shahshahani for their comments. Also, they are very grateful for the financial support of the Institute for Studies in Theoretical Physics and Mathematics (IPM).

References

[1] M. O. ALBERTSON AND K. L. COLLINS, *Homomorphisms of 3-chromatic graphs*, Disc. Math., **54** (1985), 127–132.

[2] D. ALDOUS AND J. FILL, *Preliminary version of a book on finite Markov chains*, 1996. http://www.stat.berkeley.edu/users/aldous.

[3] N. ALON, *Eigenvalues and expanders*, Combinatorica, **6** (1986), 83–96.

[4] L. BABAI, *Automorphism groups, isomorphism, reconstruction*, in Handbook of Combinatorics I, R. L. Graham, M. Grötschel, and L. Lovaśz, eds., North–Holland, Amsterdam, 1995, ch. 27, 1447–1540.

[5] E. BELSLEY, *Rates of convergence of Markov chains related to Association schemes*, PhD thesis, Dept. Math., Harvard University, 1993.

[6] ——, *Rates of convergence of random walk on distance regular graphs*, Probab. Theory Relat. Fields, **112** (1998), 493–533.

[7] J. A. BONDY AND P. HELL, *A note on the star chromatic number*, J. Graph Theory, **14** (1990), 479–482.

[8] P. BRÉMAUD, *Markov Chains, Gibbs fields, Monte Carlo simulation and queues*, vol. 31 of TAM, Springer–Verlag, New York, 1999.

[9] M. W. BUCK, *Expanders and diffusers*, SIAM J. Alg. Disc. Math., **7** (1986), 282–304.

[10] D. M. CVETKOVIĆ, M. DOOB, AND H. SACHS, *Spectra of graphs*, Academic Press, New York, 1980.

[11] D. M. CVETKOVIĆ, P. ROWLINSON, AND S. SIMIĆ, *Eigenspaces of graphs*, Cambridge Univ. Press, Cambridge, U.K., 1997.

[12] A. DANESHGAR AND H. HAJIABOLHASSAN, *Graph homomorphisms through random walks*, J. Graph Theory **44** (2003), no. 1, 15–38.

[13] P. DIACONIS, *Group representations in probability and statistics*, vol. 11 of Lecture Notes–Monograph Series, Institute of Mathematicsl Statistics, Hayward, California, 1988.

[14] P. DIACONIS AND L. SALOFF-COSTE, *Comparison techniques for random walk on finite groups*, Ann. Probab., **21** (1993), 2131–2156.

[15] ——, *Comparison theorems for reversible Markov chains*, Ann. Appl. Probab., **3** (1993), 696–730.

[16] ——, *Logarithmic sobolev inequalities for finite Markov chains*, Ann. Appl. Probab., **6** (1996), 695–750.

[17] ——, *Nash inequalities for finite Markov chains*, J. Theoret. Probab., **9** (1996), 495–510.

[18] P. DIACONIS AND M. SHAHSHAHANI, *Generating a random permutation with random transpositions*, Z. Wahrscheinlichkeitstheorie verw. Gebriete, **57** (1981), 159–179.

[19] ——, *Time to reach stationarity in the Bernoulli–Laplace diffusion model*, SIAM J. Math. Analysis, **18** (1987), 208–218.

[20] P. DIACONIS AND D. STROOCK, *Geometric bounds for eigenvalues of Markov chains*, Ann. Appl. Probab., **1** (1991), 36–61.

[21] W. D. FELLNER, *On minimal graphs*, Theoret. Comput. Sci., **17** (1982), 103–110.

[22] J. FULMAN AND E. L. WILMER, *Comparing eigenvalue bounds for Markov chains: when does Poincaré beat Cheeger?*, Ann. Appl. Probab., **9** (1999), 1–13.

[23] C. D. GODSIL AND G. ROYLE, *Algebraic Graph Theory*. (Preprint), http://bilby.uwaterloo.ca/books.html.

[24] ——, *Algebraic Combinatorics*, Chapman and Hall, New York, 1993.

[25] W. HAEMERS, *Eigenvalue methods in packing and covering in combinatorics*, no. 106 in Math. Centre Tract, Mathematisch Centrum, Amsterdam, 1979, 15–38.

[26] ——, *Eigenvalue techniques in design and graph theory*, PhD thesis, Eindhoven University of Technology, 1979.

[27] ——, *Interlacing eigenvalues and graphs*, Linear Algebra Appl., **227/228** (1995), 593–616.

[28] G. HAHN AND C. TARDIF, *Graph homomorphisms: structure and symmetry*, in Graph Symmetry, G. Hahn and G. Sabidussi, eds., no. 497 in NATO Adv. Sci. Inst. Ser. C Math. Phys. Sci., Kluwer, Dordrecht, 1997, 107–167.

[29] P. HELL AND J. NEŠETŘIL, *Cohomomorphisms of graphs and hypergraphs*, Math. Nachr., **87** (1979), 53–61.

[30] ——, *On the complexity of H–coloring*, J. Combin. Theory (B), **48** (1990), 92–110.

[31] R. HORN AND C. JOHNSON, *Matrix Analysis*, Cambridge Univ. Press, Cambridge, U.K., 1985.

[32] B. LAROSE, F. LAVIOLETTE, AND C. TARDIF, *On normal Cayley graphs and hom–idempotent graphs*, European J. Combin., **19** (1998), 867–881.

[33] A. LUBOTZKY, *Discrete groups, expanding graphs and invariant measures*, vol. 125 of Progress in Mathematics, Birkhäuser Verlag, Basel;Boston;Berlin, 1994. Appendix by J. D. Rogawski.

[34] A. LUBOTZKY, R. PHILLIPS, AND P. SARNAK, *Ramanujan graphs*, Combinatorica, **8** (1988), 261–277.

[35] B. MOHAR, *A domain monotonicity theorem for graphs and Hamiltonicity*, Disc. Appl. Math., **36** (1992), 169–177.

[36] ——, *Laplace eigenvalues of graphs – a survey*, Disc. Appl. Math., **109** (1992), 171–183.

[37] L. SALOFF-COSTE, *Lectures on finite Markov chains*, in Lectures on probability theory and statistics X (1996), P. Bernard, ed., no. 1665 in Lecture Notes In Mathematics, Springer–Verlag, New York, 1997, ch. 3, 304–413.

[38] P. M. SOARDI, *Potential theory on infinite networks*, vol. 1590 of Lecture Notes in Mathematics, Springer–Verlag, Berlin, 1994.

[39] W. WOESS, *Random walks on infinite graphs and groups*, vol. 138 of Cambridge Tracts In Mathematics, Cambridge University Press, Edinburgh, UK, 2000.

DEPARTMENT OF MATHEMATICAL SCIENCES, SHARIF UNIVERSITY OF TECHNOLOGY, P.O. BOX 11365–9415, TEHRAN, IRAN

E-mail address: `daneshgar@sharif.ac.ir`

INSTITUTE FOR STUDIES IN THEORETICAL, PHYSICS AND MATHEMATICS (IPM), P.O. BOX 19395–5746, TEHRAN, IRAN

E-mail address: `hhaji@sbu.ir`

DIMACS Series in Discrete Mathematics
and Theoretical Computer Science
Volume **63**, 2004

Recent results on Parameterized H-Coloring

Josep Díaz, Maria Serna, and Dimitrios M. Thilikos

ABSTRACT. We survey recent results on the complexity of several versions of
the H-coloring and the list-H-coloring problems that are amenable to parameterization.

1. Introduction

The notion of homomorphism between graphs is a natural algebraic characteristic which has been used to study the structural properties of several combinatorial problems on graphs. Recall that given two graphs $G = (V(G), E(G))$ and
$G' = (V(G'), E(G'))$ an *homomorphism* of G into G' is a map $\theta : V(G) \rightarrow V(G')$
with the property that $\{v, w\} \in E(G) \Rightarrow \{\theta(v), \theta(w)\} \in E(G')$. For a fixed graph
H, possibly with loops but without multiple edges, we say that a graph G has an
H-coloring if there exists a homomorphism θ from G to H. In Figure 1 we have an
example of an H-coloring of a graph. Given any graph G as input, the *H-coloring
problem* asks whether there exists an H-coloring of G. Notice that in the particular
case when H is the complete graph K_c, the H-coloring problem is the problem of
deciding if G is c-colorable. The complexity of the H-coloring problem is well known
(see [**42**] for a survey). Hell and Nešetřil [**31**] proved a dichotomy theorem stating
that if H is bipartite or it has a loop, then the H-coloring problem can be trivially
solved in polynomial time, otherwise the H-coloring problem is NP-complete. In
the particular case where G has bounded degree such a dichotomy result seems
difficult [**28**].

An interesting extension of the H-coloring problem is *list H-coloring*. Given a
graph G and, for every $v \in V(G)$, a set $L(v) \subseteq V(H)$, the *list H-coloring* problem
asks whether there is an H-coloring χ of G such that for every $u \in V(G)$ we have
$\chi(u) \in L(u)$. Notice that when every list $L(u) = V(H)$, the list H-coloring is the
H-coloring problem. Moreover, the list K_c-coloring is the *list-coloring problem* (see
for [**18, 38**]). From the previous remarks, we know that if the H-coloring problem
is NP-complete, the list H-coloring problem is also NP-complete. However, the
dichotomy for the list H-coloring problem is different and harder to achieve. The

2000 *Mathematics Subject Classification.* Primary 05C15; Secondary 68W99.

Key words and phrases. H-coloring, parameterized complexity, FPT, counting, bounded
trcewidth.

The work of all the authors was supported by the Future Emerging Technologies Program of
the EU under contract number IST-99-14186 (ALCOM-FT) and by the spanish CYCIT projects
TIC-2000-1970-CE and TIC1999-0754-C03 (MALLBA). The work of the third author was also
supported by the Ministry of Education and Culture of Spain (Resolución 31/7/00 - BOE 16/8/00).

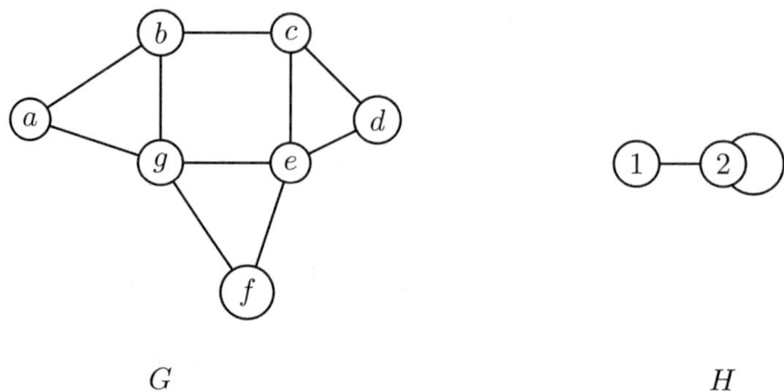

FIGURE 1. An example of an H-coloring of G. The homomorphism is given by: $\theta(a) = \theta(d) = \theta(f) = 1$ and $\theta(b) = \theta(c) = \theta(e) = \theta(g) = 2$

first result showed that when H is a *reflexive* (every vertex is looped) interval graph, the list H-coloring problem can be solved in polynomial time, otherwise the problem is NP-complete [**25**]. The second result states that if H is *irreflexive* (no vertex is looped) and the complement of H is a circular arc graph of clique covering number two, the H-coloring problem can be solved in polynomial time, otherwise the problem is NP-complete [**26**]. The third and last result establishes the dichotomy: when H is a *general* (with looped and unlooped vertices) bi-arc graph, the list H-coloring problem can be solved in polynomial time, otherwise the problem is NP-complete [**27**].

An interesting issue is *counting* the number of solutions to the above mentioned problems. Most counting versions of problems, whose decision versions are NP-complete, are known to be #P-complete. Moreover, the search for efficient approximations to counting problems, known to be #P-complete, has become one of the exciting areas of research in the last decade [**35**]. It should be noted that many counting problems on a graph G can be restated as counting the number of homomorphisms from G to a particular fixed graph H. For instance, the classical problem of counting the number of proper c-colorings of a given graph G is equivalent to counting the number of K_c-colorings of G. As a second example, the problem of counting the number of independent sets corresponds to the problem of counting the number of H-colorings, where H consists of a single edge with a loop in one of its two vertices (see Figure 1).

Given an input graph G, the *#H-coloring problem* is the problem of evaluating the number of H-colorings in G. In the same manner, given the input $(G, \{L(v)\})$, the *list #H-coloring problem* is the problem of evaluating the number of list H-colorings of $(G, \{L(v)\})$.

The following dichotomy result is due to Dyer and Greenhill: the #H-coloring problem is #P-complete if H has a connected component which is not a complete reflexive graph or a complete irreflexive bipartite graph, otherwise it is in P. This result holds even in the case when G has degree bounded by a suitable large constant [**23**].

Regarding the complexity of the problem of counting list H-colorings, it is not difficult to prove that the number of list H-colorings can be computed in polynomial time whenever H is a reflexive complete graph or an irreflexive complete bipartite graph. This observation together with the #P-completenes result for the #H-coloring problem implies the following dichotomy result [16, 32]: the list #H-coloring problem is #P-complete if H has a connected component which is not a complete reflexive graph or a complete irreflexive bipartite graph, otherwise the counting problem is in P.

The aim of this paper is to survey the use of ideas from parameterized complexity, to further explore the complexity of the H-coloring and the list H-coloring problems. In defining a parameterization, the issue is not whether a problem is hard, but what makes the problem hard or easy to compute. To study the *structural hardness* of a difficult problem, the approach is to split the input into two parts: a *difficult* part (the non-parameterizided) and an *easy* part (the parameterized), where we impose some restrictions. For several problems, it is known that the parameterization of the input does not break the NP-completeness barrier. A classical example is the *coloring problem* parameterized by the number k of colors that can be used. It is well known that the problem is NP-complete, for $k \geq 3$. On the other hand, problems like the *maximum independent set* or the *minimum vertex cover* of a graph become polynomially solvable when we parameterize them by the size of the independent set or by the size of the vertex cover. Even in the cases where a parameterization of a NP-complete problem leads to a polynomial time algorithm, there are different upper bounds of the running time for the best known algorithm. For instance, the running time could be $O(n^{f(k)})$ or it could be $O(f(k)n^{\alpha})$, where $f(k)$ is a function of the parameter k, and α is a positive constant. A parameterized problem is said to be *Fixed Parameter Tractable* if there exist an $O(f(k)n^{\alpha})$-algorithm that solves the problem. The class FPT is the class of all fixed parameter tractable problems. In a series of papers on parameterized complexity, Downey and Fellows defined a parameterized complexity hierarchy, the W-*hierarchy*, that falls between FPT and W[P], FPT $\subseteq$ W[1] $\subseteq$ W[2] $\subseteq \cdots \subseteq$ W[P], along with suitable notions of reducibility and completeness (see [21] for a nice exposition of parameterized complexity). It is an open problem whether the inclusion in the hierarchy is strict. Moreover, there is evidence that if a problem is complete for some level of the W-hierarchy, then it is not expected to have an $O(f(k)n^{\alpha})$-algorithm. For example, the parameterized *vertex cover problem* belongs to FPT [4], while the the parameterized *independent set problem* is known to be W[1]-complete [19].

We survey the parameterized complexity of different variants of the H-coloring and the list H-coloring problems, by considering two different types of parameterizations. The first approach is to set as parameter the *treewidth* of the input graph. Intuitively, the parameter of treeewidth is a measure of the global connectivity of a graph (see the formal definitions in Section 2). We survey recent results proving that several extensions and variants of the H-coloring problem, can be solved in time linear to the size of the input graph. We stress that the algorithms are easy to implement and remain polynomial even in the general case where we consider H to be a part of the input.

The second approach is to parameterize the problems by restricting the number of vertices in G that are mapped into a specific subset of vertices of H. We present characterizations of H that allow the classification of the decisional parameterized problems as in FPT, in P, or as NP-complete.

Notice that the counting versions, the list $\#H$-coloring and the $\#H$-coloring, are functional problems and therefore cannot fit into the current framework of the parameterized complexity theory. To study the complexity of the counting versions of the parameterized problems we define the class #PPT (#P Parameter Tractable problems) as the class of parameterized counting problems, which can be solved by an algorithm within time $O(f(k)n^\alpha)$, where f is some function of the parameter k, and α is a positive integer. We investigate characterizations for which the parameterized counting problems remain #P-complete, as well as conditions for which the parameterization classifies the problem in the classes P or #PPT. Notice that the solution to the list H-coloring and the H-coloring problems, can be considered as a particular case of their counting versions, therefore, when possible, we describe algorithms for solving the counting versions of the problems and we state their decision versions as corollaries to those algorithms.

In the last section, we present some comments and conjectures on the parameterized approach for the counting and decisional versions of the H-coloring problem, and propose some alternative lines of research.

Throughout the paper, we use some standard notation. For the input graph G, we set $n = |V(G)|$ and $m = |E(G)|$. For the fixed graph H, $h = |V(H)|$ and $e = |E(H)|$. For a given graph G and a vertex subset $S \subseteq V(G)$, let $G[S]$ be the subgraph induced by S, and let $G - S$ denote the subgraph $G[V(G) - S]$. As usual K_k denotes a complete graph on k vertices and $K_{k,l}$ a complete bibartite graph with parts containing k and l vertices. As usual for a functiuon θ we use $\theta|_S$ to denote its restrictioin to the set S.

2. Parameterizing by treewidth

Treewidth was first defined by Halin in [**30**] and was reintroduced independently in [**3**] and [**48**]. It plays a key role in many proofs in structural graph theory [**49, 44**] and served as one of the cornerstone concepts for the lengthy proof of the Wagner's conjecture developed by Robertson and Seymour in their Graph Minor Series (see [**47**] for a survey). Formally,

DEFINITION 2.1. A *tree decomposition* of a graph G is a pair (X, U) where U is a tree whose vertices we will call *nodes* and $X = (\{X_i \mid i \in V(U)\})$ is a collection of subsets of $V(G)$ such that

    (1) $\bigcup_{i \in V(U)} X_i = V(G)$,

    (2) for each edge $\{v, w\} \in E(G)$, there is an $i \in V(U)$ such that $v, w \in X_i$, and

    (3) for each $v \in V(G)$ the set of nodes $\{i \mid v \in X_i\}$ forms a subtree of U.

The *width* of a tree decomposition $(\{X_i \mid i \in V(U)\}, U)$ equals $\max_{i \in V(U)}\{|X_i| - 1\}$. The *treewidth* of a graph G is the minimum width over all tree decompositions of G.

In Figure 2, we present a tree decomposition with optimal width of the graph G given in Figure 1.

As a graph parameter, treewidth has many algorithmic applications. A wide range of combinatorial optimization problems are polynomially solvable when restricted to graphs with bounded treewidth. Unfortunately, computing the treewidth

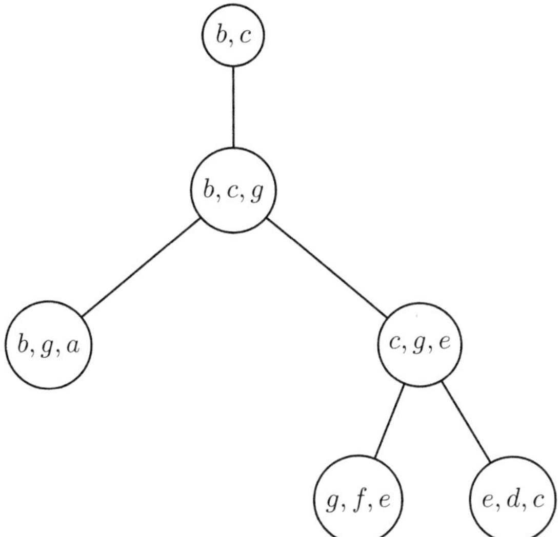

FIGURE 2. A tree decomposition of the graph G in Figure 1

of a graph is a NP-complete problem [**3**]. However, we can fix our attention only to graphs where the treewidth is bounded by a constant k. Such graphs are alternatively called *partial k-trees*. For any constant k, Bodlaender [**6**] presented a linear time algorithm that, given a graph G, checks whether G is a partial k-tree and, if so, outputs a minimum width tree decomposition.

The canonical methodology to get polynomial solutions to difficult problems, consists of a two step procedure: First find a constant width tree-decomposition of the input graph, not necessarily optimal. Then, use *dynamic programming* to get a solution, taking advantage of the bounded treewidth decomposition of the graph (see [**2, 5**]). For the purpose of solving the first canonical step, the algorithm of Bodlaender [**6**] does not seem to be feasible to implement, as it uses as a subroutine the algorithm in [**7**], which involves an enormous hidden constant. However, we can make use of earlier algorithms that output a tree decomposition of G, with width bounded by a linear function of k [**46, 39, 41**]. In particular, the deterministic algorithm by Reed [**46**] runs in $O(n \log n)$ time and, for constant k, it either returns that the treewidth of G is more than k, or it constructs a tree decomposition of width $4k$ or less. This last algorithm allows us to assume that any partial k-tree is always given together with a tree decomposition of constant width.

2.1. The main algorithm. For some problems on partial k-trees, where k is a fixed constant, Courcelle [**13**] associated the existence of a polynomial time algorithm with their expressibility by Monadic Second Order Logic, see also [**12**]. As a consequence of these results, it is possible to construct a polynomial time algorithm solving the H-coloring and the $\#H$-coloring problem for partial k-trees, when k and the size of H are fixed constants. However, the results of Courcelle do not provide implementable algorithms because of the very large hidden constants in their complexity.

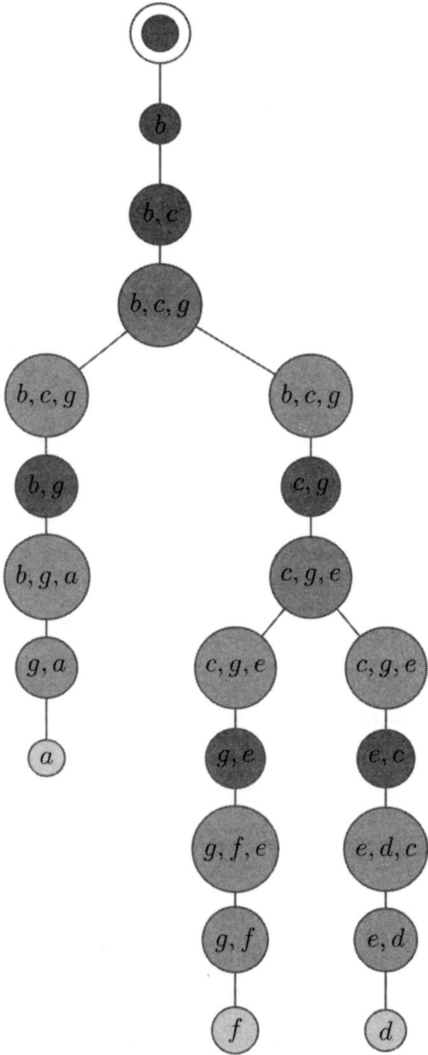

FIGURE 3. A nice tree decomposition of the graph G in Figure 1

The H-coloring problem is one of the many problems, where the dynamic programming technique has yield a polynomial-time solution for partial k-trees [50]. In terms of parameterized complexity this means that, for any H, the H-coloring problem parameterized by the treewidth of the input graph is a problem in FPT. In the remaining of the section, we present a polynomial time algorithm for counting H-colorings, in the case that the input graph G has constant treewidth and we survey its consequences.

Although the methodology follows the two cannonical steps, we shall remark that we present an easy to be implemented algorithm for the second step, asumming the aforementioned result on the existence of fast and implementable algorithms for obtaining a bounded width decomposition for a partial k-tree. Moreover, the algorithm remains polynomial even when there are no restricions on the size of H. Once more, this is the first positive result on counting H-colorings, when H is generic and only the input graph is restricted.

Instead of working with the general definition of tree decomposition, we shall consider a more refined version of decomposition, the *nice tree decomposition* defined in [**6, 7, 36**], which simplifies the exploration of the tree structure.

DEFINITION 2.2. A *rooted* tree decomposition is a triple $D = (X, U, r)$ in which U is a tree rooted at r and (X, U) is a tree decomposition.

Let $D = (X, U, r)$ be a rooted tree decomposition of a graph G where $X = \{X_i \mid i \in V(U)\}$. D is called a *nice* tree decomposition if the following are satisfied:

(1) Every node of U has at most two children,
(2) if a node i has two children j and h, then $X_i = X_j = X_h$,
(3) if a node i has one child, then either $|X_i| = |X_j| + 1$ and $X_j \subset X_i$ or $|X_i| = |X_j| - 1$ and $X_i \subset X_j$.
(4) if a node i is a leaf, then $|X_i| = 1$.

Without lost of generality, we can assume that the root r has $X_r = \emptyset$. In Figure 3 we present a nice tree decomposition for the graph G in Figure 1. It is known that for any constant $k \geq 1$, given a tree decomposition of a graph G of width at most k and $O(n)$ nodes, there exists a linear time algorithm that constructs a nice tree decomposition of G, with $O(n)$ nodes and width at most k [**7**].

Given a nice tree decomposition $D = (X, U, r)$ of G, for any $p \in X$ of D, let U_p denote the subtree of U, rooted at node p. We set $V_p = \cup_{v \in V(U_p)} X_v$ and, for any $p \in V(U)$, we define $G_p = G[V_p]$. Therefore we associate to each node a subgraph, notice that $G_r = G$. A key observation is the fact that a nice tree decomposition $D = (\{X_p \mid p \in V(U)\}, U, r)$ contains four possible node types: *Start node*, if the node is a leaf; *Join node*, if the node has two children $q_i, i = 1, 2$; *Forget node*, if the node p has only one child q and $|X_p| = |X_q| - 1$; and *Introduce node*, if the node p has only one child q and $|X_p| = |X_q| + 1$. Observe that the root r is a *forget* node.

The optimization of the space used in the counting part of the algorithm is attained by the construction of a special topological ordering of $V(U)$, called *stingy*. This ordering has the property that for any j, the number of nodes in the set $\{u_1, \ldots, u_j\}$ whose parent appears at a position $k > j$ is at most $\log n$. The stingy ordering can be computed in linear time, performing an adequate tree traversal.

Given a nice tree decomposition $D = (X, U, r)$ of a graph G, for each $p \in V(U)$, define the set $F_p = \{\varphi : X_p \to V(H)\}$. Notice that if G has treewidth k, then for any $p \in V(U)$, $|F_p| \leq h^{k+1}$. Moreover, for the root we have $F_r = \{\emptyset\}$, where $\emptyset$ represents the empty function. The table associated to a node $p \in V(U)$ will have an entry for each $\varphi \in F_p$, holding the value $I_p(\varphi) = |\{\theta \mid \theta$ is an H-colorings of G_p with $\theta|_{X_p} = \varphi\}|$. As we always have $\theta|_\emptyset = \emptyset$, we get that the number of H-colorings of G is equal to $I_r(\emptyset)$.

THEOREM 2.1 ([**15**]). Given a partial k-tree G, together with a nice tree decomposition $D = (X, U, r)$ with width k, and a stingy ordering of the nodes in D, the algorithm Count-H computes the number of H-colorings of G in $O(nh^{k+1} \min\{k, h\})$ steps, using $O(h^{k+1} \log n)$ additional space.

If we use an $O(f(k)n)$ algorithm to solve the first canonical step, we have the following corollary to Theorem 2.1.

function count-H(G, H, D, S)

$D = (X, U, r)$ is a nice tree decomposition of G with width k
$n = |V(G)|$, $h = |V(H)|$, $s = |V(U)|$
$S = (p_1, \ldots, p_s = r)$ is a stingy ordering of $V(U)$
begin
    for $p := p_1$ **to** p_s **do**
        if p is a *start* node, with $X_p = \{v\}$ **then**
            for all $a \in V(H)$ **do** $I_p((v, a)) := 1$
        end
        if p is a *introduce* node **then**
            let q be its unique child
            $v := X_p - X_q$
            $S_q := \{u \in X_q \mid \{u, v\} \in E(G_p)\}$
            for all $\varphi \in F_q$ **and** $a \in V(H)$ **do**
                if $\forall_{u \in S_q} \{\varphi(u), a\} \in E(H)$ **then**
                $I_p(\varphi \cup \{(v, a)\}) := I_q(\varphi)$
                else $I_p(\varphi \cup \{(v, a)\}) := 0$
                end
            end for
            erase the information on node q
        end
        if p is a *forget* node **then**
            let q be its unique child
            $v := X_q - X_p$
            for all $\varphi \in F_p$ **do**
                $I_p(\varphi) := \sum_{a \in V(H)} I_q(\varphi \cup \{(v, a)\})$
            end for
            erase the information on node q
        end
        if p is a *join* node with children q_1, and q_2 **then**
            for all $\varphi \in F_p$ **do**
                $I_p(\varphi) := I_{q_1}(\varphi) \cdot I_{q_2}(\varphi)$
            end for
            erase the information on nodes q_1 and q_2
        end
    end for
    return $I_r(\varnothing)$
end

FIGURE 4. An algorithm for counting H-colorings

COROLLARY 2.1. Let k be fixed. The $\#H$-coloring problem, for graphs with treewidth bounded by k, can be solved in $O(n(h^{k+1} \min\{k, h\}) + f(k)))$ steps.

2.2. Other versions of H-coloring. An earlier version of the H-coloring problem is the *exact H-coloring problem*: given a graph G, decide if there is an H-coloring θ of G such that $\theta(V(G)) = V(H)$ and $\theta(E(G)) = E(H)$. This problem appeared in [**29**] as problem GT-52 and it is known to be NP-complete, even for the case where H is a triangle [**40**]. An intermediate problem is the one of asking whether there is an H-coloring θ of G such that $\theta(V(G)) = V(H)$ and we call it *vertex exact H-coloring*. The *list exact H-coloring* and the *list vertex exact H-coloring* are defined in the obvious way and we call their counting versions *list exact #H-coloring* and *list vertex exact #H-coloring*. Algorithm count-H can be adapted to solve all these counting variations of the $\#H$-coloring problem.

THEOREM 2.2 ([**15**]). *The exact $\#H$-coloring and the vertex exact $\#H$-coloring problems, parameterized by the treewidth of the input graph, are all in $\#$PPT.*

Theorem 2.2 yields corollaries analogous to Corollary 2.1 in the obvious way.

2.3. Counting list H-colorings. It is easy to adapt the algorithm count-H to obtain an algorithm for the list $\#H$-coloring problem for graphs of bounded treewidth. The unique changes are the redefinition of the set F_p as $\{\varphi : X_p \to V(H)$ and $\forall_{x \in X_p} \varphi(x) \in L(x)\}$, and the replacement of the requirement $a \in V(H)$ by $a \in L(v)$, in the treatment of an *start* node, in function count-H. Applying similar changes to the algorithms used to prove Theorem 2.2 we can rewrite them in their "list" versions with the same complexity bounds (for further details see [**15**]).

2.4. The decision version of H-coloring problems. Notice that a graph G is H-colorable if the number of H-colorings of G is non zero. This implies that the algorithms involved in the Theorems 2.1, and 2.2 can solve the corresponding decision version of the problems examined with the same time bounds, as well as their "list" extensions. In other words the decision version of all the variations of the H-coloring introduced in this section, belong in FPT. We mention that for the case of H-coloring of partial k-trees, the result described improves the best known algorithm, which runs in $O(nh^{2(k+1)})$ steps and is due to Telle and Proskurowski in [**50**].

2.5. The directed case. Given two directed graphs $\vec{H}$ and $\vec{G}$, an $\vec{H}$-*coloring of $\vec{G}$* is any function $\theta : V(\vec{G}) \to V(\vec{H})$ with the property that $(v, w) \in E(\vec{G}) \Rightarrow (\theta(v), \theta(w)) \in E(\vec{H})$. If $\vec{H}$ is a fixed directed graph, the $\vec{H}$-*coloring problem* asks whether, a given directed graph $\vec{G}$ has an $\vec{H}$-coloring. We define the $\#\vec{H}$-*coloring problem* as the problem of, given a directed graph $\vec{G}$, counting all the $\vec{H}$-colorings of $\vec{G}$. In an analougous way, we can extend the definition for the list, exact and vertex exact variations.

It is easy to adapt the algorithms of Theorems 2.1, and 2.2 as well as their "list" extensions for the directed case. The complexity of the corresponding algorithms is the same (for details, see [**15**]).

2.6. Enumeration. Notice that for each of the algorithms described so far, if we retain the information of all the tables, it is possible to use it in order to enumerate all the homomorphisms. The storing of all the tables implies a burden of $O(n/\log n)$ to the space reported in Theorems 2.1 and 2.2. In particular, setting up a suitable bookkeeping of the enumerated homomorphisms, a top-down traversal of the table information can pop-up each of them in $O(n)$ steps.

2.7. The general homomorphism problem. Consider the *homomorphism* problem of deciding, given two input graphs G and H, whether there is an homomorphism from G to H. From Theorem 2.1 we have that the homomorphism problem can be solved in polynomial time if we parameterize it by the treewidth of G. Therefore, the H-coloring problem for partial k-trees remains polynomially solvable even if we consider H to be a part of the input. Similar observations hold for the corresponding list and counting versions.

Notice that if we parameterize the homomorphism problem by considering the treewidth of H, instead of by the treewidth of G, we do not affect the generality of the dichotomy theorem of Hell and Nešetřil [**42**] as, for any k, graphs with treewidth k can be bipartite or non-bipartite. However treewidth related restrictions, for a target graph H, the so called *bounded treewidth duality*, lead to a polynomial solution to the H-coloring problem [**33**], even for the case when H is directed. Similar results are known for the general algebraic homomorphism problem: given two finite relational structures A and B, is there an homomorphisms $H : A \rightarrow B$?. This generic formulation include the graph homomorphism problem as well as *constraint satisfaction* problems and *conjunctive-query containment* problems [**37**].

2.8. Coloring and chromatic polynomial. Several researchers have considered the problem of counting the number of proper c-colorings in a graph G. The problem is #P-hard for $c \geq 3$ and maximum degree Δ of G at least 3 [**34**]. In the same paper, it is proved that there exists a *Fully Polynomial time Randomized Aproximation Scheme* (FPRAS) for the number of colorings in the case when $c \geq 2\Delta + 1$. Bubley et al. [**9**] proved that the problem is #P-hard for fixed Δ, but there is a FPRAS for $c = 5$ and $\Delta = 3$. Edwards [**24**] proved that if $c \geq 3$ and the minimum degree where $\delta \geq \alpha n$, the counting problem is #P-complete if $\alpha < \frac{c-2}{c-1}$, but it is in P for $\alpha > \frac{c-2}{c-1}$. Recall that the problem of counting the number of c-colorings of a graph G is the #K_c-coloring problem. Therefore, we conclude that

COROLLARY 2.2. *An algorithm can be constructed to compute the number of c-colorings of a partial k-tree graph, in $O(nc^{k+1}\min\{k,c\})$ steps.*

Andrzejak [**1**] has given an $O(n^{2+7\log_2 c})$-time algorithm to compute the Tutte polynomial of a partial k-tree graph on n vertices, where c is twice the number of partitions of a set with $3(k+1)$ elements. Andrzejak's algorithm gave the best procedure to compute the chromatic polymonial for a partial k-tree. As a consequence of the results on counting H-colorings for partial k-trees, the following result is proved in [**15**].

COROLLARY 2.3. *The chromatic polynomial of a partial k-tree can be constructed in $O(kn^{k+3})$ steps.*

Notice the the goal of the previous result is not the evaluation of the chromatic polynomial, for which better bouns can be obtained through the evaluation of the Tutte polynomial of a bounded treewidth graph [**43**].

2.9. Other extensions. Recall that in the case where H is an edge with only one looped verex, as in Figure 1, the #H-coloring problem is equivalent to the problem of counting independet sets. Using our previous results, we get

COROLLARY 2.4. *The number of independent sets of a partial k-tree can be obtained in $O(n2^{k+1})$ steps.*

It has been observed that, by specializing the structure of H, we can generate an arbitrary number of counting problems that are, in general, #P-complete (see [**23, 22**]). This includes, for example, the problem of counting the q-particle Widom-Rowlinson configurations in bounded treewidth graphs. For further applications of the results described in this section see [**15**].

FIGURE 5. The graphs (H, C, K) for the parameterized independent set and vertex cover as parameterized colorings (the big vertices represent the labeled vertices of H)

Finally, notice that, for the counting and decision version of the problems, Theorems 2.1 and 2.2 provide polynomial time algorithms even in the case where $k = O(\log n)$.

3. Parameterizing H-coloring by image restriction

Let us introduce the second type of *parameterization* for the H-coloring problems. Let $C = \{a_1, \ldots, a_r\}$ be a set of r vertices in $V(H)$ and let $K = (k_1, \ldots, k_r)$ be an r-tuple of non negative integers, where each k_i is associated to a_i. The triple (H, C, K) is called a *partial weighted assignment* on H and a mapping $\chi : V(G) \to V(H)$ is an (H, C, K)-*coloring* of G if χ is an H-coloring of G such that, for all $a_i \in C$, $|\chi^{-1}(a_i)| = k_i$. In this section, we let k denote $\sum_{i=1,\ldots,r} k_i$, for any given partial weighted assignment (H, C, K). We say that a partial weighted assignment (H, C, K) is a *weighted extension* of a graph F if $H - C = F$. We call a partial weighted assignment (H, C, K) *positive* if all the integers in K are positive.

For fixed H, C and K, given an input graph G, the (H, C, K)-*coloring* problem asks whether there exists an (H, C, K)-coloring of G. Notice, that the (H, C, K)-coloring can express well known parameterized problems, like the the *independent set* and the *vertex cover* problems (see Figure 5).

The new parameterization can be extended in a straightforward manner to define the list (H, C, K)-coloring problem. We define the $\#(H, C, K)$-*coloring* as the problem of counting the number of (H, C, K)-colorings of an input graph G, and the *list* $\#(H, C, K)$-*coloring* as the problem of counting the number of list (H, C, K)-colorings of an input $(G, \{L(v)\})$.

3.1. Easy cases. In this subsection, we survey results from [16], giving necessary conditions for the $\#(H, C, K)$-coloring problem and the list $\#(H, C, K)$-coloring, to be solvable in linear time. A sufficient condition, for the list $\#(H, C, K)$-coloring problem to be solved in linear time, is that the set C of parameterized vertices forms a vertex cover. The core of the proof is the Algorithm Count-List-Colorings in Figure 6, which given a graph G, counts the number of list (H, C, K)-coloring, using $O(kn + f(k, h))$ steps, for some function f.

THEOREM 3.1. If (H, C, K) is a partial weighted assignment, where $H - C$ has no edges, there exists an $O(kn + f(k, h))$ time algorithm that solves the list $\#(H, C, K)$-coloring problem.

Algorithm Count-List-Colorings(G, L, H, C, K).

Input: Two graphs G, H, a function $L : V(G) \to 2^{V(H)}$,
           and a partial weighted assignment
           (H, C, K) on H such that $E(H - C) = \emptyset$.

Output: The number of list (H, C, K)-colorings of G.

1:   Let R_1 be the set of vertices in G of degree $> k$.

2:   If $|R_1| > k$ or $|E(G)| > kn$ then return 0.

3:   Set $G' = G - R_1$.

4:   Let R_2 be the non isolated vertices in G' and
      let R_3 be the isolated vertices in G'.

5:   If $|R_2| > k^2 + k$ then return 0.

6:   Setup a partition $\mathcal{R} = (P_1, \ldots, P_m), m \le q \cdot 2^h$ of R_3 where
      q is the number of different neighborhoods of vertices in R_3 and
      $\forall_{v,u \in R_3} \ (\exists_{1 \le i \le m} \ \{v, u\} \in P_i \Leftrightarrow (N_G(v) = N_G(u) \wedge (L(v) = L(u))))$.

7:   Let $Q = \emptyset$.

8:   For $i = 1, \ldots, m$,
      if $|P_i| \le k + 1$, then set $F_i = P_i$,
         otherwise let F_i be any subset of P_i where $|F_i| = k + 1$.
      Set $Q = Q \cup F_i$ and $P_i = P_i - F_i$.

9:   Let $\mathcal{H}$ be the set of all the list (H, C, K)-colorings of $G[R_1 \cup R_2 \cup Q]$.

10:   Set $\eta = 0$.

11:   For any $\chi \in \mathcal{H}$, do
      Set $\beta = 1$.
      For any nonempty $P_i \in \mathcal{R}$, do
         Set $\sigma = \sum_{a \in S} |\chi^{-1}(a) \cap P_i|, \ \tau = \sum_{a \in C} |\chi^{-1}(a) \cap P_i|$
         Set $\beta = \beta \cdot \binom{|P_i|}{\tau} \frac{\tau!}{\prod_{a \in C} |\chi^{-1}(a) \cap P_i|!} \binom{|P_i| - \tau}{\sigma} \sigma! \ \sigma^{|P_i| - \tau - \sigma}$.
      Set $\eta = \eta + \beta$.

12:   return η.

13:   End.

FIGURE 6. Algorithm Count-List-Colorings

Recall that any H-coloring can be seen as a list H-coloring in which the list for each vertex is $V(H)$. Therefore, as a corollary to the previous lemma, we get a necessary condition for the $\#(H, C, K)$-coloring problem to be in the class $\#\mathsf{PPT}$,

COROLLARY 3.1. *If (H, C, K) is a partial weighted assignment, where $H - C$ has no edges, then there exists an $O(kn + f(k, h))$ time algorithm that solves the $\#(H, C, K)$-coloring problem.*

It is easy to observe that, when $E(H - C) = \emptyset$, if an (H, C, K)-coloring of G exists then G has a vertex cover of size k. Therefore G must have bounded treewidth, and both Theorem 3.1 and Corollary 3.1 could be obtained as a variation of Algorithm Count-H. However both results improve the complexity bound, as they do not require the computation of a tree decomposition.

3.2. Cases in P. In this section we present the cases of the $\#(H, C, K)$-coloring and list $\#(H, C, K)$-coloring problems which are known to be solved by an algorithm with time bound $O(n^{k+c})$, and therefore can be solved in polynomial time.

The main result relates the problem of counting list (H, C, K)-colorings with the problem of counting list $(H - C)$-colorings. This result is the key for all the positive results in this subsection and it appears in [16].

THEOREM 3.2. The list $\#(H, C, K)$-coloring problem can be solved in n^{k+c} steps, whenever the list $\#(H - C)$-coloring problem can be solved in $O(n^c)$ steps.

Recalling, once more, that the number of H-colorings is the same as the number of list H-coloring for an adequate list selection, we get

COROLLARY 3.2. The $\#(H, C, K)$-coloring problem can be solved in n^{k+c} steps, if the list $\#(H - C)$-coloring problem can be solved in $O(n^c)$ steps.

As mentioned in the introduction, the dichotomy for counting list H-colorings is the same as for counting H-colorings. Therefore, if we can solve $\#H$-coloring problem in polynomial time, we can also solve the list $\#H$-coloring problem in polynomial time.

COROLLARY 3.3. The list $\#(H, C, K)$-coloring and the $\#(H, C, K)$-coloring problems can be solved in n^{k+c} steps, if the $\#(H - C)$-coloring problem can be solved in polynomial time. Where c is a constant independent of k.

3.3. Hardness results. Let us consider negative results giving conditions under which the list $\#(H, C, K)$-coloring remains $\#$P-complete. The following result has not been published previously and we provide here a complete proof.

THEOREM 3.3. For any parameter assignment (H, C, K), the list $\#(H, C, K)$-coloring problem is $\#$P-complete whenever the $\#(H - C)$-coloring problem is $\#$P-complete.

PROOF. Let $H' = H - C$. We reduce the $\#H'$-coloring problem to the list $\#(H, C, K)$-coloring problem. Let G be an instance of the H'-coloring problem. We construct an instance G' of the (H, C, K)-coloring problem as follows: Take G' as the disjoint union of G and a graph F, that consists of k isolated vertices. We assign to every $v \in V(G)$, the list $L(v) = V(H')$ and to every $f \in V(F)$, we assign the list $L(f) = C$.

Notice that, from the way we have defined the lists, the additional vertices can be only mapped to the set of parameterized vertices, but any arrangement of images is possible. The size of F guarantees that every parameterized vertex receives the prescribed number of images. Furthermore, combining an arrangement of the vertices of F with an H'-coloring of G we obtain a valid list H-coloring of $(G', \{L(v)\})$.

Therefore, for any H'-colorings of G there are $k!$ list H-colorings of $(G', \{L(v)\})$ and the statement of the theorem follows. $\square$

Putting together Theorems 3.2, 3.3 and the results in [16, 32, 23] we get the following dichotomy.

THEOREM 3.4. The list $\#(H, C, K)$-coloring is $\#$P-complete if $H - C$ has a connected component which is not a complete reflexive graph or a complete irreflexive bipartite graph. Otherwise the counting problem is in P.

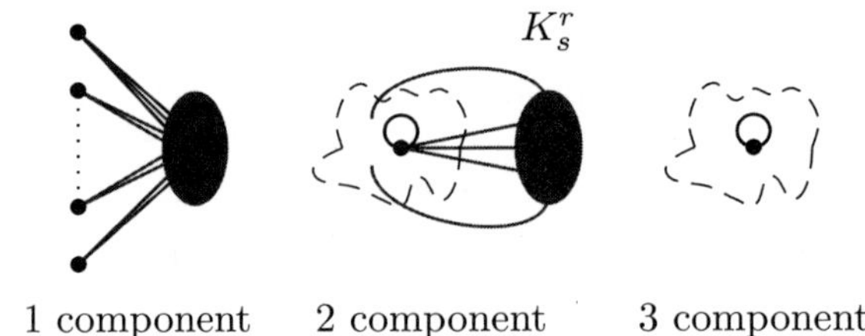

FIGURE 7. Components for a compact partial weighted assignment

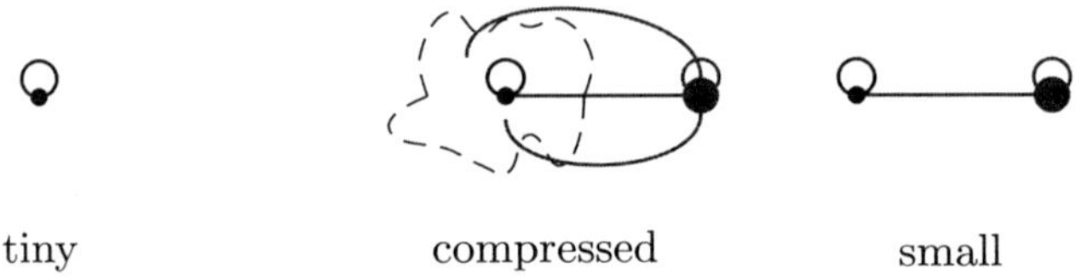

FIGURE 8. Special cases of components for a compact partial weighted assignment

3.4. The decision version. All the positive results for the counting versions can be translated to their decision versions, however, polynomial time and hardness results can be reinforced. In [**17**] another necesary condition for the (H, C, K)-coloring problem to be in the class FPT was presented. First we need a definition.

We say that a partial weighted assignment (H, C, K) is *compact* when each connected component H_i of H satisfies one of the following exclusive conditions:

(1) $E(H_i - C) = \emptyset$,
(2) $H_i[C]$ is a non-empty reflexive clique with all its vertices adjacent with one looped vertex of $H_i - C$, or
(3) $V(H_i) \cap C = \emptyset$ and H_i contains at least one looped vertex.

Figure 7 gives a representation of the types of components allowed in a compact weighted assignment. Let (H, C, K) be a compact partial weighted assignment. Notice that if $\chi : V(G) \to V(H)$ is a (H, C, K)-coloring of G, the total size of the connected components of G that are mapped to the vertices of some component H_i of type (2), is at least k_i, where a_i is H_i contribution to the total weight.

The next theorem was proved in [**17**]. The crunch of the proof is an $O(n(k + h)) + \gamma(G)f(k, h))$ time algorithm, where $\gamma(G)$ is the number of the connected components of the input graph G. The basic ideas of the algorithm are the following: first, reduce the problem to a compact partial weighted assignment for the particular components depicted in Figure 8. When G has "many" connected components an adaptation of the algorithm Count-List-Colorings is used to check whether a "small" number of them are mapped to the 1-components of H, if this is possible plenty of components are left unmapped, which will allow an extension of the mapping to the remainning of G. When G has few connected components, the number of components of G that can be mapped to the components of H does not depend on the size of G, and a exhaustive search algorithm checks the existence of an (H, C, K)-coloring.

THEOREM 3.5. *If (H, C, K) is compact, then the (H, C, K)-coloring problem, parameterized by K, belongs to the class* FPT.

It is not known any equivalent result for the list (H, C, K)-coloring problem. For the hardness results, an extension of the argument used in the proof of Theorem 3.3 yields the following result on the complexity of the list (H, C, K)-coloring problem.

COROLLARY 3.4. *The list (H, C, K)-coloring problem is* NP-*complete, whenever the list $(H - C)$-coloring problem is* NP-*complete.*

On the other hand, a much more involved argument is presented in [**17**], which provides a similar hardness result for the (H, C, K)-coloring problem.

THEOREM 3.6. *The (H, C, K)-coloring problem is* NP-*complete if $(H - C)$ is not bipartite and it does not contain a loop.*

On the positive side, Theorem 3.2 can be extended to get both problems in the class P.

COROLLARY 3.5. *The (H, C, K)-coloring and the list (H, C, K)-coloring problems can be solved in n^{k+c} steps, whenever the list $(H - C)$-coloring can be solved in $O(n^c)$ steps.*

The previous result include the particular cases where $H - C$ is a reflexive interval graph [**25**] and where $H - C$ is an irreflexive graph whose complement is a circular arc graph of clique covering number two [**26**]. Furthermore, Theorem 3.6 and Corollary 3.5 provide a dichotomy theorem for the list (H, C, K)-coloring problem.

THEOREM 3.7. *The list (H, C, K)-coloring problem is* NP-*complete if $(H - C)$ is not a bi-arc graph, otherwise it is in* P.

For the (H, C, K)-coloring problem, there is a gap, because the positive results relate to the dichotomy for the list H-coloring problem, while the negative results relate to the dichotomy for the H-coloring problem. For some graphs in the gap, there are negative and positive results depending on the weigthed extension selected.

The next result proved in [**17**] shows that if F is a graph, for which the list F-coloring problem is NP-complete, but the F-coloring problem is in P, then F has weighted extensions for which the problem falls in P.

THEOREM 3.8. *Let F_1 be either a bipartite graph or a graph with at least one loop, and let F_2 be any graph. Then, the $(F_1 \oplus F_2, V(F_2), K)$-coloring can be solved in polynomial time for any partial weighted assignment K.*

In the previous theorem, $F_1 \oplus F_2$ denotes the graph obtained from F_1 and F_2 adding all the edges between vertices in different graphs. It is also worth mentioning that for any weighted extension of a graph F for which the F-coloring problem is in P, the problem is also polynomially solvable in the trivial case $K = (0, \ldots, 0)$.

The condition of Theorem 3.6 is not necessary. The next result presents weighted assignments (H, C, K) for which the $(H - C)$-coloring problem is in P, but the (H, C, K)-coloring problem is NP-compete.

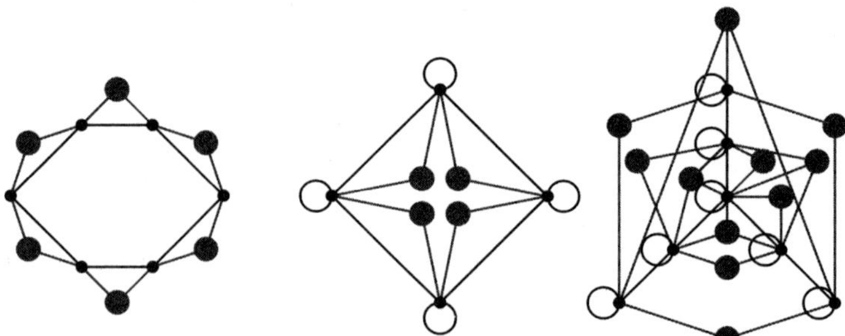

FIGURE 9. Some hard partial weighted assignments(the big vertices represent the labeled vertices of H)

THEOREM 3.9. *The (H, C, K)-coloring problem is NP-complete for the partial weighted assignments depicted in Figure 9, provided that (H, C, K) is positive.*

Finally, notice that the subgraph induced by the non parameterized vertices in each of the 3 examples given in Figure 9 have, by Theorem 3.8, another weighted extension, so that the corresponding parameterized coloring problem belongs to the class P.

4. Remarks and open problems

The results for all the versions of the H-coloring problem are sumarized in Tables 1 and 2. Notice that, all the hardness results presented in this paper, can be extended to the case in which the input graph has degree bounded by some constant. Several problems remain unsolved, we present some of them together with some conjectures. We propose other interesting ways to parameterize the H-coloring problem, and finish by citing some open problems related to lines of attack #P-hard problems.

4.1. Dichotomy conjectures. The hardness result for the $\#(H, C, K)$-coloring problem seems more difficult to obtain. The results in [16] provide a partial answer, by showing #P-completeness when H has the property that no homomorphism from H to H sends vertices in C to vertices in $V(H) - C$.

CONJECTURE 4.1. *The $\#(H, C, K)$-coloring is #P-complete if $H - C$ has a connected component which is not a complete reflexive graph or a complete irreflexive bipartite graph. Otherwise the counting problem is in P.*

The results in Theorems 3.6 and 3.8 are sharp in the following sense: If F is a graph where the F-coloring is NP-complete then, for any weighted extension (H, C, K) of F, the (H, C, K)-coloring is also NP-complete. On the other hand, if the F-coloring problem is in P then *there exist* a weighted extension (H, C, K) of F so that the (H, C, K)-coloring problem is in P. Moreover, the results in Theorem 3.9 and Corollary 3.5 are also sharp, in the sense that if F is a graph where the list F-coloring problem is in P then, for any weighted extension (H, C, K) of F, the (H, C, K)-coloring problem is also in P. We provided some examples where the list F-coloring problem is NP-complete and F has a weighted extension (H, C, K) such that the (H, C, K)-coloring problem is also NP-complete.

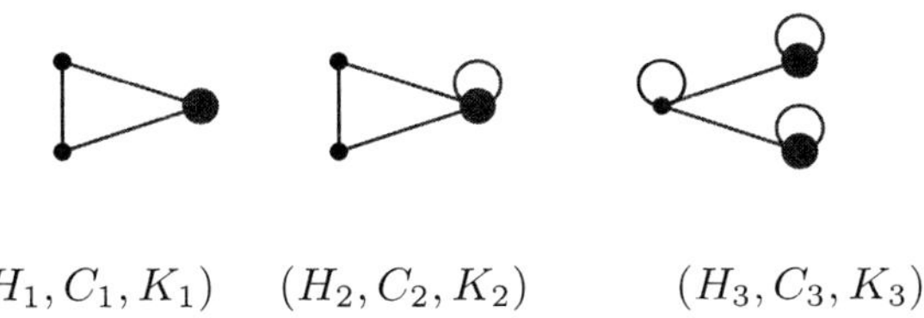

$$(H_1, C_1, K_1) \qquad (H_2, C_2, K_2) \qquad (H_3, C_3, K_3)$$

FIGURE 10. Three weighted extensions conjectured to be W[1]-hard when K is parameterized (the big vertices represent the labeled vertices of H)

The above observations indicate that different weighted extensions of some graphs produce parameterized coloring problems with different complexities. It seems to be a hard problem to achieve a dichotomy discriminating those parameterized assignments (H, C, K) of a given graph F for which the (H, C, K)-coloring problem is NP-complete or in P. However we conjecture the following.

CONJECTURE 4.2. For any graph F such that the list F-coloring problem is NP-complete, there is a weighted extension (H, C, K) of F such that the (H, C, K)-coloring problem is also NP-complete.

The results on NP-completeness indicate, that the frontier depends not only on the structure of the graph $H - C$, but also on the structure imposed by the weighted vertices. We observe that in all the complexity results described in this survey, positive or negative, the dichotomy does not depend on the choice of the numbers in K (when K is positive).

We now fix our attention on conditions classifying the (H, C, K)-coloring problem in FPT. We conjecture that the condition of Theorem 3.5 is also necessary.

CONJECTURE 4.3. For any partial weighted assignment (H, C, K), the (H, C, K)-coloring problem is in FPT if (H, C, K) is compact, otherwise the problem is W[1]-hard.

In support of this conjecture we recall that the parameterized *independent set* problem, known to be W[1]-complete [20], satisfies the conjecture.

Conjecture 4.3 is sufficiently general to express several open problems in parameterized complexity such as

(1) (Parameter: k) Does G contain an independent set S, where $|S| = k$, and such that $G[V(G) - S]$ is bipartite? (This problem can be seen as a parameterization of 3-coloring where some color should be used exactly k times.)

(2) (Parameter: k) Is there a set $S \subseteq V(G)$, with $|S| = k$, such that $G[V(G) - S]$ is bipartite?

(3) (Parameters: k, l) Does G contain $K_{k,l}$ as subgraph?

These three problems correspond to the parameterized colorings given in Figure 10. Observe that the (H_3, C_2, K_3)-problem is equivalent to ask if the complement of G contains K_{k_1, k_2} as a subgraph.

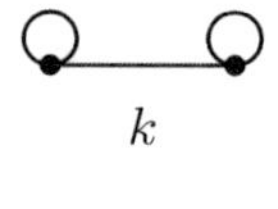

FIGURE 11. The graphs (H, C, K) for the parameterized cut as parameterized colorings

4.2. Other parameterizations of H-coloring. Other interesting parameterizations of the H-coloring problem are obtained by modifying the equality condition on K. We can consider the $(H, C, \geq K)$-coloring and the $(H, C, \leq K)$-coloring problems, defined like the (H, C, K)-coloring problem with the difference that, for any $a_i \in C$, the number of preimages of a_i is at most k_i or at least k_i, respectively. If we consider the first partial weighted assignment in Figure 5, the $(H, C, \geq K)$-coloring problem is equivalent to the (H, C, K)-coloring problem and the $(H, C, \leq K)$-coloring problem is trivial. For the second partial weighted assignment in Figure 5, the $(H, C, \leq K)$-coloring problem is equivalent to the (H, C, K)-coloring problem and the $(H, C, \geq K)$-coloring problem is trivial.

In the parameterization described in Section 3, we fixed the number of preimages of a set of vertices in an H-coloring. As an H-coloring of G maps also the edges of G to edges of H, one could enhance the definition of a (H, C, K)-coloring by including also restrictions on the number of preimages of a subset of edges. This can be done by allowing C to be a subset of $V(H) \cup E(H)$. As usual, if θ is an H-coloring, we define the preimages of an edge $e = \{a, b\}$ as $\theta^{-1}(e) = \{\{v, u\} \in E(G) \mid \theta(v) = a$ and $\theta(u) = b\}$. Both the $(H, C, \leq K)$-coloring and the $(H, C, \geq K)$-coloring can be generalized to allow edge subsets in the obvious way.

An interesting case is the parameterization (H, C, K) (see Figure 11), where H is the graph with two looped vertices connected by an edge e, $C = \{e\}$ and $K = (k)$. This particular problem is equivalent to the problem of deciding whether G has an edge cut of size equal to k. Accordingly, the "$\leq$"-version and the "$\geq$"-version of the same problem asks for an edge cut of size $\geq k$ (the decision version of the *min-cut* problem, which is in P) and an edge cut of size $\leq k$ (the decision version of the *max-cut* problem, which is NP-complete) respectively. Moreover, the *max-cut* problem is known to be in the class FPT. Recall, that all the problems in the class MAXNP are also in FPT [**10**], and furthermore it is known that the *max-cut* problem is MAXSNP-complete [**45**].

On the other hand, if **1** denotes the vector with all components set to one, the $(H, E(H), \geq \mathbf{1})$-coloring is the exact H-coloring problem, which is NP-complete [**40**].

Another interesting parameterization for the list H-coloring, is to set as parameter the maximum allowed size for each list.

4.3. Other directions. Given the hardness results of this survey, a natural question is to approximate the counting versions of the H-coloring that are known to be in #P. As the problem is self reducible, a candidate technique would be

-coloring	FPT	P	NP-complete
list H	bounded trewidth [15]	dichotomy [27]	
H	bounded trewidth [15]	dichotomy [31]	
list (H, C, K)	$E(V - C) = \emptyset$ [17]	dichotomy [17]	
(H, C, K)	(H, C, K) compact [17]	list $(H - C)$-coloring in P [17]	$(H - C)$-coloring NP-hard [17]

TABLE 1. Complexity of H-coloring problems

-coloring	#PPT	P	#P-complete
list $\#H$	bounded trewidth [15]	dichotomy [16, 32]	
$\#H$	bounded trewidth [15]	dichotomy [23]	
list $\#(H, C, K)$	$E(V - C) = \emptyset$ [16]	dichotomy [16]	
$\#(H, C, K)$	$E(V - C) = \emptyset$ [16]	list $\#(H - C)$-coloring in P [16]	(H, C, K) irreducible [16]

TABLE 2. Complexity of the counting H-coloring problems

the almost uniform generation, via the Markov chain method, to generate an almost uniform sampling. In [11, 22], there are given negative results for uniform sampling in the case of some particular selection of H and one positive result for weighted sampling when H is a tree. It remains a challenging problem, to get more general results to approximate the $\#H$-coloring and the list $\#H$-coloring and their parameterizations.

Other model of research consider that H has weights on all the vertices. This weighting of the vertices of H is interpreted as a product measure on the set of H-colorings [8]. It will be of interest to know if it is posible to have a polynomial time algorithm for sampling H-colorings accordingly to the probability distribution induced by the product measure.

Positive and negative complexity results concerning the special case of the coloring problem in which the degree of the input graph is large were studied in [24]. Edwards results on colorings were partially extended to the H-coloring problem in [14]. It will be of interest to extend the results on counting colorings in [24] to the H-coloring problem.

Acknowledgement. We thank an anonymous referee for improving the quality of the reading.

References

[1] A. Andrzejak. An algorithm for the Tutte polynomials of graphs of bounded treewidth. *Discrete Mathematics*, 190:39–54, 1998.

[2] S. Arnborg. Efficient algorithms for combinatorial problems on graphs with bounded decomposability – A survey. *BIT*, 25:2–23, 1985.

[3] S. Arnborg, D. G. Corneil, and A. Proskurowski. Complexity of finding embeddings in a k-tree. *SIAM Journal on Algebraic and Discrete Methods*, 8:277–284, 1987.

[4] R. Balasubramanian, M. Fellows, and V. Raman. An improved fixed-parameter algorithm for vertex cover. *Information Proccessing Letters*, 65:163–168, 1998.

[5] H. Bodlaender. Treewidth: algorithmic techniques and results. In *Mathematical Foundations of Computer Science*, Lectures Notes in Computer Science, pages 19–36. Springer-Verlag, 1997.

[6] H. L. Bodlaender. A linear time algorithm for finding tree-decompositions of small treewidth. *SIAM Journal on Computing*, 25:1305–1317, 1996.

[7] H. L. Bodlaender and T. Kloks. Efficient and constructive algorithms for the pathwidth and treewidth of graphs. *Journal of Algorithms*, 21:358–402, 1996.

[8] G. Brightwell and P. Winkler. Graph homomorphism and phase transitions. *Journal of Combinatorial Theory (series B)*, 77:415–435, 1999.

[9] R. Bubley, M. Dyer, C. Greenhill, and M. Jerrum. On approximately counting colorings of small degree graphs. *SIAM Journal on Computing*, 29(2):387–400, 1999.

[10] L. Cai and J. Chen. On fixed-parameter tractability and approximability of NP optimization problems. *Journal of Computer and System Sciences*, 54(3):465–474, 1997.

[11] C. Cooper, M. Dyer, and A. Frieze. On Markov chains for randomly H-coloring a graph. *Journal of Algorithms*, 39:117–134, 2001.

[12] B. Courcelle, J. Makowski, and U.Rotics. On the fixed parameter complexity of graph enumeration problems definable in monadic second order logic. *Discrete Applied Mathematics*, 108(1-2):23–52, 2001.

[13] B. Courcelle and M. Mosbah. Monadic second-order evaluations on tree-decomposable graphs. *Theoretical Computer Science*, 109:49–82, 1993.

[14] J. Díaz, J. Nešetřil, M. Serna, and D. M. Thilikos. H-coloring of large degree graphs. In *Advances in Information and Communication Technology (EURASIA-ICT-2002)*, volume 2510 of *Lecture Notes in Computer Science*, pages 850–857. Springer-Verlag, 2002.

[15] J. Díaz, M. Serna, and D. M. Thilikos. Counting H-colorings of partial k-trees. In *Computing and Combinatorics, COCOON 2001*, volume 2108 of *Lectures Notes in Computer Science*, pages 298–307. Springer-Verlag, 2001. To appear in *Theoretical Computer Science*.

[16] J. Díaz, M. Serna, and D. M. Thilikos. Fixed parameter algorithms for counting and deciding bounded restrictive list H-colorings. Technical Report, Departament de Llenguatges i Sistemes Informàtics, Universitat Politècnica de Catalunya, Barcelona, 2004.

[17] J. Díaz, M. Serna, and D. M. Thilikos. (H, C, K)-colorings: Fast, easy and hard cases. In *Mathematical Foundations of Computer Science 2001, MFCS-2001*, volume 2136 of *Lecture Notes in Computer Science*, pages 304–315. Springer-Verlag, 2001.

[18] Q. Donner. On the number of list-colorings. *Journal of Graph Theory*, 16(3):239–245, 1992.

[19] R. G. Downey and M. R. Fellows. Fixed-parameter tractability and completeness. II. On completeness for $W[1]$. *Theoretical Computer Science*, 141(1-2):109–131, 1995.

[20] R. G. Downey and M. R. Fellows. *Parameterized complexity*. Springer-Verlag, New York, 1999.

[21] R. G. Downey and M. R. Fellows. Parameterized complexity after (almost) ten years: review and open questions. In *Combinatorics, computation & logic '99 (Auckland)*, pages 1–33. Springer, Singapore, 1999.

[22] M. Dyer, A. Frieze, and M. Jerrum. On counting independent sets in sparse graphs. In *Proceedings of the 40th Annual IEEE Symposium on Foundations of Computer Science*, pages 210–217, 1999.

[23] M. Dyer and C. Greenhill. The complexity of counting graph homomorphisms. *Random Structures Algorithms*, 17:260–289, 2000.

[24] K. Edwards. The complexity of coloring problems on dense graphs. *Theoretical Computer Science*, 16:337–343, 1986.

[25] T. Feder and P. Hell. List homomorphisms to reflexive graphs. *Journal of Combinatorial Theory (series B)*, 72(2):236–250, 1998.

[26] T. Feder, P. Hell, and J. Huang. List homomorphisms and circular arc graphs. *Combinatorica*, 19:487–505, 1999.

[27] T. Feder, P. Hell, and J. Huang. Bi-arc graphs and the complexity of list homomorphism. manuscript, 2001.

[28] A. Galluccio, P. Hell, and J. Nešetřil. The complexity of H-colouring of bounded degree graphs. *Discrete Mathematics*, 222(1-3):101–109, 2000.

[29] M. R. Garey and D. S. Johnson. *Computers and Intractability: A Guide to the Theory of NP-Completeness*. Freeman, San Francisco, 1979.

[30] R. Halin. S-functions for graphs. *Journal of Geometry*, 8:171–186, 1976.

[31] P. Hell and J. Nešetřil. On the complexity of *H*-coloring. *Journal of Combinatorial Theory (series B)*, 48:92–110, 1990.

[32] P. Hell and J. Nešetřil. Counting list homomorphisms and graphs with bounded degrees. this procedings, 2001.

[33] P. Hell, J. Nešetřil, and X. Zhu. Duality and polynomial testing of tree homomorphisms. *Transactions of the American Mathematical Society*, 348(4):1281–1297, 1996.

[34] M. Jerrum. A very simple algorithm for estimating the number of k-colorings of a low degree graph. *Random Structures and Algorithms*, 7:157–165, 1995.

[35] M. Jerrum and A. Sinclair. The Markov chain Monte Carlo method: An approach to approximate counting and integration. In D. S. Hochbaum, editor, *Approximation Algorithms for NP-hard problems*, pages 482–520. PWS, Boston, 1995.

[36] Z. Kloks. *Treewidth. Computations and Approximations*, volume 842 of *Lecture Notes in Computer Science*. Springer-Verlag, 1994.

[37] Ph. G. Kolaitis and M. Y. Vardi. Conjunctive-Query containment and Constraint Satisfaction. Journal of Computer and System Sciences, 61(2):302–332 2000.

[38] J. Kratochvíl and Z. Tuza. Algorithmic complexity of list colorings. *Discrete Applied Mathematics*, 50(3):297–302, 1994.

[39] J. Lagergren. Efficient parallel algorithms for graph with bounded tree-width. *Journal of Algorithms*, 20:20–44, 1996.

[40] L. Levin. Universal sequential search problems. *Problems of Information Transmissions*, 9:265–266, 1973.

[41] J. Matoušek and R. Thomas. Algorithms finding tree-decompositions of graphs. *Journal of Algorithms*, 12:1–22, 1991.

[42] J. Nešetřil. Aspects of structural combinatorics (graph homomorphisms and their use). *Taiwanese Journal of Mathematics*, 3(4):381–423, 1999.

[43] S. D. Noble. Evaluating the Tutte polynomial for graphs of bounded treewidth. *Combinatorics, Probability and Computing*, 7:307–321, 1998.

[44] B. Oporowski, J. Oxley, and R. Thomas. Typical subgraphs of 3- and 4-connected graphs. *Journal of Combinatorial Theory (series B)*, 57(2):239–257, 1993.

[45] C. Papadimitriou and M. Yannakakis. Optimization, approximation, and complexity classes. *Journal of Computer and System Sciences*, 43:425–440, 1991.

[46] B. Reed. Finding approximate separators and computing tree-width quickly. In *Proceedings of the 24th ACM Symposium on Theory of Computing*, pages 221–228, 1992.

[47] N. Robertson and P. D. Seymour. Graph minors - a survey. In I. Anderson, editor, *Surveys in Combinatorics*, pages 153–171. Cambridge University Press, 1985.

[48] N. Robertson and P. D. Seymour. Graph minors. II. algorithmic aspects of tree-width. *Journal of Algorithms*, 7:309–322, 1986.

[49] N. Robertson and P. D. Seymour. Graph minors. V. Excluding a planar graph. *Journal of Combinatorial Theory (series B)*, 41:92–114, 1986.

[50] J. A. Telle and A. Proskurowski. Algorithms for vertex partitioning problems on partial k-trees. *SIAM Journal on Discrete Mathematics*, 10(4):529–550, 1997.

DEPARTAMENT DE LLENGUATGES I SISTEMES INFORMÀTICS. UNIVERSITAT POLITÈCNICA DE CATALUNYA. CAMPUS NORD MÒDUL C6. C/ JORDI GIRONA SALGADO 1-3, 08034, BARCELONA, SPAIN.
E-mail address: `diaz@lsi.upc.es`

DEPARTAMENT DE LLENGUATGES I SISTEMES INFORMÀTICS. UNIVERSITAT POLITÈCNICA DE CATALUNYA. CAMPUS NORD MÒDUL C6. C/ JORDI GIRONA SALGADO 1-3, 08034, BARCELONA, SPAIN.
E-mail address: `mjserna@lsi.upc.es`

DEPARTAMENT DE LLENGUATGES I SISTEMES INFORMÀTICS. UNIVERSITAT POLITÈCNICA DE CATALUNYA. CAMPUS NORD MÒDUL C6. C/ JORDI GIRONA SALGADO 1-3, 08034, BARCELONA, SPAIN.
E-mail address: `sedthilk@lsi.upc.es`

DIMACS Series in Discrete Mathematics
and Theoretical Computer Science
Volume **63**, 2004

Rapidly mixing Markov chains for dismantleable constraint graphs

Martin Dyer, Mark Jerrum, and Eric Vigoda

ABSTRACT. If $G = (V_G, E_G)$ is an input graph, and $H = (V_H, E_H)$ a fixed constraint graph, we study the set Ω of homomorphisms (or colorings) from V_G to V_H, i.e., functions that preserve adjacency. Brightwell and Winkler introduced the notion of *dismantleable* constraint graph to characterize those H whose associated set Ω of homomorphisms is, for every G, connected under single vertex recolorings. Given fugacities $\lambda(c) > 0$ ($c \in V_H$) our focus is on sampling a coloring $\omega \in \Omega$ according to the *Gibbs distribution*, i.e., with probability proportional to $\prod_{v \in V_G} \lambda(\omega(v))$. The *Glauber dynamics* is a Markov chain on Ω which recolors a single vertex at each step, and leaves invariant the Gibbs distribution. We prove that, for each dismantleable H and degree bound Δ, there exist positive constant fugacities on V_H such that the Glauber dynamics has mixing time $O(n^2)$, for all graphs G whose vertex degrees are bounded by Δ.

1. Introduction

Graph homomorphisms provide a natural generalization of many well-studied combinatorial problems, including independent sets and colorings. Our focus is the computational complexity of randomly generating a homomorphism and computing the number of homomorphisms.

We consider an input graph $G = (V_G, E_G)$ with maximum degree Δ and a constraint graph $H = (V_H, E_H)$, where the latter may have loops. Let $n = |V_G|$, $h = |V_H|$, and let $x \sim y$ denote adjacency of a pair of vertices in G or H. Our interest is in the set Ω of H-*colorings* (or homomorphisms) $\sigma : V_G \to V_H$ where $\sigma(v) \sim \sigma(w)$ for all $v \sim w$. Just two of the numerous combinatorial structures that can be expressed as H-colorings are illustrated in Figure 1. It is easily verified that when $H = H_{\mathrm{IS}}$ the set Ω of H-colorings of a graph G is in direct correspondence with the set of all independent sets in G; while when $H = H_{3\mathrm{Col}}$, the set Ω consists of all proper 3-colorings of G.

Each color $c \in V_H$ is assigned a *fugacity* $\lambda(c)$. For convenience, we extend λ to colorings $\sigma \in \Omega$ by defining $\lambda(\sigma) = \prod_{v \in V_G} \lambda(\sigma(v))$. We are interested in the *Gibbs* (or "hard-core") probability distribution π over Ω given by $\pi(\sigma) = \lambda(\sigma)/Z$ for all

This work was supported by EPSRC Research grant "Sharper Analysis of Randomised Algorithms: A Computational Approach" and in part by the ESPRIT Project RAND-APX. The last author was partially supported by a Koshland Scholar award from the Weizmann Institute of Science.

87

FIGURE 1. Independent sets (left) and proper 3-colorings (right).

$\sigma \in \Omega$, where the normalising factor $Z = \sum_{\sigma \in \Omega} \lambda(\sigma)$ is the *partition function* for H-colorings of G. The complexity of exactly computing Z has been investigated by Dyer and Greenhill [6] and is well-understood; in fact the problem is #P-complete for every non-trivial H even when we restrict attention to low-degree graphs G and uniform fugacity 1.

The picture appears (at the moment) more vague when we consider approximate computation of Z, though Dyer, Goldberg, Greenhill and Jerrum [7] discuss the case where all fugacities are equal, and uncover a part of the picture. In this paper, we examine the closely related problem of sampling colorings from the Gibbs distribution, or at least a close approximation to that distribution. We identify a class of constraint graphs H for which sampling can be done in polynomial time, for appropriately chosen non-zero fugacities. Since approximate counting is efficiently reducible to sampling for this class of constraint graphs, we obtain a polynomial-time approximation algorithm (technically an FPRAS) for this class of constraint graphs with their associated fugacities.

The typical approach to efficient sampling is to set up an ergodic Markov chain (X_t) on Ω whose stationary distribution is the desired distribution π. By simulating this Markov chain for sufficiently many steps, samples from a distribution arbitrarily close to π may be obtained. There are a number of ways of specifying appropriate transition probabilities for this Markov chain; a basic one is provided by the *Glauber dynamics*, which changes one vertex color at a time, according to the following experiment:

G1. Choose a vertex $v \in V_G$ uniformly at random.

G2. Let $X_{t+1}(w) := X_t(w)$ for all $w \neq v$.

G3. Let $S = \{c \in V_H : c \sim X_t(w) \text{ for all } w \sim v\}$ denote the set of valid colors for v.

G4. Choose the color $X_{t+1}(v)$ randomly from S with probability proportional to its fugacity.

In what circumstances might the Glauber dynamics provide an effective solution to the problem of sampling H-colorings? Certainly we require the state space Ω to be connected in some sense. A Markov chain with finite state space Ω and transition matrix $P : \Omega \times \Omega \to [0,1]$ is called *ergodic* if the following conditions hold:

- *Irreducibility*: for all states (colorings) $\sigma, \tau \in \Omega$, there exists a time $t = t(\sigma, \tau)$ such that $P^t(\sigma, \tau) > 0$;
- *Aperiodicity*: for all states σ, $\gcd\{t : P^t(\sigma, \sigma) > 0\} = 1$.

It is a classical theorem from stochastic processes that a finite, ergodic Markov chain has a unique stationary distribution.

In many situations it is easy to verify that a candidate distribution is stationary. Specifically, a distribution π' on Ω that satisfies the so-called detailed balance conditions,

$$(1) \qquad \pi'(\sigma)\,P(\sigma,\tau) = \pi'(\tau)\,P(\tau,\sigma) \quad \text{for all } \sigma,\tau \in \Omega,$$

is a stationary distribution. It is straightforward to verify that the Gibbs distribution π satisfies the detailed balance conditions for the Glauber dynamics on H-colorings. Nevertheless, we still need to check that π is the *unique* stationary distribution, i.e., that the Glauber dynamics meets the conditions of irreducibility and aperiodicity.

Since the Glauber dynamics satisfies $P(\sigma,\sigma) > 0$ for all colorings σ (it is always valid to recolor a vertex with the same color as before) it is immediate that the Markov chain specified by the Glauber dynamics is aperiodic. The question of irreducibility is more subtle. Brightwell and Winkler [1] characterized the constraint graphs H for which the Glauber dynamics on H-colorings is irreducible for every input graph G. They call such constraint graphs "dismantleable" and prove a host of equivalent conditions for dismantleability.

For efficient sampling, however, it is not sufficient that the Glauber dynamics converges eventually to the stationary (Gibbs) distribution; we need to know that the rate of convergence is rapid. Our main result is that for every dismantleable H and degree bound Δ there exists a set of non-zero fugacities such that the Glauber dynamics has polynomial mixing time[1] uniformly over graphs G of maximum degree at most Δ. The term *rapidly mixing* is often applied to a Markov chain, such as this one, whose mixing time is polynomial in some natural measure of input size, in this case the order of G.

If H is a complete graph with loops on all vertices then the Glauber dynamics is trivially rapidly mixing for any input graph G. Cooper, Dyer and Frieze [3] proved that, for any other H, there exists a set of fugacities (actually, one may set all fugacities equal to 1) and a sufficiently large degree Δ such that the Glauber dynamics on H-colorings has exponential mixing time, for some infinite family of Δ-regular input graphs G. In contrast, we prove here that, provided H is dismantleable, there is another set of fugacities such that the Glauber dynamics on H-colorings has polynomial mixing time, on every bounded degree input graph G.

2. Definitions and results

In graph-theoretic terms, a constraint graph H is said to be *dismantleable* if there exists an ordering $<$ of V_H for which the following holds:

> Let c^* be the smallest vertex (color) in the ordering $<$. Then,
> for all $c \in V_H \setminus \{c^*\}$, there exists $p(c) < c$ such that $c' \leq c$ and
> $c' \sim c$ entail $c' \sim p(c)$.

It is quite easy to show that the above condition is sufficient to ensure irreducibility. Fix an ordering on colors (i.e., vertices of H) for which the dismantleability condition holds, with c^* being the least color in the ordering. Starting from an arbitrary coloring $\sigma \in \Omega$, repeating the following procedure reaches the monochromatic coloring $(c^*)^{V_G}$: let c denote the greatest color appearing in σ; recolor all occurrences

---

[1] A precise definition of mixing time is given in §2, but roughly it is the time t at which the t-step distribution of a Markov chain comes sufficiently close to the stationary distribution π in l_1 distance.

of c by $p(c)$. Note that the steps can be reversed to get from $(c^*)^{V_G}$ back to the original coloring.[2] That dismantleability is a necessary condition is a little trickier, and we refer the reader to [1].

We are interested in the asymptotic distance of the Glauber dynamics from stationarity. The traditional measure of distance in this context is (total) variation distance, defined as:

$$d_{\mathrm{TV}}(P^t(\sigma, \cdot), \pi) = \frac{1}{2} \sum_{\tau \in \Omega} |P^t(\sigma, \tau) - \pi(\tau)|,$$

where σ is the initial state (coloring). Our focus is the time to get close to stationarity, known as the mixing time. For an initial state $\sigma \in \Omega$ and $0 \leq \varepsilon < 1$, let

$$T_\sigma(\varepsilon) = \min\{t : d_{\mathrm{TV}}(P^t(\sigma, \cdot), \pi) \leq \varepsilon\}.$$

The *mixing time* is defined as

$$T(\varepsilon) = \max_{\sigma \in \Omega} T_\sigma(\varepsilon).$$

For the purposes of the analysis we do not consider the Glauber dynamics directly, but consider instead a variant dynamics with a more restricted set of possible transitions. (It is perhaps counterintuitive that the proof of rapid mixing might be simplified by limiting the available transitions!) In brief, we allow only transitions in which a color c is replaced by its "parent" $p(c)$ or vice versa. If we were simply to restrict the available transitions in this way, while retaining the original fugacities $\lambda(c)$, then we would alter the stationary distribution. To correct for this, we assign to each color $c \in V_H$ a weight $\mu(c)$, which is in general different from $\lambda(c)$. The weight function μ influences transition probabilities just as λ does, but we avoid calling $\mu(c)$ a "fugacity", since the relation of the weights μ to the stationary distribution of the Markov chain is not quite the same as that of the fugacities λ to the Gibbs distribution.

Fix an ordering $<$ and parents $p(\cdot)$ consistent with the definition of dismantleability. Then our new Markov chain (X_t) has transitions $X_t \to X_{t+1}$ defined by the following experiment:

V1. Choose a vertex $v \in V_G$ uniformly at random. Let c denote the current color of v.

V2. Let $X_{t+1}(w) := X_t(w)$ for all $w \neq v$.

V3. Let $R = R(c) = \{c' \in V_H : c' = p(c), c' = c, \text{ or } c = p(c')\}$ denote the set of "relatives" of color c.

V4. Choose the color c' randomly from R with probability proportional to its weight $\mu(c')$. Provided it leads to a valid H-coloring, set $X_{t+1}(v) := c'$; otherwise set $X_{t+1}(v) := c$.

As before, the variant dynamics is an ergodic Markov chain provided H is dismantleable. The stationary distribution under the variant dynamics has probabilities $\pi(\sigma)$ proportional to $\prod_{v \in V_G} \mu(\sigma(v)) \, \mu(R(\sigma(v)))$, where $\mu(R(c)) = \sum_{c' \in R(c)} \mu(c')$. This may be verified from the detailed balance conditions (1). Suppose σ and τ agree at all vertices except v, at which $c = \sigma(v)$ and $c' = \tau(v)$. (Note that σ

---

[2]Incidentally, the procedure just described provides a demonstration that $(c^*)^{V_G}$ is a valid H-coloring of G, provided *some* valid H-coloring exists. Thus, except in some trivial situations, the vertex c^* must be looped in H.

and τ must be of this form if $P(\sigma, \tau)$ is to be non-zero.) Then we have $P(\sigma, \tau) = \mu(c')/\mu(R(c))$ and $P(\tau, \sigma) = \mu(c)/\mu(R(c'))$. Hence

$$\frac{P(\sigma, \tau)}{P(\tau, \sigma)} = \frac{\mu(c')\,\mu(R(c'))}{\mu(c)\,\mu(R(c))} = \frac{\pi(\tau)}{\pi(\sigma)},$$

as required. Thus the stationary distribution is the Gibbs distribution with fugacities $\lambda(c) = \mu(c)\,\mu(R(c))$. We can now state our main theorem.

THEOREM 1. *For every input graph $G = (V_G, E_G)$ with maximum degree Δ and dismantleable constraint graph $H = (V_H, E_H)$, there exists a set of weights (depending only on Δ and $h = |V_H|$) such that the variant dynamics (as defined in V1–V4 above) has mixing time $O(n \log n)$. Specifically, $T(\varepsilon) = O\big(n(\log n + \log \varepsilon^{-1})\big)$.*

COROLLARY 2. *Under the conditions of Theorem 1, there exists a set of fugacities (depending only on Δ and h) such that the Glauber dynamics (as defined in G1–G4 above) has mixing time $O(n^2)$. Specifically, $T(\varepsilon) = O\big(n(n + \log \varepsilon^{-1})\big)$.*

The corollary follows easily from the Diaconis and Saloff-Coste technique [4] for comparing the associated Dirichlet forms of the Markov chains. Indeed a slightly weaker bound on mixing time, with $n \log n + \log \varepsilon^{-1}$ replacing $n + \log \varepsilon^{-1}$, can be obtained simply by substituting the mixing time bound from Theorem 1 into a general comparison theorem of Randall and Tetali [9, Prop. 4]. However, by applying exactly the same method, but working from first principles, we can avoid the factor $\log n$. The argument is this. Applying Sinclair's [10, Prop. 1(ii)] to the mixing time bound of Theorem 1 with $\varepsilon = n^{-1}$, we discover that the spectral gap of the variant dynamics is $\Omega(n^{-1})$. Now the spectral gaps of the Glauber and variant dynamics are the same to within a constant factor: this can be seen by inspecting the variational characterisation of the second largest eigenvalue. Corollary 2 then follows by plugging the $\Omega(n^{-1})$ bound on spectral gap into [10, Prop. 1(i)].

Since the above comparison argument relies only on each possible transition in the variant dynamics being matched (in general with different probability) in the Markov chain under comparison, Corollary 2 holds for other single-site update rules, e.g., Metropolis. We conjecture that the true mixing time here is also $O\big(n(\log n + \log \varepsilon^{-1})\big)$ but the proof of this (if true) appears rather more complex than our proofs of Theorem 1 and Corollary 2.

As mentioned in the introduction, the existence of an efficient sampling procedure for certain structures usually entails the existence of an efficient approximation algorithm for the partition (or generating) function for those structures. The current situation is no exception.

COROLLARY 3. *Under the same conditions as Theorem 1, there is a fully polynomial randomized approximation scheme (FPRAS) for the partition function Z of H-colorings of G.*

For the definition of FPRAS, and also the reduction from approximate counting to sampling required to establish Corollary 3, see for example Jerrum's survey article [8], specifically §2. The reduction is given there in the context of usual (proper) colorings, but it works almost without change for H-colorings, when H

is dismantleable.[3] To verify the reduction in the current context it is necessary to check that the ratios ρ_i appearing in that reduction are bounded away from 0. This can be done using a straightforward extension of the argument used to prove ergodicity of the Glauber dynamics.

Finally, note that many constraint graphs H are covered both by Theorem 1 and by the result of Cooper et al. [**3**]. In other words, there are graphs H — the simplest being K_2 with a single loop, which corresponds to independent sets in the input graph G — for which the mixing time of the Glauber dynamics is either polynomial or exponential, depending on the fugacities.

3. Coupling

We prove the main theorem via coupling. A coupling of a Markov chain is a joint evolution of two copies of the chain, designed to minimize the time till the copies coalesce. In order to be a valid coupling we need that individually each copy behaves faithfully, and once the pair coalesce they evolve together. To be precise, a (Markovian) coupling is a Markov chain P' on $\Omega \times \Omega$ which satisfies the following conditions:

$$\sum_{\tau' \in \Omega} P'((\sigma, \tau), (\sigma', \tau')) = P(\sigma, \sigma') \qquad \text{for all } \sigma, \sigma', \tau \in \Omega;$$

$$\sum_{\sigma' \in \Omega} P'((\sigma, \tau), (\sigma', \tau')) = P(\tau, \tau') \qquad \text{for all } \sigma, \tau, \tau' \in \Omega;$$

and

$$P'((\sigma, \sigma), (\sigma', \sigma')) = P(\sigma, \sigma') \qquad \text{for all } \sigma, \sigma' \in \Omega.$$

Our goal is to define a coupling which, in expectation, makes progress with respect to an appropriately defined distance metric after every coupled transition. Defining and analyzing a coupling for an arbitrary pair of states is typically a complicated task; however, the path coupling lemma of Bubley and Dyer simplifies matters. In particular, it suffices to focus on colorings which differ by a single transition of the dynamics, referred to as *adjacent* colorings. We write $\sigma \sim_v \tau$ to indicate that the pair of colorings σ, τ differ only at vertex v, and $\sigma \sim \tau$ if $\sigma \sim_v \tau$ for some vertex v.

We analyze this set of adjacent colorings with respect to the following distance metric. Each color $c \in V_H$ will be assigned a distance weight $d(c)$. For colorings $\sigma \sim_v \tau$, where without loss of generality $\tau(v) = p(\sigma(v))$, we will assign distance $d(\sigma, \tau) = d(\sigma(v))$. For any $\sigma, \tau \in \Omega$, let $\rho(\sigma, \tau)$ denote the collection of paths η such that $\sigma = \eta_0 \sim \eta_1 \sim \cdots \sim \eta_\ell = \tau$, where $\ell(\eta)$ is the length (in terms of number of transitions) of η. We define the distance between an arbitrary pair of states as the total distance along a shortest path:

$$d(\sigma, \tau) = \min_{\eta \in \rho(\sigma, \tau)} \sum_{0 \leq i < \ell(\eta)} d(\eta_i, \eta_{i+1}).$$

For a pair of colorings σ, τ and a coupling (σ, τ), let σ', τ' denote the resulting pair of colorings after the coupled transition. The specialization of the path coupling lemma to our setting is as follows.

---

[3]It is in fact possible to reduce approximate counting to sampling without restriction on the graph H, but the reduction is then substantially more involved and, crucially, in the context of Corollary 3, does not preserve the bound Δ on vertex degrees [**5**].

LEMMA 4 (Bubley and Dyer [2]). *If there exists a $\beta < 1$ and a coupling for all $\sigma \sim \tau$ such that*

$$\mathbb{E}[d(\sigma', \tau')] < \beta\, d(\sigma, \tau),$$

then the mixing time is bounded by

$$T(\varepsilon) \leq \frac{\log(D/\varepsilon)}{1 - \beta},$$

where $D = \max_{\sigma, \hat{\sigma} \in \Omega} d(\sigma, \hat{\sigma})$.

In our application, we shall discover that $\beta = 1 - \Theta(1/n)$ and $D = \Theta(n)$, leading easily to Theorem 1.

4. Proof of Theorem 1

Fix an ordering on the colors $c_1 < c_2 < \cdots < c_h$ for which the dismantleability condition holds. Set $d(c_i) = 2(\Delta + 2)^{i+1}$ and let $d_{\max} = d(c_h)$. Now set $\mu_i = \mu(c_i) = (d_{\max} + 1)^{-i}$, for $i = 1, \ldots, h$. Observe that the inequality

$$\frac{\sum_{k > i} \mu_k}{\mu_i} < \frac{1}{d_{\max}}$$

holds for all i.

Fix a pair of colorings $\sigma \sim_v \tau$ and once again let σ', τ' denote the resulting colorings after our coupled transition. To prove the theorem we need to demonstrate a coupling for which $\mathbb{E}[d(\sigma', \tau')] - d(\sigma, \tau)$ is negative and bounded away from zero.

We couple the two chains so that both chains attempt to modify the same vertex at every step. Therefore, for a vertex x, it is well defined to let

$$\mathbb{E}_x[d(\sigma', \tau')] = \mathbb{E}\left[d(\sigma', \tau') \mid \text{vertex } x \text{ is selected by the coupled process}\right].$$

Observe that the distance metric does not change when we recolor a vertex sufficiently far from v. Specifically, we only need to consider recolorings of v or a neighbor w of v. In summary, we have

$$n\big(\mathbb{E}[d(\sigma', \tau')] - d(\sigma, \tau)\big) = \big(\mathbb{E}_v[d(\sigma', \tau')] - d(\sigma, \tau)\big)$$

(2)
$$+ \sum_{w : w \sim v} \big(\mathbb{E}_w[d(\sigma', \tau')] - d(\sigma, \tau)\big).$$

Of all the colors that are valid for $\sigma'(x)$, let $c_\sigma^*(x)$ denote the one of greatest weight. Note that either $c_\sigma^*(x) = \sigma(x)$ or $c_\sigma^*(x) = p(\sigma(x))$, and that, in the latter case, $c_\sigma^*(x)$ has the greatest weight among the various colors that may be proposed. Either way, if vertex x is selected, then x will acquire color $c_\sigma^*(x)$ unless a color of lower weight than $c_\sigma^*(x)$ is proposed. Thus, conditioned on vertex x being selected, we have

(3)
$$\Pr[\sigma'(x) \neq c_\sigma^*(x)] \leq \frac{\sum_{c > c_\sigma^*(x)} \mu(c)}{\mu(c_\sigma^*(x))} < \frac{1}{d_{\max}}.$$

Defining $c_\tau^*(x)$ similarly, the analogous inequality holds with respect to τ as well.

In the light of inequality (3) and its analogue for τ we may complete the definition of the coupled process as follows. With probability $1 - 1/d_{\max}$ we couple $\sigma'(x) = c_\sigma^*(x)$ with $\tau'(x) = c_\tau^*(x)$. We call such a coupled move a *type A transition*. With the remaining probability of $1/d_{\max}$ the two chains independently recolor x

(using the residual distributions), a *type B transition*. In the latter case, it is clear that $d(\sigma', \sigma) \le d_{\max}$ and $d(\tau', \tau) \le d_{\max}$ which implies

$$(4) \qquad \mathbb{E}_x[d(\sigma', \tau')|\ \text{type B transition}] \le 2d_{\max} + d(\sigma, \tau).$$

We now proceed to bound the expected change in distance from a type A transition. It is clear that the distance after the type A transition on x is maximized when $c_\sigma^*(x) \ne c_\tau^*(x)$.

Recall that, by convention, $\tau(v) = p(\sigma(v))$, rather than vice versa. Consider the recoloring of a neighbor w of v. We begin by proving that $c_\sigma^*(w) \ne c_\tau^*(w)$ implies $\sigma(w) < \sigma(v)$. Suppose $\sigma(w) \ge \sigma(v)$. Vertex w has the same color in both chains, therefore colors $c_\sigma^*(w)$ and $c_\tau^*(w)$ are either $\sigma(w)$ or $p(\sigma(w))$. Since $\sigma(w) \sim \sigma(v)$ and $\sigma(w) \sim \tau(v)$, from the definition of dismantleability and our ordering on V_H we know $p(\sigma(w)) \sim \sigma(v)$ and $p(\sigma(w)) \sim \tau(v)$. Similarly if $u \ne v$ is any other neighbor of w, then dismantleability implies $p(\sigma(w)) \sim \sigma(u) = \tau(u)$. Therefore, recoloring vertex w to color $p(\sigma(w))$ is valid in both chains or neither and hence $c_\sigma^*(w) = c_\tau^*(w)$.

Now suppose $c_\sigma^*(w) \ne c_\tau^*(w)$. In addition to the fact $\sigma(w) < \sigma(v)$, it is clear that $c_\sigma^*(w) = \sigma(w)$ and $c_\tau^*(w) = p(\sigma(w))$. Following the type A transition for w we have $\sigma' = \sigma$ and $d(\tau', \tau) = d(\sigma(w)) \le d(\sigma(v))/(\Delta + 2)$.

In summary, we have proven

$$(5) \qquad \mathbb{E}_w[d(\sigma', \tau')|\ \text{type A transition}] \le d(\sigma, \tau)\left[1 + \frac{1}{\Delta + 2}\right].$$

Combining (4) and (5) — recalling that a type B transition occurs with probability at most $1/d_{\max}$ — yields

$$(6) \qquad \mathbb{E}_w[d(\sigma', \tau')] - d(\sigma, \tau) \le 2 + \frac{d(\sigma, \tau)}{\Delta + 2}.$$

We complete the proof by considering the effect of recoloring v by a type A transition. Since $p(\sigma(v)) = \tau(v)$, it is clear that $p(\sigma(v))$ is a valid color for both $\sigma'(v)$ and $\tau'(v)$. Moreover, $c_\sigma^*(v) = p(\sigma(v))$ and $c_\tau^*(v)$ is either $p(\sigma(v))$ or $p(p(\sigma(v)))$. In other words, $d(\sigma', \tau') > 0$ implies $c_\tau^*(v) = p(p(\sigma(v)))$. In which case we have $d(\sigma', \tau') \le d(p(\sigma(v))) \le d(\sigma, \tau)/(\Delta + 2)$. Restating our bound:

$$\mathbb{E}_v[d(\sigma', \tau')|\ \text{type A transition}] \le \frac{d(\sigma, \tau)}{\Delta + 2}.$$

The above inequality together with (4) implies

$$
\begin{aligned}
\mathbb{E}_v[d(\sigma', \tau')] - d(\sigma, \tau) \ &\le\ 2 + d(\sigma, \tau)\left[\frac{1}{d_{\max}} + \frac{1}{\Delta + 2} - 1\right] \\
(7) \qquad\qquad\qquad &\le\ 3 - d(\sigma, \tau)\left[1 - \frac{1}{\Delta + 2}\right].
\end{aligned}
$$

Let $\Delta(v)$ denote the degree of vertex v. Putting inequalities (6) and (7) into (2), and recalling $d(\sigma, \tau) = d(\sigma(v))$, we obtain

$$n\big(\mathbb{E}[d(\sigma', \tau')] - d(\sigma, \tau)\big) \leq \Delta(v)\left[2 + \frac{d(\sigma, \tau)}{\Delta + 2}\right] + \left[3 - \frac{(\Delta + 1)\, d(\sigma, \tau)}{\Delta + 2}\right]$$
$$\leq (2\Delta + 3) - \frac{d(\sigma(v))}{\Delta + 2}$$
$$\leq (2\Delta + 3) - 2(\Delta + 2)$$
$$= -1.$$

Since $\sigma \sim \tau$ implies $d(\sigma, \tau) \leq d_{\max}$, we conclude

$$\mathbb{E}[d(\sigma', \tau')] \leq (1 - 1/nd_{\max})\, d(\sigma, \tau),$$

and hence we may take $\beta = 1 - 1/nd_{\max}$ in Lemma 4. If $\hat\sigma = (c_1)^{V_G}$, we know that $d(\sigma, \hat\sigma) \leq nhd_{\max}$ for all $\sigma \in \Omega$, by the path of transitions described in §2. Therefore, we may take $D = 2nhd_{\max}$ in Lemma 4. Theorem 1 now follows.

References

[1] G. R. Brightwell and P. Winkler. Gibbs measures and dismantleable graphs. *J. Combin. Theory Ser. B*, 78(1):141–166, 2000.

[2] R. Bubley and M. Dyer. Path coupling, Dobrushin uniqueness, and approximate counting. In *38th Annual Symposium on Foundations of Computer Science*, pages 223–231, Miami Beach, FL, October 1997. IEEE.

[3] C. Cooper, M. Dyer, and A. Frieze. On Markov chains for randomly H-coloring a graph. *J. Algorithms*, 39(1):117–134, 2001.

[4] P. Diaconis and L. Saloff-Coste. Comparison theorems for reversible Markov chains. *The Annals of Applied Probability*, 3(3):696–730, 1993.

[5] M. Dyer, L.A. Goldberg and M. Jerrum. Counting and Sampling H-colourings. To appear in *Proceedings of the 6th. International Workshop on Randomization and Approximation Techniques in Computer Science*, Springer-Verlag LNCS, September 2002.

[6] M. Dyer and C. Greenhill. The complexity of counting graph homomorphisms. *Random Structures Algorithms*, 17(3-4):260–289, 2000.

[7] M. E. Dyer, L. A. Goldberg, C. S. Greenhill, and M. R. Jerrum. On the relative complexity of approximate counting problems. In *Proceedings of APPROX 2000*, Lecture Notes in Computer Science vol. 1913, pages 108–119, Springer Verlag, 2000.

[8] M. Jerrum. Mathematical foundations of the Markov chain Monte Carlo method. In *Probabilistic Methods for Algorithmic Discrete Mathematics* (M. Habib, C. McDiarmid, J. Ramirez-Alfonsin & B. Reed, eds), Algorithms and Combinatorics vol. 16, Springer-Verlag, 1998, 116–165.

[9] D. Randall and P. Tetali. Analyzing Glauber dynamics by comparison of Markov chains. *J. Math. Phys.*, 41(3):1598–1615, 2000.

[10] A. Sinclair. Improved bounds for mixing rates of Markov chains and multicommodity flow. *Combinatorics, Probability and Computing*, 1(4):351–370, 1992.

SCHOOL OF COMPUTING, UNIVERSITY OF LEEDS, LEEDS LS2 9JT, UNITED KINGDOM
E-mail address: dyer@comp.leeds.ac.uk

LABORATORY FOR FOUNDATIONS OF COMPUTER SCIENCE, UNIVERSITY OF EDINBURGH, JCMB, THE KING'S BUILDINGS, EDINBURGH EH9 3JZ, UNITED KINGDOM
E-mail address: mrj@dcs.ed.ac.uk

DEPARTMENT OF COMPUTER SCIENCE, UNIVERSITY OF CHICAGO, 1100 E 58TH STREET, CHICAGO, IL 60637, USA
E-mail address: vigoda@cs.uchicago.edu

DIMACS Series in Discrete Mathematics
and Theoretical Computer Science
Volume **63**, 2004

On weighted graph homomorphisms

David Galvin and Prasad Tetali

ABSTRACT. For given graphs G and H, let $|Hom(G, H)|$ denote the set of graph homomorphisms from G to H. We show that for any finite, n-regular, bipartite graph G and any finite graph H (perhaps with loops), $|Hom(G, H)|$ is maximum when G is a disjoint union of $K_{n,n}$'s. This generalizes a result of J. Kahn on the number of independent sets in a regular bipartite graph. We also give the asymptotics of the logarithm of $|Hom(G, H)|$ in terms of a simply expressed parameter of H.

We also consider weighted versions of these results which may be viewed as statements about the partition functions of certain models of physical systems with hard constraints.

1. Introduction

Let G be an n-regular, N-vertex bipartite graph on vertex set $V(G)$, and let H be a fixed graph on vertex set $V(H)$ (perhaps with loops). We will always use u, v for the vertices of G and i,j for those of H. Set

$$Hom(G, H) = \{f : V(G) \to V(H) \ : \ u \sim v \Rightarrow f(u) \sim f(v)\}.$$

That is, $Hom(G, H)$ is the set of graph homomorphisms from G to H. (For graph theory basics, see e.g. [2], [5]).

When $H = H_{ind}$ consists of one looped and one unlooped vertex connected by an edge, an element of $Hom(G, H_{ind})$ can be thought of as a specification of an independent set (a set of vertices spanning no edges) in G. Our point of departure is the following result of Kahn [7], bounding the number of independent sets in regular bipartite graphs. For any graph G, write $\mathcal{I}(G)$ for the set of independent sets of G.

THEOREM 1.1. *For any n-regular, N-vertex bipartite graph G,*

$$|\mathcal{I}(G)| \le (2^{n+1} - 1)^{N/2n}.$$

An approximate version of Theorem 1.1 — $\log |\mathcal{I}(G)| \le (1/2 + o(1))N$, where $o(1) \to 0$ as $n \to \infty$ — for general n-regular, N-vertex G was earlier proved by Alon [1]. Note that $|Hom(K_{n,n}, H_{ind})| = 2^{n+1} - 1$ (where $K_{n,n}$ is the complete

---

2000 *Mathematics Subject Classification.* 05A16.
Key words and phrases. Graph homomorphisms, hard constraint models, graph coloring.
Second author's research supported in part by NSF grant DMS-0100298.

bipartite graph with n vertices on each side), so we may paraphrase Theorem 1.1 by saying that $|Hom(G, H_{ind})|$ is maximum when G is a disjoint union of $K_{n,n}$'s. Our main result is a generalization of this statement (and our proof is a generalization of Kahn's).

PROPOSITION 1.2. *For any n-regular, N-vertex bipartite G, and any H,*

$$|Hom(G, H)| \leq |Hom(K_{n,n}, H)|^{N/2n}.$$

Somewhat surprisingly, we can also exhibit a lower bound that is good enough to allow us to obtain the asymptotics of $\log|Hom(G, H)|$ for fixed H as $n \to \infty$ (here, and throughout the rest of the paper, we use log for the base 2 logarithm). To state the result, it is convenient to introduce a parameter of H that is very closely related to $|Hom(K_{n,n}, H)|$, but is easier to work with. Set

$$\eta(H) = \max\{|A||B| : A, B \subseteq V(H), i \sim j \ \forall i \in A, j \in B\}.$$

(When H is loopless, this is the maximum number of edges in a complete bipartite subgraph of H. Peeters [10] has recently shown that determining $\eta(H)$, even when H is bipartite, is NP-complete.)

PROPOSITION 1.3. *For any n-regular, N-vertex bipartite G, and any H,*

$$\frac{\log \eta(H)}{2} \leq \frac{\log|Hom(G, H)|}{N} \leq \frac{\log \eta(H)}{2} + \frac{|V(H)|}{2n}.$$

We use the example of $H = K_k$, the complete graph on k vertices, to illustrate the definition of η. It is easy to see that for any $A, B \subseteq V(K_k)$, we have $i \sim j \ \forall i \in A, j \in B$ iff A and B are disjoint, and so $|A||B|$ is maximum when $|A|$ and $|B|$ are as close as possible to $k/2$. Hence $\eta(K_k) = \lfloor k/2 \rfloor \lceil k/2 \rceil$. Since an element of $Hom(G, K_k)$ is exactly a proper k coloring of G, we get as a corollary of Proposition 1.3 an approximate count of the number of k-colorings of a regular bipartite graph.

COROLLARY 1.4. *For any n-regular, N-vertex bipartite G,*

$$|Hom(G, K_k)| = (\lfloor k/2 \rfloor \lceil k/2 \rceil)^{N(1/2 + o(1))}.$$

We now consider weighted versions of Propositions 1.2 and 1.3. Following [3], we put a measure on $Hom(G, H)$ as follows. To each $i \in V(H)$ assign a positive "activity" λ_i, and write Λ for the set of activities. Give each $f \in Hom(G, H)$ weight

$$w^\Lambda(f) = \prod_{v \in V(G)} \lambda_{f(v)}.$$

The constant that turns this assignment of weights on $Hom(G, H)$ into a probability distribution is

$$Z^\Lambda(G, H) = \sum_{f \in Hom(G,H)} w^\Lambda(f).$$

When all activities are 1, we have $Z^\Lambda(G, H) = |Hom(G, H)|$, and so the following is a generalization of Proposition 1.2.

PROPOSITION 1.5. *For any n-regular, N-vertex bipartite G, any H, and any system Λ of positive activities on $V(H)$,*

$$Z^\Lambda(G, H) \leq \left(Z^\Lambda(K_{n,n}, H)\right)^{N/2n}.$$

It was observed in [3] that $Z^\Lambda(G, H)$ may be related to $|Hom(G, H')|$ for an appropriate modification H' of H. That observation (which will be discussed in more detail in Section 3) is central to the proof of Proposition 1.5.

Proposition 1.3 also generalizes. For a set of activities Λ on $V(H)$, set

$$\eta^\Lambda(H) = \max\left\{\left(\sum_{i \in A} \lambda_i\right)\left(\sum_{j \in B} \lambda_j\right) : A, B \subseteq V(H), i \sim j \ \forall i \in A, j \in B\right\}.$$

PROPOSITION 1.6. *For any n-regular, N-vertex bipartite G, any H, and any system Λ of positive activities on $V(H)$,*

$$\frac{\log \eta^\Lambda(H)}{2} \leq \frac{\log Z^\Lambda(G, H)}{N} \leq \frac{\log \eta^\Lambda(H)}{2} + \frac{|V(H)|}{2n}.$$

We may put these results in the framework of a well-known mathematical model of physical systems with "hard constraints" (see [3]). These are systems with strictly forbidden configurations. An example is the hard-core lattice gas model, in which a legal configuration of particles on a lattice is precisely one in which no two adjacent lattice sites are occupied. (By way of contrast, consider the ferromagnetic Ising model, where adjacent particles are discouraged from having opposing spins, but not forbidden — this is a "soft constraint".)

We think of the vertices of G as particles and the edges as bonds between pairs of particles, and we think of the vertices of H as possible "spins" that particles may take. Pairs of vertices of G joined by a bond may have spins i and j only when i and j are adjacent in H (in particular, they may both have spin i only when i has a loop in H). Thus the legal spin configurations on the vertices of G are precisely the homomorphisms from G to H. We think of the activities on the vertices of H as a measure of the likelihood of seeing the different spins; the probability of a particular spin configuration is proportional to the product over the vertices of G of the activities of the spins. Propositions 1.5 and 1.6 concern the "partition function" of this model — the normalizing constant that turns the above-described system of weights on the set of legal configurations into a probability measure.

The results we actually prove are in a slightly more general weighted model. Write $\mathcal{E}_G$ and $\mathcal{O}_G$ for the partition classes of G, and to each $i \in V(H)$ assign a positive *pair* of activities (λ_i, μ_i). Write (Λ, M) for the set of activities. Give each $f \in Hom(G, H)$ weight

$$w^{(\Lambda, \mathrm{M})}(f) = \prod_{v \in \mathcal{E}_G} \lambda_{f(v)} \prod_{v \in \mathcal{O}_G} \mu_{f(v)}.$$

The constant that turns this assignment of weights on $Hom(G, H)$ into a probability distribution is

$$(1) \qquad Z^{(\Lambda, \mathrm{M})}(G, H) = \sum_{f \in Hom(G, H)} w^{(\Lambda, \mathrm{M})}(f).$$

A special case of this model was considered by Kahn [8] (see also [6]), where Theorem 1.1 was extended to

THEOREM 1.7. *For any n-regular, N-vertex bipartite G, and any $\lambda, \mu \geq 1$,*

$$\sum_{I \in \mathcal{I}(G)} \prod_{v \in \mathcal{E}_G} \lambda^{|I \cap \mathcal{E}_G|} \prod_{v \in \mathcal{O}_G} \mu^{|I \cap \mathcal{O}_G|} \leq ((1 + \lambda)^n + (1 + \mu)^n - 1)^{N/2n}.$$

It was conjectured in [**8**] that the assumption $\lambda, \mu \geq 1$ may be relaxed to $\lambda, \mu \geq 0$. We show that this is indeed true, by generalizing Proposition 1.5 to:

PROPOSITION 1.8. *For any n-regular, N-vertex bipartite G, any H, and any system (Λ, M) of positive activities on $V(H)$,*

$$Z^{(\Lambda,\mathrm{M})}(G, H) \leq \left(Z^{(\Lambda,\mathrm{M})}(K_{n,n}, H) \right)^{N/2n}.$$

We also generalize Proposition 1.6 to this setting. Set

$$\eta^{(\Lambda,\mathrm{M})}(H) = \max \left\{ \left(\sum_{i \in A} \lambda_i \right) \left(\sum_{j \in B} \mu_j \right) : A, B \subseteq V(H), i \sim j \ \forall i \in A, j \in B \right\}.$$

PROPOSITION 1.9. *For any n-regular, N-vertex bipartite G, any H, and any system (Λ, M) of positive activities on $V(H)$,*

$$\frac{\log \eta^{(\Lambda,\mathrm{M})}(H)}{2} \leq \frac{\log Z^{(\Lambda,\mathrm{M})}(G, H)}{N} \leq \frac{\log \eta^{(\Lambda,\mathrm{M})}(H)}{2} + \frac{|V(H)|}{2n}.$$

Proposition 1.8 generalizes to the case of biregular G (a bipartite graph G with partition classes $\mathcal{E}_G$ and $\mathcal{O}_G$ is (a, b)-*biregular* if all vertices in $\mathcal{E}_G$ have degree a and all in $\mathcal{O}_G$ have degree b). The proof of the following proposition, which is a straightforward modification of the proof of Proposition 1.8, is omitted.

PROPOSITION 1.10. *For any (a, b)-biregular, N-vertex, bipartite G, any H, and any system (Λ, M) of positive activities on $V(H)$,*

$$Z^{(\Lambda,\mathrm{M})}(G, H) \leq \left(Z^{(\Lambda,\mathrm{M})}(K_{a,b}, H) \right)^{N/(a+b)}.$$

It was conjectured in [**7**] that Theorem 1.1 remains true without the assumption that G is bipartite. We similarly conjecture that biparticity is unnecessary in Proposition 1.8, and hence also in Propositions 1.2 and 1.5. (Proposition 1.3, and hence also Propositions 1.6 and 1.9, is easily seen to fail for non-bipartite G.)

The proof of Proposition 1.8 requires entropy considerations; these are reviewed in Section 2. The proofs are then given in Section 3.

2. Entropy

Here we briefly review the relevant entropy material. Our treatment is mostly copied from [**7**]. For a more thorough discussion, see e.g. [**9**]. In what follows $\mathbf{X}$, $\mathbf{Y}$ etc. are discrete random variables, which in our usage are allowed to take values in any finite set.

The *entropy* of the random variable $\mathbf{X}$ is

$$H(\mathbf{X}) = \sum_x p(x) \log \frac{1}{p(x)},$$

where we write $p(x)$ for $\mathbf{P}(\mathbf{X} = x)$ (and extend this convention in natural ways below). The *conditional entropy* of $\mathbf{X}$ given $\mathbf{Y}$ is

$$H(\mathbf{X}|\mathbf{Y}) = \mathbf{E} H(\mathbf{X}|\{\mathbf{Y} = y\}) = \sum_y p(y) \sum_x p(x|y) \log \frac{1}{p(x|y)}.$$

Notice that we are also writing $H(\mathbf{X}|Q)$ with Q an event (in this case $Q = \{\mathbf{Y} = y\}$):

$$H(\mathbf{X}|Q) = \sum p(x|Q) \log \frac{1}{p(x|Q)}.$$

When we condition on a random variable and an event simultaneously, we use ";" to separate the two.

For a random vector $\mathbf{X} = (\mathbf{X}_1, \ldots, \mathbf{X}_n)$ (note this is also a random variable), we have

$$(2) \qquad H(\mathbf{X}) = H(\mathbf{X}_1) + H(\mathbf{X}_2|\mathbf{X}_1) + \cdots + H(\mathbf{X}_n|\mathbf{X}_1, \ldots, \mathbf{X}_{n-1}).$$

We will make repeated use of the inequalities

$$(3) \qquad H(\mathbf{X}) \leq \log |\mathrm{range}(\mathbf{X})| \text{ (with equality if } \mathbf{X} \text{ is uniform)},$$

$$H(\mathbf{X}|\mathbf{Y}) \leq H(\mathbf{X}),$$

and more generally,

$$(4) \qquad \text{if } \mathbf{Y} \text{ determines } \mathbf{Z} \text{ then } H(\mathbf{X}|\mathbf{Y}) \leq H(\mathbf{X}|\mathbf{Z}).$$

Note that (2) and (4) imply

$$H(\mathbf{X}) \leq H(\mathbf{Y}) + H(\mathbf{X}|\mathbf{Y})$$

and

$$(5) \qquad H(\mathbf{X}_1, \ldots, \mathbf{X}_n) \leq \sum H(\mathbf{X}_i)$$

We also have a conditional version of (5):

$$H(\mathbf{X}_1, \ldots, \mathbf{X}_n|\mathbf{Y}) \leq \sum H(\mathbf{X}_i|\mathbf{Y}).$$

We will also need the following lemma of Shearer (see [4, p. 33]). For a random vector $\mathbf{X} = (\mathbf{X}_1, \ldots, \mathbf{X}_m)$ and $A \subseteq [m]$, set $\mathbf{X}_A = (\mathbf{X}_i : i \in A)$.

LEMMA 2.1. *Let* $\mathbf{X} = (\mathbf{X}_1, \ldots, \mathbf{X}_m)$ *be a random vector and* $\mathcal{A}$ *a collection of subsets (possibly with repeats) of* $[m]$, *with each element of* $[m]$ *contained in at least* t *members of* $\mathcal{A}$. *Then*

$$H(\mathbf{X}) \leq \frac{1}{t} \sum_{A \in \mathcal{A}} H(\mathbf{X}_A).$$

3. Proofs

For a regular bipartite graph G, we write $\mathcal{E}_G$ and $\mathcal{O}_G$ for the partition classes. For ease of notation, we write $\mathcal{E}_n$ for $\mathcal{E}_{K_{n,n}}$ and $\mathcal{O}_n$ for $\mathcal{O}_{K_{n,n}}$. For a partition $U \cup L$ of $V(H)$, set

$$Hom^{U,L}(G, H) = \{f \in Hom(G, H) : f(\mathcal{E}_G) \subseteq U, f(\mathcal{O}_G) \subseteq L\}.$$

(For a set X we write $f(X)$ for $\{f(x) : x \in X\}$.)

We begin by deriving an expression for $|Hom^{U,L}(K_{n,n}, H)|$. For $A \subseteq L$ set

$$\mathcal{H}(A) = \{f \in Hom^{U,L}(K_{n,n}, H) : f(\mathcal{O}_n) = A\},$$

$$T(A) = \{g : [n] \rightarrow A \ : \ g \text{ surjective}\}$$

and

$$C^U(A) = \{j \in U : j \sim i \ \forall i \in A\}.$$

(Note that $C^U(A)$ is the set of possible images of $v \in \mathcal{E}_G$ under a member of $Hom^{U,L}(G, H)$, given that the image of $N(v)$ is A.) It is easy to see that $\{\mathcal{H}(A) :$

$A \subseteq L\}$ forms a partition of $Hom^{U,L}(K_{n,n}, H)$, and also that for each A, $|\mathcal{H}(A)| = |T(A)||C^U(A)|^n$. Thus we have

$$(6) \qquad |Hom^{U,L}(K_{n,n}, H)| = \sum_{A \subseteq L} |T(A)||C^U(A)|^n.$$

The following is the central lemma in the proofs of Propositions 1.8 and 1.9.

LEMMA 3.1. *For any n-regular, N-vertex bipartite G, and any H with $U \cup L$ a partition of $V(H)$,*

$$|Hom^{U,L}(G, H)| \leq |Hom^{U,L}(K_{n,n}, H)|^{N/2n}.$$

Proof: The proof is based on [7, Thm. 1.9]. Let $\mathbf{f}$ be chosen uniformly from $Hom^{U,L}(G, H)$. For $v \in V(G)$, write $\mathbf{f}_v$ for $\mathbf{f}(v)$, $\mathbf{N}_v$ for $\mathbf{f}|_{N(v)}$ and $\mathbf{M}_v$ for $\{\mathbf{f}_w : w \in N(v)\}$. For $v \in \mathcal{E}_G$ and $A \subseteq L$, write $m_v(A)$ for $\mathbf{P}(\mathbf{M}_v = A)$. (Note that $\sum_A m_v(A) = 1$.) We have (with the main inequalities justified below; the remaining steps follow in a straightforward way from the material of Section 2)

$$
\begin{aligned}
\log |Hom^{U,L}(G, H)| &= H(\mathbf{f}) \\
&= H(\mathbf{f}|_{\mathcal{O}_G}) + H(\mathbf{f}|_{\mathcal{E}_G} \mid \mathbf{f}|_{\mathcal{O}_G}) \\
&\leq H(\mathbf{f}|_{\mathcal{O}_G}) + \sum_{v \in \mathcal{E}_G} H(\mathbf{f}_v \mid \mathbf{f}|_{\mathcal{O}_G}) \\
&\leq H(\mathbf{f}|_{\mathcal{O}_G}) + \sum_{v \in \mathcal{E}_G} H(\mathbf{f}_v \mid \mathbf{N}_v)
\end{aligned}
$$

$$(7) \qquad \leq \frac{1}{n} \sum_{v \in \mathcal{E}_G} H(\mathbf{N}_v) + \sum_{v \in \mathcal{E}_G} H(\mathbf{f}_v \mid \mathbf{N}_v)$$

$$\leq \frac{1}{n} \sum_{v \in \mathcal{E}_G} [H(\mathbf{M}_v) + H(\mathbf{N}_v | \mathbf{M}_v)] + \sum_{v \in \mathcal{E}_G} H(\mathbf{f}_v \mid \mathbf{N}_v)$$

$$\leq \frac{1}{n} \sum_{v \in \mathcal{E}_G} [H(\mathbf{M}_v) + H(\mathbf{N}_v | \mathbf{M}_v) + nH(\mathbf{f}_v | \mathbf{N}_v)]$$

$$\leq \frac{1}{n} \sum_{v \in \mathcal{E}_G} \sum_{A \subseteq L} \Big[m_v(A) \log \frac{1}{m_v(A)} +$$
$$m_v(A) H(\mathbf{N}_v | \{\mathbf{M}_v = A\}) +$$
$$n m_v(A) H(\mathbf{f}_v | \mathbf{N}_v; \{\mathbf{M}_v = A\}) \Big]$$

$$\leq \frac{1}{n} \sum_{v \in \mathcal{E}_G} \sum_{A \subseteq L} \Big[m_v(A) \log \frac{1}{m_v(A)} +$$
$$(8) \qquad m_v(A) \log |T(A)| + n m_v(A) \log |C^U(A)| \Big]$$

$$= \frac{1}{n} \sum_{v \in \mathcal{E}_G} \sum_{A \subseteq L} m_v(A) \log \frac{|T(A)||C^U(A)|^n}{m_v(A)}$$

$$(9) \qquad \leq \frac{1}{n} \sum_{v \in \mathcal{E}_G} \log \left[\sum_{A \subseteq L} |T(A)||C^U(A)|^n \right]$$

$$(10) \qquad = \frac{N}{2n} \log |Hom^{U,L}(K_{n,n}, H)|.$$

The main inequality (7) involves an application of Lemma 2.1, with $\mathcal{A} = \{N(v) : v \in \mathcal{E}_G\}$, and (9) is an application of Jensen's inequality. In (8), we use (3), noting that conditioning on the event $\{\mathbf{M}_v = A\}$ there are $|T(A)|$ possible values for $\mathbf{N}_v$, and $|C^U(A)|$ possible values for $\mathbf{f}_v$. Finally, (10) follows from (6). $\qquad\square$

It is worth noting at this point that Lemma 3.1 easily implies Proposition 1.2. Let H' be the graph on vertex set $\cup_{i \in V(H)}\{v_i, w_i\}$ with v_i and w_j adjacent exactly when i and j are adjacent in H. Set $U = \{v_1, \ldots, v_{|V(H)|}\}$ and $L = \{w_1, \ldots, w_{|V(H)|}\}$. It is easy to check that $|Hom(G, H)| = |Hom^{U,L}(G, H')|$, from which Proposition 1.2 follows via an application of Lemma 3.1.

This idea of "doubling" H, combined with the construction of [**3**] that relates $Z^\Lambda(G, H)$ to $|Hom(G, H')|$ for an appropriate modification H' of H, allows us to pass from Proposition 1.2 to Proposition 1.8. The details are as follows.

Recall that our aim is to upper bound the partition function $Z^{(\Lambda,\mathrm{M})}(G, H)$ (see (1)). By continuity, we may assume that all activities are rational. Let C be the least positive integer such that $C\lambda_i$ and $C\mu_i$ are integers for each $i \in V(H)$. Let $H^{(\Lambda,\mathrm{M})}$ be the graph whose vertex set is obtained from H by replacing each $i \in V(H)$ by two sets, $D_i^U = \{i_1^U, \ldots, i_{C\lambda_i}^U\}$ and $D_i^L = \{i_1^L, \ldots, i_{C\mu_i}^L\}$ of $C\lambda_i$ and $C\mu_i$ vertices. For each $i, j \in V(H)$ (not necessarily distinct), $i' \in D_i^U$ and $j' \in D_j^L$, join i' to j' exactly when i and j are adjacent in H. Set $U = U(H^{(\Lambda,\mathrm{M})}) = \cup_{i \in V(H)}D_i^U$ and $L = L(H^{(\Lambda,\mathrm{M})}) = \cup_{i \in V(H)}D_i^L$.

We now show that $Z^{(\Lambda,\mathrm{M})}(G, H)$ is related to $|Hom^{U,L}(G, H^{(\Lambda,\mathrm{M})})|$. Say that $g \in Hom^{U,L}(G, H^{(\Lambda,\mathrm{M})})$ is a *lift* of $f \in Hom(G, H)$ if for all $v \in V(G)$,

$$g(v) = \begin{cases} f(v)_k^U, & \text{some } 1 \le k \le C\lambda_{f(v)} & \text{if } v \in \mathcal{E}_G, \\ f(v)_k^L, & \text{some } 1 \le k \le C\mu_{f(v)} & \text{if } v \in \mathcal{O}_G. \end{cases}$$

Set

$$\mathcal{G}(f) = \{g \in Hom^{U,L}(G, H^{(\Lambda,\mathrm{M})}) : g \text{ is a lift of } f\}.$$

It is easy to check that $|\mathcal{G}(f)| = w^{(\Lambda,\mathrm{M})}(f)C^N$ for each $f \in Hom(G, H)$, and that $\{\mathcal{G}(f) : f \in Hom(G, H)\}$ forms a partition of $Hom^{U,L}(G, H^{(\Lambda,\mathrm{M})})$. It follows that

$$(11) \qquad Z^{(\Lambda,\mathrm{M})}(G, H) = \frac{|Hom^{U,L}(G, H^{(\Lambda,\mathrm{M})})|}{C^N}.$$

We now have all we need to prove Propositions 1.8 and 1.9.
Proof of Proposition 1.8: Applying (11) with $G = K_{n,n}$ we get

$$(12) \qquad Z^{(\Lambda,\mathrm{M})}(K_{n,n}, H) = \frac{|Hom^{U,L}(K_{n,n}, H^{(\Lambda,\mathrm{M})})|}{C^{2n}}.$$

Proposition 1.8 now follows from (11), (12) and Lemma 3.1. $\qquad\square$

Proof of Proposition 1.9: For each $A \subseteq V(H)$, set $C(A) = \{j \in H : j \sim i \ \forall i \in A\}$ and

$$\mathcal{D}(A) = \{f \in Hom(K_{n,n}, H) : f(\mathcal{E}_n) \subseteq A, \ f(\mathcal{O}_n) \subseteq C(A)\}.$$

104   DAVID GALVIN AND PRASAD TETALI

By Proposition 1.8 we have

$$\left(Z^{(\Lambda,\mathrm{M})}(G,H)\right)^{2n/N} \leq Z^{(\Lambda,\mathrm{M})}(K_{n,n},H)$$

$$\leq \sum_{A\subseteq V(H)}\sum_{f\in\mathcal{D}(A)} w^{(\Lambda,\mathrm{M})}(f)$$

$$= \sum_{A\subseteq V(H)}\left(\sum_{i\in A}\lambda_i\right)^n\left(\sum_{j\in C(A)}\mu_j\right)^n$$

$$(13) \qquad\qquad \leq 2^{|V(H)|}\left(\eta^{(\Lambda,\mathrm{M})}(H)\right)^n.$$

This gives the upper bound. For the lower bound, let $A, B \subseteq V(H)$ satisfying $i \sim j \;\forall i \in A, j \in B$ be such that $\eta^{(\Lambda,\mathrm{M})}(H) = \left(\sum_{i\in A}\lambda_i\right)\left(\sum_{j\in B}\mu_j\right)$. We have

$$Z^{(\Lambda,\mathrm{M})}(G,H) \geq \sum\{w^{(\Lambda,\mathrm{M})}(f) : f(\mathcal{E}_G)\subseteq A, f(\mathcal{O}_G)\subseteq B\}$$

$$= \left(\sum_{i\in A}\lambda_i\right)^{N/2}\left(\sum_{j\in B}\mu_j\right)^{N/2}$$

$$= \eta^{(\Lambda,\mathrm{M})}(H)^{N/2}.$$

$$\square$$

Acknowledgment This research was done while the second author was visiting Microsoft Research. He would like to thank Microsoft Research, and especially the theory group, for providing him with this opportunity.

References

[1] N. Alon, Independent sets in regular graphs and sum-free subsets of finite groups, *Israel J. Math.* **73** (1991), 247–256.

[2] B. Bollobás, *Modern Graph Theory*, Springer, New York, 1998.

[3] G. Brightwell and P. Winkler, Graph homomorphisms and phase transitions, *J. Combin. Theory Ser. B* **77** (1999), 221–262.

[4] F.R.K. Chung, P. Frankl, R. Graham and J.B. Shearer, Some intersection theorems for ordered sets and graphs, *J. Combin. Theory Ser. A.* **48** (1986), 23–37.

[5] R. Diestel, *Graph Theory*, Springer, New York, 1997.

[6] O. Häggström, Ergodicity of the hard-core model on $\mathbf{Z}^2$ with parity-dependent activities, *Ark. Mat.* **35** (1997), 171–184.

[7] J. Kahn, An entropy approach to the hard-core model on bipartite graphs, *Combin. Prob. Comp.* **10** (2001), 219–237.

[8] J. Kahn, Entropy, independent sets and antichains: a new approach to Dedekind's problem, *Proc. Amer. Math. Soc.* **130** (2002), 371–378

[9] R.J. McEliece, *The Theory of Information and Coding*, Addison-Wesley, London, 1977.

[10] R. Peeters, The maximum edge biclique problem is NP-complete, *Research Memorandum 789*, Faculty of Economics and Business Administration, Tilberg University (2000).

MICROSOFT RESEARCH, 1 MICROSOFT WAY, REDMOND WA 98052, USA

E-mail address: `galvin@microsoft.com`

SCHOOL OF MATHEMATICS AND COLLEGE OF COMPUTING, GEORGIA TECH, ATLANTA GA 30332-0160, USA

E-mail address: `tetali@math.gatech.edu`

DIMACS Series in Discrete Mathematics
and Theoretical Computer Science
Volume **63**, 2004

Counting List Homomorphisms for Graphs with Bounded Degrees

Pavol Hell and Jaroslav Nešetřil

ABSTRACT. We discuss a number of variants of the homomorphism problem, where input graphs G are to be mapped homomorphically to a fixed target graph H. We set up a structure unifying the work on finding graph homomorphisms, due to the authors, of counting graph homomorphisms, due to Dyer and Greenhill, and of finding list homomorphisms, due to Feder, Hell, and Huang.

We also consider the effect of restricting the maximum degree of the input graph. We identify a number of interesting problems in this context. Our main result is a complete classification of the complexity (as polynomial time solvable or $\#P$-complete) of the problem of counting list homomorphisms of graphs with bounded degrees.

1. Introduction

Our graphs are undirected and without multiple edges, but may have loops. A graph without loops is called *irreflexive*, and a graph in which each vertex has a loop is called *reflexive*.

Let G and H be graphs. A *homomorphism* f of G to H is a mapping $f : V(G) \to V(H)$ such that $f(g)f(g')$ is an edge of H whenever gg' is an edge of G. Each fixed graph H gives rise to a decision problem in which one is to decide whether or not a given input graph G admits a homomorphism to H. This basic decision problem has been studied by several authors, and a complete classification of its complexity has been given in [**11**]. A number of variants of this basic problem has been considered since. In each of the ones discussed in this paper, the graph H is fixed. Instead of deciding the existence of a homomorphism of G to H, we may want to count the total number of such homomorphisms. This problem arises in statistical physics [**1**], and a complete classification of its complexity has been given in [**4**]. In another direction, one may ask whether the problem of deciding the existence of a homomorphism of G to H is any easier when G is restricted to have bounded degrees. It turns out that this is so in certain cases, although the full answer is not yet known, [**9**]. Finally, there is a version in which the input consists of a graph G together with lists, $L(g) \subseteq V(H), g \in V(G)$, and the question

---

Supported by NSERC.
Partially supported by the Project LN00A056 of the Czech Ministery of Education.

to decide is whether or not there exists a homomorphism f of G to H in which each $g \in V(G)$ has $f(g) \in L(g)$. The complexity of deciding the existence of a list homomorphism has been completely classified in [**5, 6, 7**].

Clearly, these variants can be combined in various ways - we may want to count list homomorphisms, or decide the existence of list homomorphisms for graphs with bounded degrees, and so on.

To organize these variants, we introduce the following notation: HOM_H will denote the basic problem (for a fixed graph H) – given an input graph G is there a homomorphism of G to H ? The variants will be distinguished by introducing superscripts: The superscript $+L$ will mean we are considering lists, the superscript $+C$ will mean we are interested in counting the number of homomorphisms, or list homomorphisms, rather than just deciding their existence, and the superscript $+\Delta$ will mean the degrees of the input graphs are restricted to be at most Δ. Moreover, we shall omit the reference to H if we want to discuss the problem in general - over all possible graphs H. Thus, for example, HOM^{+C+L} is the family of problems HOM_H^{+C+L} of counting list homomorphisms to H, over all possible target graphs H.

To be precise, here are the definitions of the individual problems:

- HOM_H

  Instance: A graph G

  Goal: Decide if there is a homomorphism of G to H
- HOM_H^{+C}

  Instance: A graph G

  Goal: Count the number of homomorphisms of G to H
- HOM_H^{+L}

  Instance: A graph G together with lists $L(g) \subseteq V(H), g \in V(G)$

  Goal: Decide if there is a homomorphism of G to H such that $f(g) \in L(g)$ for each $g \in V(G)$
- HOM_H^{+C+L}

  Instance: A graph G together with lists $L(g)$ as above

  Goal: Count the number of homomorphisms of G to H such that each $f(g) \in L(g)$
- $\mathrm{HOM}_H^{+\Delta}$, $\mathrm{HOM}_H^{+C+\Delta}$, $\mathrm{HOM}_H^{+L+\Delta}$, $\mathrm{HOM}_H^{+C+L+\Delta}$, are defined analogously, except the instance graphs G are restricted to have all degrees less than or equal to Δ.

A summary of all the results and open problems is presented in a table at the end of the article.

2. Without Degree Constraints

The following is the classification of the **basic problem**:

THEOREM 2.1. [11] HOM_H *is polynomial time solvable when H has a loop or is bipartite, and is NP-complete otherwise.*

When lists are added, we have the following classification:

THEOREM 2.2. [7] HOM_H^{+L} *is polynomial time solvable when H is a bi-arc graph, and is NP-complete otherwise.*

Bi-arc graphs are defined as follows: Let C be a fixed circle, with two specified points n and s. A *bi-arc* is an ordered pair of arcs (N, S) on C such that N contains n but not s, and S contains s but not n. A graph H is a *bi-arc graph* if there exists a family of bi-arcs $(N_h, S_h), h \in V(H)$, such that for any $h, h' \in V(H)$ one of the following two alternatives must happen: Either h and h' are adjacent in H, N_h intersects $S_{h'}$, and $N_{h'}$ intersects S_h; or h and h' are not adjacent in H, N_h does not intersect $S_{h'}$, and $N_{h'}$ does not intersect S_h. (In both cases $h = h'$ is possible.) We note, cf. [5, 6], that a reflexive graph is a bi-arc graph if and only if it is an *interval graph*, and an irreflexive graph is a bi-arc graph if and only if it is bipartite and its complement is a *circular arc graph*.

The complexity of **counting** homomorphisms has been classified by M. Dyer and C. Greenhill:

THEOREM 2.3. [4] HOM_H^{+C} *is polynomial time solvable when each component of H is either a reflexive complete graph or an irreflexive complete bipartite graph, and is $\#P$-complete otherwise.*

We now complete the picture by classifying the complexity of **counting list homomorphisms** (the same result has been independently proved in [2]):

THEOREM 2.4. HOM_H^{+C+L} *is polynomial time solvable when each component of H is either a reflexive complete graph or an irreflexive complete bipartite graph, and is $\#P$-complete otherwise.*

Proof: When H is not of the kind described, we have seen that even the counting of ordinary homomorphisms is $\#P$-complete. Since homomorphisms are just list homomorphisms for inputs G with each $L(g) = V(H)$, the $\#P$-completeness follows. To count the number of list homomorphisms of G to H, when H is as described above, it will suffice to count the number of list homomorphisms of a connected G to a connected H (which is then either a reflexive complete graph or an irreflexive complete bipartite graph). Indeed, the number of list homomorphisms of a connected G to a disconnected H is the sum of the counts of the list homomorphisms of G to H' all the components H' of H, and the number of list homomorphisms of a disconnected G to H is the product of the counts of list homomorphisms of all components G' of G to H. But counting list homomorphisms of a connected G to such a connected H is easy: If H is a reflexive complete graph, then any mapping conforming to the lists is a homomorphism, thus the count is the product of the sizes of the *reduced* lists. (The reduced list of a vertex consists of all those members of the original list which belong to the component of H under consideration.) If H is an irreflexive bipartite graph, then the count is 0 when G

contains an odd cycle of any size, including size one (i.e., a loop). Otherwise G is also a connected irreflexive bipartite graph and the count is the sum of the counts assuming the first part of G maps to the first part of H and the second part of G to the second part of H, or conversely. Each of these counts is easy to evaluate by taking a product of the appropriately reduced lists.

3. With Degree Constraints

In [9] we have investigated the effect of **adding degree constraints to the basic problem** - introducing the family of problems $\mathrm{HOM}^{+\Delta}$. It turns out there are graphs H such that $\mathrm{HOM}_H^{+\Delta}$ is polynomial time solvable, even though without the degree constraints, the corresponding problem HOM_H is NP-complete, i.e., H is nonbipartite. An immediate example of this phenomenon is the problem of three-colouring:

THEOREM 3.1. *Let* $H = K_3$ *(the irreflexive triangle):*
If $\Delta = 3$ *then* $\mathrm{HOM}_H^{+\Delta}$ *is polynomially solvable.*
If $\Delta \geq 4$ *then* $\mathrm{HOM}_H^{+\Delta}$ *is* NP-*complete.*

(Both these observations are well known, cf. [9]; the first one is due to the theorem of Brooks which implies that a graph with all degrees at most three is three-colourable if and only if it contains no component isomorphic to K_4.)

There are more interesting examples [9]. Let $\Delta \geq 3$. It is shown in [10] that for any connected graph A there exists a graph H such that, for each graph G with degrees at most Δ, G admits a homomorphism to H if and only if A admits a homomorphism to G. Since A is fixed, the existence of a homomorphism of A to G can be tested in polynomial time (in terms of the size of G), and therefore we have a polynomial algorithm for $\mathrm{HOM}_H^{+\Delta}$.

Here is a concrete nonbipartite example of such a construction: Let X be any set. We construct a graph $H(X)$ as follows: The vertices are ordered pairs (x, T), where T is a three-element subset of X, and x is an element of $X \setminus T$. Two vertices $(x, T), (x', T')$ are adjacent in $H(X)$ just if T, T' are disjoint, and $x \in T', x' \in T$. It is proved in [10] that a cubic graph G admits a homomorphism to $H(X)$ if and only if G is triangle-free, as long as the set X is large enough.

THEOREM 3.2. [10] *Let* $H = H(X)$, *where* $|X| \geq 22$, *and let* $\Delta = 3$. *Then* $\mathrm{HOM}_H^{+\Delta}$ *is polynomial time solvable.*

It is easy to see that $H = H(X)$ is irreflexive and nonbipartite (in fact, it has a fairly big chromatic number [9]), thus HOM_H is NP-complete. Additional results concerning the possible chromatic numbers of constructs related to $H(X)$ can be found in [3].

There are, on the other hand, many nonbipartite graphs H for which $\mathrm{HOM}_H^{+\Delta}$ remains NP-complete, even for $\Delta = 3$. This is the case, for instance, for irreflexive odd cycles, other than the triangle:

THEOREM 3.3. [9] *Let* $H = C_{2k+1}, k \geq 2$, *and* $\Delta \geq 3$. *Then* $\mathrm{HOM}_H^{+\Delta}$ *is* NP-*complete.*

In fact, the same conclusion ($\mathrm{HOM}_H^{+\Delta}$ NP-complete) applies [9] whenever H is a triangle free graph in which each vertex belongs to a pentagon and in which no two pentagons share more than one edge. (This does not include the Petersen graph, and we do not know the complexity of $\mathrm{HOM}_H^{+\Delta}$ when H is the Petersen graph and $\Delta = 3$.) We also know [**9**] there exist graphs H of arbitrarily high girth and chromatic number, for which $\mathrm{HOM}_H^{+\Delta}$ is NP-complete (whenever $\Delta \geq 3$). The dependence of the problem $\mathrm{HOM}_H^{+\Delta}$ on the girth of the input graphs G, in the case when H is an irreflexive cycle, is related to the work in [**12, 13, 15**].

Based on the types of reductions used in proving completeness results, T. Feder et al. made the following (meta-)conjecture:

CONJECTURE 3.1. [**8**] *Any homomorphism type problem (including list homomorphisms, and homomorphisms of more general structures) which is NP-complete without degree constraints is also NP-complete with degree constraints, provided the degree bound is high enough.*

According to Conjecture 3.1, each nonbipartite graph H admits a minimum value of Δ such that $\mathrm{HOM}_H^{+\Delta}$ is NP-complete. Determining this value of Δ (i.e., classifying the complexity of all problems in $\mathrm{HOM}^{+\Delta}$) is an interesting open problem.

M. Dyer and C. Greenhill [**4**] have also considered the effect on the complexity of **counting the number of homomorphisms of graphs with bounded degrees** - yielding the family of problems $\mathrm{HOM}^{+C+\Delta}$. They made the following conjecture, asserting that restricting the degrees of input graphs does not change the classification of **any** basic counting problem:

CONJECTURE 3.2. [**4**] *If HOM_H^{+C} is #P-complete (i.e. if H has a component that is not a reflexive complete graph or an irreflexive complete bipartite graph), and $\Delta \geq 3$, then $HOM_H^{+C+\Delta}$ is also #P-complete.*

They have verified their conjecture in the following special case:

THEOREM 3.4. [**4**] *Suppose H is a graph whose adjacency matrix is non-singular and which contains a component that is not a reflexive complete graph or an irreflexive complete bipartite graph, and let $\Delta \geq 3$. Then $\mathrm{HOM}_H^{+C+\Delta}$ is #P-complete.*

It follows (as shown in [**4**]) that for each graph H for which HOM_H^{+C} is #P-complete there exists a Δ such that also the problem $\mathrm{HOM}_H^{+C+\Delta}$ is #P-complete. This lends further support for Conjecture 3.1.

To complete our catalogue of the variants of the homomorphism problems, we only have to deal with **lists in conjunction with degree bounds**, i.e., with the problem families $\mathrm{HOM}^{+L+\Delta}$ and $\mathrm{HOM}^{+C+L+\Delta}$. In the first version of this paper we have posed the classification of the complexity of $\mathrm{HOM}^{+L+\Delta}$ as an open problem. That has now been solved in [**8**].

THEOREM 3.5. [**8**] *Let $\Delta \geq 3$. Then $HOM_H^{+L+\Delta}$ is polynomial time solvable when H is a bi-arc graph, and is NP-complete otherwise.*

We note that since $\mathrm{HOM}_H^{+L+\Delta}$ is a subproblem of HOM_H^{+L}, the polynomial time solvability follows from Theorem 2.2. The difficulty of Theorem 3.5 is in adapting the NP-completeness proofs to graphs with bounded degrees [**8**], and it can be seen as additional evidence for Conjecture 3.1.

Thus adding degree restrictions (with $\Delta \geq 3$) does not change the classification of tractable and intractable list homomorphism problems. On the other hand, there are other list type problems which are intractable but become tractable when degrees are restricted. Particularly nice examples given in [8] include the case of *extension problems*, i.e., list homomorphism problems where the lists are restricted to be either singletons (corresponding to *pre-coloured vertices*) or the whole set $V(H)$ (corresponding to unrestricted vertices). It is, for instance, shown in [8] that when H is the reflexive k-cycle, then this extension problem without degree constraints is polynomial time solvable when $k = 3$ (any mapping extending the given pre-colouring is a homomorphism), and is NP-complete otherwise. Restricting the degrees of the input graphs to any Δ greater than four has no effect on this classification [8]. With degree constraint $\Delta = 4$, the problem becomes polynomial time solvable for the reflexive four-cycle, but all longer cycles remain NP-complete. With degree constraint $\Delta = 3$ the problem becomes polynomial time solvable also for the reflexive five-cycle (and all longer cycles remain NP-complete) [8].

We close our discussion with the following complete classification of the complexity of problems in the family $\text{HOM}^{+C+L+\Delta}$:

THEOREM 3.6. *If each component of H is a reflexive complete graph or an irreflexive complete bipartite graph, then $\text{HOM}_H^{+C+L+\Delta}$ is polynomial time solvable. Otherwise $\text{HOM}_H^{+C+L+\Delta}$ is $\#P$-complete.*

Proof: In the former case, we argue that $\text{HOM}_H^{+C+L+\Delta}$ is a subproblem of HOM_H^{+C+L}, which is solvable in polynomial time by Theorem 2.4. In the latter case H contains a component G which

   (1) contains both a vertex with a loop and a vertex without a loop, or
   (2) is reflexive but not complete, or
   (3) is irreflexive but not complete bipartite.

Below we give the adjacency matrices of four small graphs A, B, C, D (for simplicity we shall denote the graphs and their adjacency matrices by the same symbol). The graph A consists of two adjacent vertices exactly one of which has a loop. Any component G satisfying condition 1 above must contain A as an induced subgraph. The graph B is the reflexive path with three vertices. Any component G satisfying condition 2 above must contain B as an induced subgraph. The graph C is the irreflexive triangle K_3, and the graph D is the irreflexive path with four vertices. Any component G satisfying condition 3 is either not bipartite, and hence contains C or D as an induced subgraph, or is bipartite but not complete bipartite, and hence contains D as an induced subgraph.

$$A = \begin{pmatrix} 1 & 1 \\ 1 & 0 \end{pmatrix}, \ B = \begin{pmatrix} 1 & 1 & 0 \\ 1 & 1 & 1 \\ 0 & 1 & 1 \end{pmatrix}, \ C = \begin{pmatrix} 0 & 1 & 1 \\ 1 & 0 & 1 \\ 1 & 1 & 0 \end{pmatrix}, \ D = \begin{pmatrix} 0 & 1 & 0 & 0 \\ 1 & 0 & 1 & 0 \\ 0 & 1 & 0 & 1 \\ 0 & 0 & 1 & 0 \end{pmatrix}$$

Thus H contains as an induced subgraph one of A, B, C, D. It is easy to verify that each of the adjacency matrices A, B, C, D is non-singular. Thus the problems $\text{HOM}_H^{+C+\Delta}$ for $H = A, B, C, D$ are $\#P$-complete, according to Theorem 3.4. Since H contains one of these as an induced subgraph, $\text{HOM}_H^{+C+L+\Delta}$ must also be NP-complete. Indeed, if, say, A is an induced subgraph of H, then any instance of

$\text{HOM}_A^{+C+\Delta}$ may be viewed as an instance of $\text{HOM}_H^{+C+L+\Delta}$ (where each list is either $V(H)$ or $V(A)$).

Our result shows that, at least for counting list homomorphisms, adding the degree constraints did not change the classification of any graph H - thus lending further support to Conjecture 3.2.

4. Conclusions

We first justify our frequent assumption that $\Delta \geq 3$:

THEOREM 4.1. *If $\Delta \leq 2$, then all problems in HOM^{Δ}, $\text{HOM}^{+C+\Delta}$, $\text{HOM}^{+L+\Delta}$, and $\text{HOM}^{+C+L+\Delta}$ are polynomial time solvable.*

Proof: Clearly, it suffices to treat $\text{HOM}_H^{+C+L+\Delta}$, since it contains all the other problems. If G is *any graph of bounded treewidth* (and graphs with maximum degree at most two have treewidth at most two), one can count the number of list homomorphisms (to any H) by the standard techniques dealing with nice tree decompositions, cf., e.g., [14, 2].

We summarize the results below. The first three columns express, in boolean values, whether or not the variant considers counting, lists, or bounded degrees. The same information is contained, in the agreed notation, in the fourth colums (name). The next two columns (source and theorem) give a reference both in the literature, and in this paper. The final column (easy graphs) gives an abbreviation for the class $\mathcal{C}$ of graphs H such that the problem is tractable for graphs in $\mathcal{C}$ and apparently intractable for graphs not in $\mathcal{C}$. We say that a graph is *-bipartite if it is bipartite or has a loop, and is *-complete if each component is either a reflexive complete graph or an irreflexive complete bipartite graph.

SUMMARY TABLE

count?	lists?	degrees?	name	source	theorem	easy graphs
0	0	0	HOM	[11]	2.1	*-bipartite
0	1	0	HOM^{+L}	[7]	2.2	bi-arc
1	0	0	HOM^{+C}	[4]	2.3	*-complete
1	1	0	HOM^{+C+L}	here	2.4	*-complete
0	0	1	$\text{HOM}^{+\Delta}$	[9]	3.1,3.2,3.3	OPEN
0	1	1	$\text{HOM}^{+L+\Delta}$	[8]	3.5	bi-arc
1	0	1	$\text{HOM}^{+C+\Delta}$	[4]	3.4	OPEN
1	1	1	$\text{HOM}^{+C+L+\Delta}$	here	3.6	*-complete

References

[1] G. Brightwell and P. Winkler, Graph homomorphisms and phase transitions, *J. Combinatorial Th. B* 77 (1999) 415-435.

[2] J. Diaz, M. Serna, and D.M. Thilikos, The complexity of parametrized H-colorings: A survey, *this volume.*

[3] P.A. Dryer, Jr., C. Malon, and J. Nešetřil, Universal H-colourable graphs without a given configuration, *Discrete Math.* 250 (2002) 245-252.

[4] M. Dyer and C. Greenhill, The complexity of counting graph homomorphisms, *Random Structures and Algorithms* 17 (2000) 260-289.

[5] T. Feder and P. Hell, List homomorphisms to reflexive graphs, *J. Combinatorial Theory B* 72 (1998) 236-250.

[6] T. Feder, P. Hell, and J. Huang, List homomorphisms and circular arc graphs, *Combinatorica* 19 (1999) 487-505.

[7] T. Feder, P. Hell, and J. Huang, Bi-arc graphs, and the complexity of list homomorphisms, *Journal of Graph Theory* to appear.

[8] T. Feder, P. Hell, and J. Huang, Extension problems and list homomorphisms for graphs with bounded degrees, manuscript 2002.

[9] A. Galluccio, P. Hell, J. Nešetřil, The complexity of H-colouring of bounded degree graphs, *Discrete Math.* 222 (2000), 101-109.

[10] R. Häggkvist and P. Hell, Universality of A–mote graphs, *European J. of Combinatorics* 14 (1993) 23 - 27.

[11] P. Hell and J. Nešetřil, On the complexity of H-colouring, *J. Combinatorial Theory B* 48 (1990) 92-110.

[12] J.H. Kim, J. Nešetřil, On colourings of bounded degree graphs, manuscript 2002.

[13] A. Kostochka, J. Nešetřil, P. Smolíková, Colorings and homomorphisms of degenerate and bounded degree graphs, *Discrete Math.* 233, 1-3 (2001), 257-276.

[14] A. Proskurowski and S. Arnborg, Linear time algorithms for NP-hard problems restricted to partial k-trees, *Discrete Appl. Math.* 23 (1989) 11-24.

[15] I.M. Wanless, N.C. Wormald, Graphs with no homomorphisms onto cycles, *J. Comb. Th. (B)* 82 (2001) 155-160.

SCHOOL OF COMPUTING SCIENCE, SIMON FRASER UNIVERSITY, BURNABY, B.C., CANADA V5A 1S6

E-mail address: pavol@cs.sfu.ca

DEPARTMENT OF APPLIED MATHEMATICS, INSTITUTE FOR THEORETICAL COMPUTER SCIENCE (ITI), CHARLES UNIVERSITY, MALOSTRANSKÉ NÁM. 25, 118 00 PRAHA 1, CZECH REPUBLIC

E-mail address: nesetril@kam.mff.cuni.cz

DIMACS Series in Discrete Mathematics
and Theoretical Computer Science
Volume **63**, 2004

On the satisfiability of random k-Horn formulae

Gabriel Istrate

ABSTRACT. We determine the satisfaction probability of a random at-most-k-Horn formula, via a probabilistic analysis of a simple version, called PUR, of positive unit resolution. We show that for $k = k(n) \to \infty$ the problem can be "reduced" to the case $k(n) = n$, that was solved in [**16**]. On the other hand, in the case $k = $ constant the behavior of PUR is modelled by a simple queuing chain, leading to a closed-form solution when $k = 2$. Our analysis predicts an "easy-hard-easy" pattern in this latter case. Under a rescaled parameter, the graphs of satisfaction probability corresponding to finite values of k converge to the one for the uniform case, a "dimension-dependent behavior" similar to the one found experimentally in [**18**] for k-SAT. The phenomenon is qualitatively explained by a threshold property for the number of iterations of PUR makes on random *satisfiable* Horn formulas.

1. Introduction

Finding the ground state (state of minimum energy) of a physical system and computing an optimal solution to a combinatorial optimization problem are intuitively two very similar tasks. This simple observation, that motivated the development of *simulated annealing* [**17**], a simple general-purpose heuristic for combinatorial optimization, lies behind the recent birth of a new field at the crossroads of Statistical Mechanics, Theoretical Computer Science and Artificial Intelligence, that studies *phase transitions in combinatorial problems* (see [**13**] for a readable introduction). The transfer of principles and methods from Physics (mainly from Spin Glass Theory [**22**]) to Computer Science has already been quite successful, and is responsible for a couple of interesting results, such as a better understanding of the factors that account for computational intractability [**23, 24**], strikingly accurate predictions of the average running time of various algorithms [**11, 19**], or of expected values of optimal solutions [**21**].

The need for a rigorous validation of these insights is quite obvious. The theory of spin glasses is a relatively young field, which still presents many heuristic, unsolved or plain controversial aspects Moreover, while physical grounds can guide the development of the theory for "physical" models, by corroborating (or falsifying) some of its predictions (e.g. see [**22**], for a discussion of the demise, on physical grounds, of the first formulation of the so-called *replica method*), such intuition is not available when applying this type of ideas

2000 *Mathematics Subject Classification.* Primary 68Q25, Secondary 82B27.
Key words and phrases. random Horn satisfiability, phase transitions.
This paper is part of the author's Ph.D. thesis at the University of Rochester. Support for this work has come from the NSF CAREER Award CCR-9701911 and the NSF Grant 9725021.

to combinatorial problems. Given that rigorous results are hard to obtain even for spin glasses (there has recently been some progress, see e.g. [27]) it is not surprising that an analysis of most interesting combinatorial problems is still out of reach. An approach that was popular in Statistical Mechanics, and we advocate for problems in Computer Science as well, was to gather intuition through the systematic study of *exactly solved models* [4]. These are "toy" versions of the original models that are simpler to deal with, but retain much of the properties of the former ones.

Horn satisfiability is arguably the most "practical" version of satisfiability. In a previous paper [16] we have analyzed under a random model that takes into account *all* Horn clauses over n variables, we have shown that it has a *coarse threshold*, and obtained a closed-form expression for its satisfaction probability. The model from [16] has the somewhat unsettling feature that in the "interesting" range the number of clauses is exponential in the number of variables.

A slightly more realistic model imposes an upper bound k on the maximum length of a Horn clause. In this paper we compute the satisfaction probability under this random model. The more interesting phenomenon revealed by our analysis is that **in a certain well-defined sense random k-Horn satisfiability[1] "converges" (as $k \to \infty$) to the "uniform" version of satisfiability from [16].This is an "exactly-solved" counterpart to a phenomenon experimentally observed for k-satisfiability by Kirkpatrick and Selman [18].**

Also, for $k = 2$, the one case where the satisfaction probability has a singularity we are able to study another phenomenon that was investigated in connection with phase transitions: in many cases the "hardest on the average" instances for Davis-Putnam-Logeman-Loveland algorithms appear at the transition point (even if we only consider satisfiable instances [15]); this feature was once believed to be robust with respect to the choice of the particular algorithm [6]. However, Vardi et al. [7, 1] have shown however that this is *not* the case: some algorithms based on binary decision diagrams have a complexity peak well before the phase transition. An understanding of what makes a class of instances hard for a given algorithm is clearly valuable.

As a consequence of our methods we have been able to analyze a *particular problem*, random at-most-2-Horn satisfiability, and rigorously show that the average running time of a *particular DPLL algorithm*, when restricted to satisfiable instances (the ones that are statistically significant on both sides of the critical point) is finite outside the critical point, and it diverges as we approach this point.

The layout of the paper is as follows: in section 2 we review the experimental results of Kirkpatrick and Selman, in particular discussing the concept of *critical behavior*, their heuristic arguments in favor of critical behavior, Wilson's results [28], that show that some of these intuitions are wrong, as well as the recent result of Frieze and Wormald [10] on random k-SAT with k increasing with n.

Our results are presented and discussed in section 4, while in section 11 we further discuss their significance.

2. Background

In this section we briefly review some of the relevant notions and the results from [18].

---

[1]for technical convenience, all over the paper *random k-Horn satisfiability* is understood as *random* **at-most-k-Horn satisfiability** (a Horn formula that contains no positive unit clauses is trivially satisfiable).

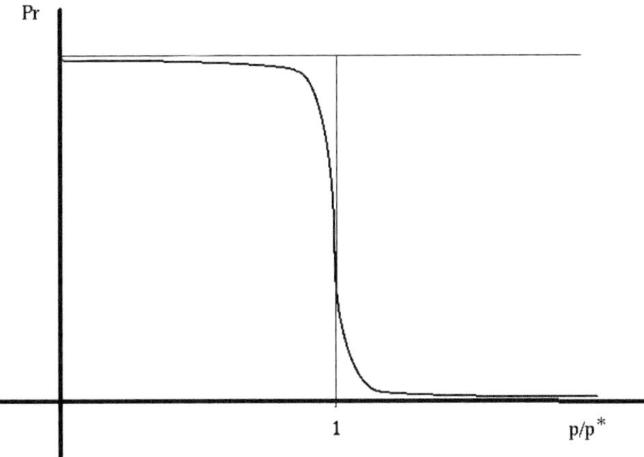

FIGURE 1. Qualitative picture of a (rescaled) sharp threshold

2.1. Threshold properties. We first discuss, briefly and limited to our interests, threshold phenomena. The best way to introduce them is through a concrete example, that of *k-CNF satisfiability*.

We generate random formulas using one control parameter, *the constraint density c*, defined as the ratio between the number of clauses m and the number of variables n of the formula. A random formula is obtained by choosing m random clauses uniformly at random among the $2^k \cdot \binom{n}{k}$ clauses of length k in the n variables.

If we plot the probability that such a random formula is satisfiable against the constraint density c, we notice the existence of a critical value c_k such that the satisfaction probability drops (as $n \to \infty$) from one to zero at c_k. Such a "sudden change" is an illustration of the mathematical concept of *sharp threshold*, qualitatively illustrated in Figure 1. The existence of a critical value c_k has not been rigorously established (except for $c_2 = 1$), even though Friedgut [8] has shown that the transition is "sharp" for every k.

Of special interest will also be the width of the so-called *scaling window (a.k.a. critical region)*. To define it consider, for $0 < \delta < 1$, $\alpha_-(n, \delta)$, the supremum over α such that for $m = \alpha n$, the probability of a random formula being satisfiable is at least $1 - \delta$. Similarly, let $\alpha_+(n, \delta)$ be the infimum over α such that for $m = \alpha n$, the probability of a random formula being satisfiable is at most δ. Then, for α within the δ-*scaling window*

$$(2.1) \qquad W(n, \delta) = (\alpha_-(n, \delta), \alpha_+(n, \delta)),$$

the probability that a random formula is satisfiable is between δ and $1 - \delta$.

We will be interested in the width of the window $W(n, \delta)$ as a function of n. It is generally believed that $|W(n)| = \theta(n^{1-1/\nu})$ for some $\nu = \nu_k \geq 1$ independent of δ, even though the existence of ν_k has only been established for $k = 2$ [5].

2.2. The mean-field approximation. An important feature that phase transitions in computational problems share with their analogues in Statistical Physics is that the various quantities of interest, such as the satisfaction probability, the ground state energy, and the location of the phase transition are hard to compute. No general-purpose methods exist, and in some cases even obtaining good non-rigorous estimates is a challenging open problem.

A technique that often provides realistic approximate values for these quantities came to be known as the *mean-field approximation*. Giving a rigorous definition of this concept is not easy, and in fact a textbook on Statistical Mechanics [2] claims: "The term "mean-field theory" conveys an impression of uniqueness, but this is false: there are many ways to generate mean-field theories".

In this paper we will deal with the following scenario, explicitly recognized as a mean-field approximation in the literature [3]: suppose we are trying to compute the average (over a certain discrete probability space) of a certain expression $f \circ (g_1, \ldots, g_n)$. If this is hard to compute we can approximate this quantity by

$$E[f(g_1(x), \ldots, g_n(x)] \sim f[E[g_1(x)], \ldots, E[g_n(x)]].$$

This technical definition does not convey a useful intuition: suppose we want to solve a combinatorial problem whose objective function depends on simultaneously satisfy several "constraints" whose effects are not independent. The mean-field approximation ignores the dependencies between various constraints, and treat them as independent.

For threshold properties this approach usually amounts to an approximation using the so-called *first-moment method*.

EXAMPLE 2.1. (k-**Satisfiability**)

The reason that the satisfiability probability of a random formula is hard to compute is that, for two assignments A, B the events $A \models \Phi$ and $B \models \Phi$ are not independent. One way to construct an mean-field theory for k-SAT is to ignore the dependencies between these events. More precisely, we have

$$1_{SAT}[\Phi] = f(g_{A_1}[\Phi], \ldots, g_{A_{2^n}}[\Phi]),$$

where

$$f(x_1, x_2, \ldots, x_{2^n}) = 1 - \prod_{i=1}^{2^n} x_i,$$

and

$$g_A[\Phi] = \begin{cases} 1, & \text{if } A \not\models \Phi, \\ 0, & \text{otherwise.} \end{cases}$$

Define $\gamma_k = 1 - 2^{-k}$. The mean-field approximation amounts to

$$\Pr[\Phi \in SAT] = E[1_{SAT}[\Phi]] \sim f(E_{g_1}[\Phi], \ldots, E_{g_{2^n}}[\Phi])$$

Since

$$E_{g_1}[\Phi] = \ldots = E_{g_{2^n}}[\Phi]) = 1 - \gamma_k^{cn}$$

this reads,

$$\Pr[\Phi \in SAT] \sim 1 - [1 - \gamma_k^{cn}]^{2^n} \sim 1 - e^{-2^n \cdot \gamma_k^{cn}} = 1 - e^{-E[\#_{SAT}[\Phi]]}$$

where $\#_{SAT}[\Phi]$ is the number of satisfying assignments for Φ. Thus (neglecting the case $E[\#_{SAT}[\Phi]] = 1$)

---

[2][**12**], pp. 105
[3]see e.g. [**25**], pp. 39–40

$$\Pr[\Phi \in SAT] \sim \begin{cases} 1, & \text{if } E[\#_{SAT}[\Phi]] \to \infty, \\ 0, & \text{if } E[\#_{SAT}[\Phi]] \to 0. \end{cases}$$

2.3. Critical exponents and behavior. A phenomenon that has been observed in various contexts is *critical behavior*. In these cases the class of problems under study has an intrinsic notion of dimensionality d, and in the limit $d \to \infty$ (or sometimes even when d is greater than a so-called *critical dimension*) "the mean-field approximation becomes exact".

A way to give precise meaning to the above quote comes from the concept of *universality*. In Statistical Mechanics one defines certain *critical exponents*, that describe the behavior of the system near the critical points; universality predicts that phase transitions with the same critical exponents are "structurally similar". A good example of this "structural similarity" is the case of random 2-satisfiability: proving unsatisfiability amounts to finding a contradictory cycle in a directed graph associated to the given formula, and the satisfiability threshold corresponds to the emergence of a giant component in this random (di)graph model. Although edges in this random graph model are *not* independent, "qualitatively" the phenomenon is very similar to the phase transition in random graphs. This intuition is made precise in [5], where it is shown that the critical exponents of the two models coincide.

Since critical exponents can be defined for the mean-field versions of the physical models too, critical behavior means that as $d \to \infty$ (or, sometimes, for d larger than the critical dimension) the critical exponents of the d-dimensional system coincide with the critical exponents of the d-dimensional mean-field model.

2.4. Rescaling. To discuss the results from [18] we another concept from Statistical Mechanics: *finite-size scaling*. The intuition behind it is that [18] "sufficiently close to a threshold or critical point, systems of all sizes are indistinguishable except for an overall change of scale." In mathematical terms this amounts to defining a new control parameter that "peeks into the *scaling window*", the region where the probability decreases from 1 to 0. It is important to note that, since a mean-field approximation yields an expression for the control parameter (in our case satisfaction probability) that will usually display a phase transition as well, a rescaled parameter can be defined for the mean-field version of the problem as well.

The definition of the rescaled parameter allows a precise formulation of the intuition that a mean-field approximation becomes exact in the limit $d \to \infty$: let P_d be a class of satisfiability problems indexed by a dimensionality parameter d, let F_d be the rescaled satisfaction probability graph of P_d, and let $F_{ann,d}$ be the rescaled graph corresponding to the mean-field approximation. The fact that the mean-field approximation becomes exact in the limit $d \to \infty$ simply means that F_d and $F_{Ann,d}$ (pointwise) converge to a common limit.

3. What did Kirkpatrick and Selman really notice ?

A recent example of critical behavior has recently been claimed, based on experimental and heuristic grounds by Kirkpatrick and Selman [18] for satisfiability problems. Their results does not explicitly mention critical exponents (although they are implicit in their discussion)[4].

---

[4]the width of the scaling window, $1 - 1/\nu_k$, is a linear combination of two critical exponents: $1 - 1/\nu_k = 2 \cdot \beta + \gamma$ (see [5])

Kirkpatrick and Selman define an (approximate) rescaled control parameter for k-SAT

$$y_k = n^{1/\nu_k}\frac{(c - c_k)}{c_k},$$

where $c = m/n$, c_k is the critical threshold for k-SAT, and ν_k is the scaling width coefficient. Also, define the "rescaled parameter"

$$y_{\infty,k} = n\frac{(c - c_k)}{c_k},$$

The rescaled limit probability graphs seem to converge (see Fig. 4 in [18]) to the "annealed limit"

$$f_\infty(y) = e^{-2^{-y}}.$$

Why did they predict critical behavior and the above form for the limit function ? The intuition presented in [18] is very simple: the major difficulty in computing the probability that a random $k - SAT$ formula is satisfiable is the fact that, for two assignments A and B, the events "$A \models \Phi$" and "$B \models \Phi$" are not generally independent, because there exist clauses of length k that are falsified by both A and B. On the other hand, qualitatively, as $k \to \infty$ clausal constraints become progressively "looser", so that in the limit we can neglect such correlations.

One can derive the exact expression for $f_\infty(y)$ by the mean-field approximation, as discussed in the previous section: for a k-CNF formula the mean-field approximation implies

$$\Pr[\Phi \in \overline{SAT}] \sim (1 - \gamma_k^{cn})^{2^n} \sim e^{-2^n \cdot \gamma_k^{cn}}.$$

But since c_k is specified (in this approximation) by $E[\#SAT] \sim 1$, i.e. $2^n \cdot \gamma_k^{c_k n} \sim 1$, or $1 + c_k \log_2 \gamma_k = 0$, this implies that as $k \to \infty$

$$\Pr[\Phi \in \overline{SAT}] \sim e^{-2^{n \cdot [1 - c/c_k]}} \sim f_\infty(y_{\infty,k}).$$

In other words, when plotted against the control parameters $y_{ann,k}$ the rescaled satisfaction probability graphs (and their mean-field counterparts) converge to the graph of f_∞.

The intuitive argument sketched in the preceding paragraph seemed to provide a beautiful explanation of the experimental results from [18]. That this intuition is, however, wrong has been shown by Wilson [28]. First note that if the previous argument were true, we would have $\nu_k \to 1$ as $k \to \infty$, since this is the width of the scaling window that the mean-field versions of $k - SAT$ predict. The paper [18] even presented experimental estimates for the critical exponents that suggested that indeed $\nu_k \to 1$ as $k \to \infty$. However these estimates are *not* accurate: Wilson proved by a simple rigorous argument that $\nu_k \geq 2$ for every constant k.

A recent result by Frieze and Wormald [10] makes the whole picture even more intriguing: they consider versions of k-SAT with k growing with n and show that if $k(n) - \log_2(n) \to \infty$ then the threshold of random k-SAT is precisely the one predicted by the mean-field approximation (the first-moment method) ! This raises the following interesting open question:

QUESTION 1: For what values of $k(n)$ is the threshold of random k-SAT *accurately predicted* by the first-moment method ?

We do not know how to answer Question 1 (we suspect the smallest values of k are those given by the result of Frieze and Wormald). More generally the question arises to investigate the intrinsic properties of random k-SAT that determine this range of values.

In this paper we answer precisely this question for a "toy" model, that of random Horn satisfiability with clauses of length *at most* k. Results in the next section show that the satisfaction probability of random k-Horn formulae converge (under a suitable rescaling, as $k \to \infty$) to a limit function. Furthermore, the limit function can be identified as a suitable "mean-field approximation that becomes exact".

We stress that we are *not* claiming that methods we used to prove this result should be of use in dealing with the infinitely more complex case of k-satisfiability, however our result is an interesting "exactly solved counterpart" to this latter one.

4. Our results

A *Horn clause* is a disjunction of literals containing *at most one positive literal*. It will be called *positive* if it contains a positive literal and *negative* otherwise. A Horn formula is a conjunction of Horn clauses. *Horn satisfiability* (denoted by HORN-SAT) is the problem of deciding whether a given Horn formula has a satisfying assignment.

Our results can be summarized as follows:

(1) For an unbounded $k = k(n)$ the threshold phenomenon is essentially the one from the "uniform case" $k(n) = n$. In particular there exists a "rescaled" parameter that makes the graphs of the limit probabilities superimpose (Theorem 4.2).

(2) For any constant k the threshold phenomenon is qualitatively described by a suitably chosen queuing model (Theorem 4.4). This yields a closed-form expression for the satisfaction probability when $k = 2$ (Theorem 4.3). This expression has a singularity (though $k = 2$ is likely the only case that does so).

(3) The rescaled limit probabilities from the cases when k is a constant converge to the one from the "infinite" case, that can in turn be seen as the result of a mean-field approximation (thus the problem displays what we have called dimension-dependent behavior).

(4) Somewhat surprisingly, the explanation for this convergence (an intrinsic feature of the problem) is a threshold property for the number of iterations of PUR (a particular algorithm) on random satisfiable Horn formulas "in the critical range."

(5) In the case when $k = 2$ PUR displays an "easy-hard-easy" pattern for the average number of iterations on satisfiable instances, peaked at the point where the limit probability has a singularity (Theorem 4.6).

Note, however, the important difference between random k-SAT and random at-most-k-HORN-SAT: for every $k \geq 2$, k-SAT has a sharp threshold [8]. All versions of HORN-SAT have coarse thresholds.

DEFINITION 4.1. Let $k = k(n) : \mathbf{N} \to \mathbf{N}$ be monotonically increasing, $1 \leq k(n) \leq n$. We define the following random model $\Omega(k, n, m)$: *formula Φ on n variables is obtained by selecting (uniformly at random and with repetition) m clauses from the set of all (non-empty) Horn clauses in the given variables of length* at most $k(n)$.

The following are our results :

THEOREM 4.2. *If $k(n) \to \infty$, $c > 0$, $H_{k(n)}$ is the number of Horn clauses on n variables having length at most $k(n)$, and $m(n) = c \cdot \dfrac{H_{k(n)}}{n}$ then*

$$(4.1) \qquad p_\infty(c) := \lim_{n \to \infty} Pr_{\Phi \in \Omega(k(n), n, m)}(\Phi \in \textit{HORN-SAT}) = 1 - F_1(e^{-c}).$$

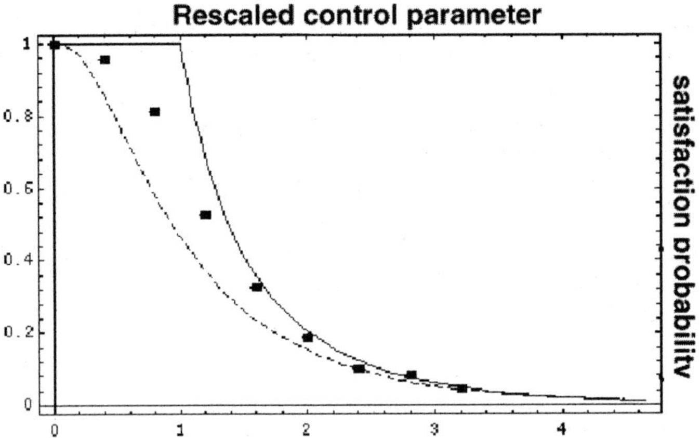

FIGURE 2. Rescaled threshold functions

THEOREM 4.3. *If $c > 0$, and $F_2 : (0,1) \to (1,\infty)$, $F_2(x) = \ln x/(x-1)$, then*

$$(4.2) \quad p_2(c) := \lim_{n\to\infty} Pr_{\Phi\in\Omega(2,n,cn)}(\Phi \in \textit{HORN-SAT}) = \begin{cases} 1, & \textit{if } c \leq \frac{3}{2}, \\ F_2^{-1}(2c/3), & \textit{otherwise.} \end{cases}$$

More generally, define $\lambda_k = \frac{k!}{k+1}$ and $S_j^i = \binom{i}{0} + \binom{i}{1} + \ldots + \binom{i}{j}$ (with the usual convention $\binom{i}{j} = 0$ for $i < j$). Then

THEOREM 4.4. *The limit probability $p_k(c) := \lim_{n\to\infty} Pr_{\Phi\in\Omega(k,n,c\cdot n^{k-1})}(\Phi \in \textit{HORN-SAT})$ is equal to the probability that the following queuing chain ever hits state zero:*

$$(4.3) \quad \begin{cases} Q_0 = 1, \\ Q_{i+1} = Q_i \dot{-} 1 + Po(c \cdot \lambda_k \cdot S_{k-2}^{i+1}), \end{cases}$$

To get a better intuition on the threshold phenomenon, as displayed by Theorems 4.2, 4.3 and 4.4, we have plotted (in Fig. 2) the limit probability functions $p_2(\cdot), p_3(\cdot), p_\infty(\cdot)$, against the "rescaled" parameter (inspired by Theorem 4.2) $\hat{c} = \frac{m\cdot n}{H_{k(n)}}$.

This rescaling has the pleasant property that it simplifies the factor λ_k from the right-hand side of (4.3), in particular mapping the critical point in Theorem 4.3 to $\hat{c} = 1$. The graphs of p_2 (continuous) and p_∞ (dashed) are obtained from their formulas in the previous results, while p_3 (dotted) is obtained via simulations. The figure makes apparent that the graphs of $p_2, p_3, \ldots, \ldots$ converge to the graph of p_∞. This statement can be proved rigorously :

THEOREM 4.5. *For every $\hat{c} > 0$, $\lim_{n\to\infty} p_n(\hat{c}) = p_\infty(\hat{c})$.*

As a bonus our analysis yields the following result:

THEOREM 4.6. *Let $q = q(c)$ be the limit of the expected number of iterations of* PUR *on a random formula $\Phi \in \Omega(2, n, cn)$, conditional on Φ being satisfiable. Then*

$$(4.4) \quad q(c) = \begin{cases} \frac{1}{1-p_2\lambda_2 c} & \textit{, if } c \neq \frac{3}{2}, \\ \infty, & \textit{otherwise.} \end{cases}$$

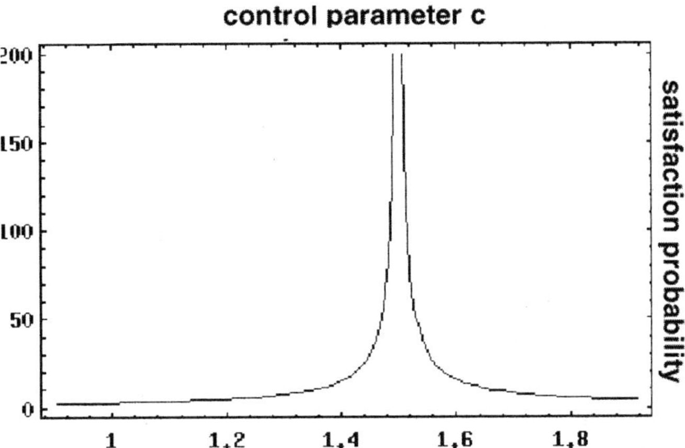

FIGURE 3. The "easy-hard-easy" pattern.

This theorem suggests (see Fig. 3) and explains the "easy-hard-easy" pattern for the average running time of PUR in this case. Experiments we performed confirm this prediction.

5. Preliminaries

Throughout this paper we use "with high probability" (w.h.p.) as a substitute for "with probability $1 - o(1)$". We denote (sometimes abusing notation) by $B(n,p)(Po(\lambda))$ a random variable having a binomial (Poisson) distribution with the corresponding parameter(s), and by $a \dot{-} b$ the value $max(a - b, 0)$. We will use the following version of the Chernoff bound

PROPOSITION 5.1. *If* $0 < \theta < 1/4$ *then* $\Pr[|B(n,p) - np| > \theta np] \leq e^{-np\frac{\theta^2}{4}}$.

as well as the related inequality from [2] :

PROPOSITION 5.2. *Let P have Poisson distribution with mean μ. For $\epsilon > 0$,*

$$\Pr[P \leq \mu \cdot (1 - \epsilon)] \leq e^{\epsilon^2 \cdot \mu/2},$$

$$\Pr[P \geq \mu \cdot (1 + \epsilon)] \leq [e^{\epsilon}(1 + \epsilon)^{-(1+\epsilon)}]^{\mu}.$$

We also use the following inequality:

PROPOSITION 5.3. *Let $k \in \mathbf{N}$ and $p \in [0, 1]$. Then for every $n \geq k$*

$$(5.1) \qquad 1 - \sum_{i=0}^{k-1} \binom{n}{i} p^i (1 - p)^{n-i} \leq \binom{n}{k} p^k.$$

Proof: Define $f : [0, 1] \to R$, $f(p) = 1 - \sum_{i=0}^{k-1} \binom{n}{i} p^i (1-p)^{n-i} - \binom{n}{k} p^k$. It is easy to see that $f'(p) = n \binom{n-1}{k-1} p^{k-1}[(1 - p)^{n-k} - 1] \leq 0$, therefore f is monotonically decreasing, and $f(0) = 0$. $\qquad\square$

We will also employ *couplings of Markov chains* (see [20]) to assert stochastic domination. The following is the definition of the type of coupling we employ in this paper:

DEFINITION 5.4. Let $(X_t)_t$ and $(Y_t)_t$ be two Markov chains on $\mathbf{Z}$. A *coupling of X and Y such that* $X_t \leq Y_t$ is a Markov chain $Z = (Z_{t,1}, Z_{t,2})$ such that:

- $Z_{t,1}$ is distributed like X_t given X_0.
- $Z_{t,2}$ is distributed like Y_t given Y_0.
- for every $i \geq 0$, $Z_{i,1} \leq Z_{i,2}$.

We use such couplings to bound the probability that a Markov chain Y_t ever decreases below a certain value a by coupling it with a chain X_t such that $X_t \leq Y_t$ and using the estimate $\Pr[\exists t : Y_t \leq a] \leq \Pr[\exists t : X_t \leq a]$ (that follows from the coupling). The couplings we construct are fairly trivial, and employ the following ideas:

- Suppose the recurrences describing ΔX_t and ΔY_t are identical, except for one term, which is $B(m_1, \tau)$ in X_t and $B(m_2, \tau)$ in Y_t, where $m_1 \leq m_2$ are positive integers and $\tau \in (0, 1)$. Obtain a coupling by identifying $B(m_1, \tau)$ with the outcome of the first m_1 Bernoulli experiments in $B(m_2, \tau)$.
- Suppose now that ΔX_t and ΔY_t differ by exactly one term which is $B(m, p)$ in ΔX_t and $B(m, q)$ in ΔY_t, $p \leq q$. Let A_i and B_i, $i = 1, m$, be independent $0/1$ experiments with success probabilities p and $\frac{q-p}{1-p}$ respectively. Define the pair $(Z_{t,1}, Z_{t,2})$ so that
  (1) $Z_{t,1}$ is the number of times A_i succeeds.
  (2) $Z_{t,2}$ is the number of times at least one of A_i and B_i succeeds.
  It is easy to see that this is a coupling with the desired properties.

We measure the distance between two probability distributions P and Q by *the total variation distance*, denoted by $d_{TV}(P, Q)$, and recall the following results, (see [26] and [3], page 2 and Remark 1.4):

LEMMA 5.5. *If $n, p, \lambda, \mu > 0$ then $d_{TV}(B(n, p), Po(np)) \leq \min\{np^2, \frac{3p}{2}\}$ and $d_{TV}(Po(\lambda), Po(\mu)) \leq |\mu - \lambda|$.*

We will also need the following simple lemma:

LEMMA 5.6. *Let c be a fixed positive integer. For every $t \in \mathbf{N}$ let ξ_t, η_t be two probability distributions. Define the Markov chains $(X_t)_t$ and $(Y_t)_t$ by recurrences*

$$(5.2) \qquad \begin{cases} X_{t+1} = X_t \dot{-} c + \xi_t, \\ Y_{t+1} = Y_t \dot{-} c + \eta_t. \end{cases}$$

Then, for every $t \geq 0$, $d_{TV}(X_t, Y_t) \leq d_{TV}(X_0, Y_0) + \sum_{i=0}^{t-1} d_{TV}(\xi_i, \eta_i)$.

Proof.
The following result gives a more convenient inequality that immediately implies Lemma 5.6

LEMMA 5.7. *Let c be a fixed positive integer. Let X, Y, ξ, η be random variables with nonnegative integer values. Define the random variables Z and T by recurrences*

$$(5.3) \qquad \begin{cases} Z = X \dot{-} c + \xi, \\ T = Y \dot{-} c + \eta. \end{cases}$$

Then, for every $d_{TV}(Z,T) \leq d_{TV}(X,Y) + d_{TV}(\xi,\eta)$.

Proof.

To prove this result, we will denote (for the "generic" r.v. A) by A_i the probability that A takes value i. We also employ the following simple inequality, valid for $a,b,c,d \geq 0$: $|ad - bc| \leq a|d - c| + |a - b|c$.

For every $a \geq 0$ we have:

$$Z_a = \sum_{i=0}^{c} X_i \xi_a + \sum_{i=c+1}^{c+a} X_i \xi_{a+c-i},$$

$$T_a = \sum_{i=0}^{c} Y_i \eta_a + \sum_{i=c+1}^{c+a} Y_i \eta_{a+c-i},$$

Applying the above-mentioned inequality and summing we get:

$$d_{TV}(Z,T)$$

$$\leq \frac{1}{2}\{\sum_{i=0}^{c}\sum_{a=0}^{\infty} X_i|\xi_a - \eta_a| + \sum_{i=0}^{c}\sum_{a=0}^{\infty}|X_i - Y_i|\eta_a +$$

$$+ \sum_{i=c+1}^{c+a}\sum_{a=0}^{\infty} X_i|\xi_{c+a-i} - \eta_{c+a-i}| + \sum_{i=c+1}^{c+a}\sum_{a=0}^{\infty}|X_i - Y_i|\eta_{c+a-i}\}.$$

Let A,B,C,D be the four terms of the sum. By simple algebraic manipulations we obtain:

$$A = (\textstyle\sum_{i=0}^{c} X_i) \cdot d_{TV}(\xi,\eta), \qquad\qquad B = \frac{1}{2}\textstyle\sum_{i=0}^{c}|X_i - Y_i|,$$
$$C = (\textstyle\sum_{i=c+1}^{\infty} X_i) \cdot d_{TV}(\xi,\eta), \qquad\qquad D = \frac{1}{2}\textstyle\sum_{i=c+1}^{\infty}|X_i - Y_i|,$$

and the result follows. $\qquad\qquad\square$

Finally, we need the following trivial occupancy property:

LEMMA 5.8. *Let a white balls and b black balls be thrown uniformly at random in n bins.*

(1) *if $r = \max(a,b) = o(n^{1/2})$ then the probability that there is a bin that contains both white and black balls is at most $\frac{4r^2}{n} = o(1)$.*

(2) *if $s = \min(a,b) = \omega(n^{1/2})$ then the probability that there is a bin that contains both white and black balls is $1 - o(1/poly)$.*

Proof. The first part is easy: the probability that two balls (of any color) end up in the same bin is at most $\binom{a+b}{2} \cdot \frac{1}{n}$. For the second part, let A be the event that no two balls of different colors end up in the same bin, and let B the event that at least $\sqrt{n}$ bins contain white balls. We have:

$$\Pr[A] \leq \Pr[A|B] + \Pr[\overline{B}].$$

But

$$\Pr[\overline{B}] \leq \binom{n}{\sqrt{n}} \cdot (\frac{1}{\sqrt{n}})^a = n^{\sqrt{n}-a/2} = o(\frac{1}{poly}),$$

and

$$\Pr[A|B] \leq (1 - \frac{1}{\sqrt{n}})^b \sim e^{-b/\sqrt{n}} = o(\frac{1}{poly}).$$

$\square$

The algorithm PUR is displayed in Figure 3. We regard PUR as working in stages, indexed by the number of variables still left unassigned; thus, the stage number decreases as PUR moves on. We say that *formula Φ survives Stage t* if PUR on input Φ does not halt at Stage t or earlier. Let Φ_i be the formula at the beginning of stage i, and let N_i denote the number of its clauses. We will also denote by $P_{i,t}(N_{i,t})$, the number of clauses of Φ_t of size i and containing one (no) positive literal. Define $\Phi_{i,t}^P$ ($\Phi_{i,t}^N$) to be the subformula of Φ_t containing the clauses counted by $P_{i,t}(N_{i,t})$.

The following lemmas were proved in [**16**], in the context of analyzing the behavior of PUR on $\Phi \in \Omega(n, n, m), m = c \cdot 2^n$.

LEMMA 5.9. (1) *Suppose* PUR *does not halt before stage t. Then, conditional on N_t, the clauses of Φ_t are random and independent.*

(2) *Suppose now that we condition on $\Gamma_t = (N_{1,t}, N_{2,t}, P_{1,t}, P_{2,t}$ and on the fact that Φ survives Stage t as well. Then we have*

$$(5.4) \qquad N_{t-1} = N_t - \Delta_{1,P}(t) - \Delta_{2,P}(t),$$

where
- $\Delta_{1,P}(t)$, *the number of positive clauses that are satisfied at stage t, has the distribution $1 + B\left(P_{1,t} - 1, \frac{1}{t}\right)$.*
- $\Delta_{2,P}(t)$, *the number of positive non-unit clauses that are satisfied at stage t, has the binomial distribution $B\left(P_{2,t}, \frac{1}{t}\right)$.*

LEMMA 5.10. *For every $c > 0$ and every $t, n - c\sqrt{n} \leq t \leq n$, the conditional probability that the inequality*

$$(5.5) \qquad N_n - (n - t)\left[1 + \frac{2(N_n - 1)}{t}\right] \leq N_j \leq N_n$$

holds for all $t \leq j \leq n$, in the event that PUR *reaches stage t, is $1 - o(1)$.*

LEMMA 5.11. *Let $X_n \in [0, n]$ be the r.v. denoting the number of iterations of* PUR *on a random satisfiable formula $\Phi \in \Omega(n, c \cdot 2^n)$. Then X_n converges in distribution to a distribution ρ on $[0, n]$ having support on the nonnegative integers, $\rho = (\rho_k)_{k \geq 0}$, $\rho_k = Prob[\rho = k]$, given by*

$$\rho_k = \frac{e^{-2^k c}}{1 - F(e^{-c})} \cdot \prod_{i=1}^{k-1}(1 - e^{-2^i c}).$$

```
Program PUR(Φ):
  if Φ (contains no positive literal as a clause)
     then return TRUE
     else
           choose such a positive unit clause x
           if (Φ contains x̄ as a clause)
              then
                    return FALSE
           else
                    let Φ' be the formula
                    obtained by setting x to 1
                    return PUR(Φ')
```

FIGURE 4. Algorithm PUR

6. The proof of Theorem 4.2

Let $c_1 < c_2 < c_3$ be arbitrary constants. Consider three random formulas $\Phi_1 \in \Omega(n, \mathbf{k(n)}, c_1 \cdot \frac{H_{k(n)}}{n}), \Phi_2 \in \Omega(n, \mathbf{n}, c_2 \cdot 2^n)$ and $\Phi_3 \in \Omega(n, \mathbf{k(n)}, c_3 \cdot \frac{H_{k(n)}}{n})$, and let Φ' be the subformula of Φ_2 consisting of the clauses of size at most $k(n)$. By the Chernoff bound, with high probability, m', the number of clauses of Φ', is in the interval $[c_1 \cdot \frac{H_{k(n)}}{n}, c_3 \cdot \frac{H_{k(n)}}{n}]$. When $n \to \infty$ the probability that $\Phi_2 \in$ HORN-SAT tends to $1 - F_1(e^{-c_2})$.

From Lemma 5.11 we infer the following easy consequence

CLAIM 6.1. *The probability that PUR accepts Φ_2 after stage $n - k(n) + 1$ is $o(1)$.*

Since in the first $k(n) - 1$ stages of PUR *only the clauses of Φ' can influence the algorithm acceptance/rejection of Φ_2 (because PUR accepts/rejects at Stage i based only on the unit clauses, and each non-simplified clause loses at most one literal at each phase),*

$$|\Pr[\Phi_2 \in \text{HORN-SAT}] - \Pr[\Phi' \in \text{HORN-SAT}]| = o(1).$$

By the monotonicity of SAT and the randomness of Φ_1, Φ_2, Φ' we have

$$\Pr[\Phi_1 \in \text{HORN-SAT}] - o(1) \leq \Pr[\Phi_2 \in \text{HORN-SAT}]$$
$$\leq \Pr[\Phi_3 \in \text{HORN-SAT}] + o(1).$$

Taking limits it follows that

$$\overline{\lim}_{n \to \infty} \Pr_{\Phi \in \Omega(n, k(n), c_1 H_{k(n)}/n)}[\Phi \in \text{HORN-SAT}] \leq 1 - F(e^{-c_2}) \leq$$
$$\underline{\lim}_{n \to \infty} \Pr_{\Phi \in \Omega(n, k(n), c_3 H_{k(n)}/n)}[\Phi \in \text{HORN-SAT}].$$

Since c_1, c_2, c_3 were chosen arbitrarily, by choosing $c_1 = c, c_2 = c+\epsilon$, and $c_2 = c-\epsilon, c_3 = c$, respectively, we infer that

$$1 - F_1(e^{-(c-\epsilon)}) \leq \underline{\lim}_{n \to \infty} \Pr_{\Phi \in \Omega(n, k(n), cH_{k(n)}/n)}[\Phi \in \text{HORN-SAT}] \leq$$
$$\overline{\lim}_{n \to \infty} \Pr_{\Phi \in \Omega(n, k(n), cH_{k(n)}/n)}[\Phi \in \text{HORN-SAT}] \leq 1 - F_1(e^{-(c+\epsilon)}).$$

As ϵ is arbitrary, we get the desired result. $\square$

OBSERVATION 1. One point about the previous proof that is intuitively clear, but gets somewhat obscured by the technical details of the proof, is that if $\Phi_2 \in \Omega(n, \mathbf{n}, c_2 \cdot 2^n)$ then Φ' behaves "for every practical purpose" as if it were a uniform formula in $\Omega(n, \mathbf{k(n)}, c_2 \cdot \frac{H_{k(n)}}{n})$. We will use a similar intuition in the proof of Proposition 4.5.

7. The uniformity lemma

The following lemma is the analog of Lemma 5.9 for the case $k = 2$, and the basis for our analysis of this case:

LEMMA 7.1. *Suppose that Φ survives up to stage t. Then, conditional on $(P_{1,t}, N_{1,t}, P_{2,t}, N_{2,t})$, the clauses in $\Phi^P_{1,t}, \Phi^N_{1,t}, \Phi^P_{2,t}, \Phi^N_{2,t}$ are chosen uniformly at random and are independent. Also, conditional on the fact that Φ survives stage t as well, the following recurrences hold:*

$$(7.1) \qquad \begin{cases} P_{1,t-1} = P_{1,t} - 1 - \Delta^P_{1,t} + \Delta^P_{12,t}, \\ N_{1,t-1} = N_{1,t} + \Delta^N_{12,t}, \\ P_{2,t-1} = P_{2,t} - \Delta^P_{12,t} - \Delta^P_{02,t}, \\ N_{2,t-1} = N_{2,t} - \Delta^N_{12,t}, \end{cases}$$

where (in distribution)

$$(7.2) \qquad \begin{cases} \Delta^P_{1,t} = B(P_{1,t} - 1, 1/t), \\ \Delta^P_{12,t} = B(P_{2,t}, 1/t), \\ \Delta^P_{02,t} = B(P_{2,t} - \Delta^P_{12,t}, 1/t), \\ \Delta^N_{12,t} = B(N_{2,t}, 2/t). \end{cases}$$

Proof. A formula will be represented by an $m \times 2$ table. The rows of the table correspond to clauses in the formula and the entries are its literals. They are gradually unveiled as the algorithm proceeds. We assume that when generating Φ we mark those clauses containing only one literal (so that we know their location, but not their content). We say that a row (or a clause) is "blocked" either if the clause is already satisfied or the clause has been turned into the empty clause. Suppose PUR arrives at stage t on Φ. Then in stages $i = n, n - 1, \ldots, t + 1$, Φ_i should contain a unit clause consisting of a positive literal but should not have contained complementary unit clauses of the same variable. To carry out the disclosure at stage i, let x be the variable set to one in this stage. We assume that the formula unveils all occurrences of x or $\overline{x}$ in Φ. For each clause we perform the following:

(1) if it contains x we unveil all its literals and block;
(2) otherwise we do nothing.

The clauses of Φ_t having size two correspond to the rows of Φ that contain no unveiled literal. The clauses of size one are either the clauses of size one in Φ that contain none of the chosen literals, or the clauses of size two that contain the negation of one chosen variable and another is yet to be chosen. Given these observations the uniformity and independence follow from the way we construct Φ.

To prove the recurrences, let x be the variable set to one in stage t (it exists since PUR does not halt at this stage). By uniformity and independence, each of the $P_{1,t} - 1$ positive unit clauses of Φ_t, other than the chosen one, is equal to x with probability $1/t$ (since there are t variables left at this stage). On the other hand, the positive unit clauses of Φ_{t-1} that are not present in Φ_t can only come from clauses of size two of Φ_t that contain $\bar{x}$ and a positive literal (therefore counted by $P_{2,t}$). Uniformity and independence imply therefore that $\Delta_1^P(t)$ has the distribution claimed in (7.2). The other relations can be justified similarly (noting that, since PUR does not reject at this stage, every negative unit clause of Φ_t is also present in Φ_{t-1}).

It will be useful to consider the Markov chain (7.1) for all values of $t = n, \ldots, 0$ (even when the algorithm halts). To accomplish that, the "minus" signs in the first equation of (7.1) and the definition of $\Delta_{1,t}^P$ should be replaced by $\dot{-}$. We also need to specify the distribution of each component of the tuple $(P_{1,n}, N_{1,n}, P_{2,n}, N_{2,n})$. Let Δ_n be a random variable having the Bernoulli distribution $B\left(cn, \frac{2n}{2n+3\binom{n}{2}}\right)$. It is easy to see that in distribution

$$(7.3) \qquad \begin{cases} P_{1,n} = B(\Delta_n, 1/2), \\ N_{1,n} = \Delta_n - P_{1,n}, \\ P_{2,n} = B(cn - \Delta_n, 2/3) \\ N_{2,n} = cn - \Delta_n - P_{2,n}. \end{cases}$$

$\square$

8. Proof of Theorem 4.3

The main intuition for the proof is that in "most interesting stages" $\Delta_{1,t}^P = 0$ and $\Delta_{12,t}^P$ is approximately Poisson distributed. Therefore, $P_{1,t}$ qualitatively behaves like the Markov Chain $(Q_t)_t$ defined by

$$(8.1) \qquad \begin{cases} Q_{n+1} = 1, \\ Q_{t-1} = Q_t \dot{-} 1 + Po(\lambda), \end{cases}$$

where $\lambda = 2c/3$. This explains the closed form of the limit probability: a well-known result states that ρ, the probability that the queuing chain Q_t reaches state 0, satisfies the equation $\rho = \Phi(\rho)$, where $\Phi(t) = e^{\lambda(t-1)}$ is the generating function of the arrival distribution $Po(\lambda)$. We will define a suitable value ω_0 such that:

(1) With high probability PUR does not reject in any of stages $n, \ldots, n - \omega_0$.
(2) PUR accepts "mostly before or at stage $n - \omega_0$" (i.e. the probability that PUR accepts after stage $n - \omega_0$, given that Φ survives this far is $o(1)$).
(3) With high probability, for every $t \in n, \ldots, n - \omega_0$, $\Delta_{1,t}^P = 0$.
(4) At stages $n, \ldots, n - \omega_0$, $P_{1,t}$ is "very close" to Q_t, with respect to total variation distance.

8.1. The case $c < \frac{3}{2}$. The program outlined before can be accomplished as described if $c < 3/2$. To prove Property 4 we make use of Lemmas 5.5 and 5.8. Property 2 is proved only implicitly: in this case (see [**14**]) the probability that $Q_i = 0$ for some i tends to one, and, in fact, by a technical result due to Frieze and Suen (Lemma 3.1 in [**9**]), $\Pr[Q_i = 0$ for some $i \geq n - \log n]$ is $1 - o(1)$. Define $\omega_0 = n^{0.1}$. The following are the main steps of the proof in this case:

LEMMA 8.1. *With probability $1 - o(1/poly)$ for every $t \in [n, \ldots, n/2]$ we have*

$$\Delta_{12,t}^P, \Delta_{02,t}^P, \Delta_{12,t}^N \leq \frac{1}{2}n^{0.1}.$$

Proof.   Use the coupling with $m_1 = P_{2,t}(N_{2,t})$, $m_2 = cn$, $\tau = 1/t$, and apply Chernoff bound to $B(cn, 1/t)$. $\square$

COROLLARY 8.2.   Consider $\omega \leq n/2$. If for every $t \in [n, \ldots, n/2]$, $\Delta^P_{12,t}, \Delta^P_{02,t}, \Delta^N_{12,t} \leq \frac{1}{2}n^{0.1}$ then, for all $t \in [n, \ldots, n-\omega]$, $P_{1,t}, N_{1,t}, |P_{2,t}-P_{2,n}|, |N_{2,t}-N_{2,n}| < (n-t)\cdot n^{0.1}$.

LEMMA 8.3.   *If for all $t \in [n, \ldots, n - \omega]$, $P_{1,t}, N_{1,t}, |P_{2,t} - P_{2,n}|, |N_{2,t} - N_{2,n}| < (n - t) \cdot n^{0.1}$ holds then w.h.p. $\Delta^P_{1,t} = 0$ for every $t \in [n, \ldots, n - \omega_0]$.*

Proof. $\Pr[B(P_{1,t} - 1, \frac{1}{t}) > 0] = 1 - \Pr[B(P_{1,t} - 1, \frac{1}{t}) = 0] = 1 - (1 - \frac{1}{t})^{P_{1,t}-1} < \frac{P_{1,t}-1}{t} < n^{-0.9}$. $\square$

LEMMA 8.4.   *W.h.p., $|P_{2,n} - \frac{2}{3}cn|, |N_{2,n} - \frac{1}{3}cn| < n^{0.6}$.*

Proof.   Directly from the Chernoff bounds on Δ_n and $P_{2,n}$. $\square$

LEMMA 8.5.   *If the events in the conclusions of Lemmas 8.2 and 8.4 hold for $\omega = \omega_0$, $\epsilon_1 = 1/6$ and $\epsilon_2 = 0.1$, then there exists a constant $r > 0$ such that for every $t = n, \ldots, n - \omega_0$, $|\frac{P_{2,t}}{t} - \frac{2}{3}c| \leq rn^{-0.4}$.*

Proof.   We have

$$|\frac{P_{2,t}}{t} - \frac{2}{3}c| \leq P_{2,t}\left|\frac{1}{t} - \frac{1}{n}\right| + \frac{|P_{2,t} - P_{2,n}|}{n} + \left|\frac{P_{2,n}}{n} - \frac{2}{3}c\right|$$

$$\leq P_{2,n}\frac{\omega_0}{n(n - \omega_0)} + \frac{n^{0.2}}{n} + n^{0.6-1},$$

by Lemma 8.5 and $n - \omega_0 \leq t \leq n$, and the result immediately follows. $\square$

LEMMA 8.6.   *If the conclusions of Lemmas 8.5 and 8.3 are true then*

$$\sum_{t=n-\omega_0}^{n} d_{TV}(P_{1,t}, Q_t) = o(1/\omega_0).$$

Proof.   By Lemma 8.5 and the inequalities on total variation distance there exist $r_1, r_2 > 0$ such that

$$d_{TV}(\Delta^P_{12,t}, Po(\lambda)) \leq d_{TV}\left(\Delta^P_{12,t}, Po\left(\frac{P_{2,t}}{t}\right)\right) + d_{TV}\left(Po\left(\frac{P_{2,t}}{t}\right), Po\left(\frac{2}{3}c\right)\right)$$

$$\leq r_1\frac{1}{t} + r_2 n^{-0.4} \leq r_3 n^{-0.4},$$

where $r_3 = r_1 + r_2$. Employing Lemma 5.6 it follows that

$$\sum_{t=n-\omega_0}^{n} d_{TV}(P_{1,t}, Q_t) \leq r_3 \sum_{t=n-\omega_0}^{n} tn^{-0.4} \leq r_3 n^{-0.4}\frac{\omega_0^2}{2},$$

and this amount is $o(1/\omega_0)$. $\square$

OBSERVATION 2.   The probability that the conditions in the previous lemma are not fulfilled is at most $\omega_0^4/n = n^{-0.6}$. Indeed, the events that ensure the applicability of the previous lemma are:

(1) for every $t \in [n, \ldots, n/2]$, $\Delta_{12,t}^{P}, \Delta_{02,t}^{P}, \Delta_{12,t}^{N} \leq \frac{1}{2}n^{0.1}$,

(2) for all $t \in [n, \ldots, n - \omega_0]$, $\Delta_{1,t}^{P} = 0$, and

(3) $|P_{2,n} - \frac{2}{3}cn|, |N_{2,n} - \frac{1}{3}cn|, < n^{0.6}$

The events in the first and the third lines have probability $1 - o(1/poly)$ (as they come from applying Chernoff bounds). The second fails (for a specific t) with probability at most $\frac{P_{1,t}}{n-t} \leq \omega_0^2/(n - \omega_0)$, so its total probability is at most $\omega_0 \cdot \omega_0^2/(n - \omega_0)$. Both terms can be absorbed into ω_0^4/n.

LEMMA 8.7. *If the event in Lemma 8.2 holds then w.h.p.* PUR *does not reject at stage t, for every t in the range $n, n - 1, \ldots, n - \omega_0$, given that Φ survives up to this stage.*

Proof. To prove Lemma 8.7 we show that, with high probability the unit clauses of each Φ_t involve different variables. This can be seen as follows: consider $P_{1,t} + N_{1,t}$ balls to be thrown into t urns. The probability that two of them arrive in the same urn is at most $\binom{P_{1,t}+N_{1,t}}{2} \cdot \frac{1}{t}$. This is upper bounded by $\frac{(\omega_0 n^{0.1})^2}{2(n-\omega_0)}$. Summing this for $t = n, \ldots, n - \omega_0$ yields an upper bound, which is $o(1)$. $\qquad\square$

The proof for the case $c < 3/2$ follows easily from these results: with probability $1 - o(1)$ all the events in Lemmas 8.1, 8.2, 8.3, 8.4, 8.6, and 8.7 take place, therefore PUR does not reject at any of the stages n to $n - \omega_0$ and $P_{1,t}$ is close to Q_t in the sense of Lemma 8.6. Therefore the probability that for some t in this range $P_{1,t} = 0$ (i.e. PUR accepts) differs by $o(1)$ from the corresponding probability for Q_t. But according to the result by Frieze and Suen [9] this latter probability is $1 - o(1)$. $\qquad\square$

8.2. The case $c > \frac{3}{2}$. Let us now concentrate on the case when $c > 3/2$ (the case when $c = 3/2$ will follow by a monotonicity argument). In the previous argument we only used the fact that $c < 3/2$ when deriving the probability that Q_t hits state 0, hence the arguments from above carry on, and the conclusion is that the probability that PUR accepts at one of the stages $n, \ldots, n - \omega_0$ differs by $o(1)$ from the probability that $Q_t = 0$ somewhere in this range. We now, however, have to consider the probability that PUR accepts at some stage later than $n - \omega_0$ and aim to prove that this probability is $o(1)$. It is conceptually simpler to divide the interval $[n - \omega_0, 0]$ into two subintervals, $[n - \omega_0, n - \omega_1]$ and its complement, such that w.h.p. $\Phi_{n-\omega_1}$ (if defined) contains two opposite unit clauses, therefore the probability that PUR accepts after stage $n - \omega_1$ is $o(1)$. In the range $[n - \omega_0, n - \omega_1]$ we would like to prove that "most of the time" $\Delta_{1,t}^{P}$ is zero and $P_{1,t}$ is "close" to Q_t and to reduce the problem to the analysis of Q_t. Unfortunately there are two problems with this approach: although the probability that each individual $\Delta_{1,t}^{P} > 0$ is fairly small, to make $\Phi_{n-\omega_1}$ unsatisfiable w.h.p., ω_1 has to be $\omega(\sqrt{n})$. This implies that we cannot sum these probabilities over $[n - \omega_0, n - \omega_1]$ and expect the sum to be $o(1)$; a similar problem arises if we want to sum the upper bounds for $d_{TV}(\Delta_{12,t}^{P}, Po(\lambda))$.

Fortunately there is a way to circumvent this, avoiding the use of total variation distance altogether: although we cannot guarantee that w.h.p. each $\Delta_{1,t}^{P} = 0$, we can arrange that w.h.p. for every sequence of p consecutive stages $t, t - 1, \ldots t - p + 1$, $\Delta_{1,t}^{P} + \Delta_{1,t-1}^{P} + \ldots + \Delta_{1,t-p+1}^{P} \leq 4$ (*). Intuitively, in any sequence of p consecutive steps at most $p + 4$ clients leave the queue, and the number of those who arrive is the sum of p approximately Poisson variables, thus approximately Poisson with parameter $p\lambda$. Choosing p large enough so that $\lambda > 1 + \frac{4}{p}$ ensures that in any p steps *the average number of customers that arrive is strictly larger than the number of customers that are served in* *this time span.* Therefore we will seek to approximate $P_{1,t}$ by a queuing chain $\overline{Q}_t$ with this property. Since $P_{1,n-\omega_0} = \overline{Q}_{n-\omega_0}$ is "large," an elementary analysis of the queuing chain implies that the probability that $\overline{Q}_t$ hits state 0 in the interval $[n - \omega_0, n - \omega_1]$ is exponentially small. So we obtain the desired result if $\overline{Q}_t$ is constructed so that it is stochastically dominated by $P_{1,t}$.

Define $\omega_1 = n^{0.51}$. The following are the auxiliary results we use in this case:

LEMMA 8.8. *Let* $A = n^{0.61}$. *For every* $k > 0$ *there exists a constant* $c_k > 0$ *such that for every* $r > 0$ *the probability that there exists* $t \in [n, n - \omega_1]$, $\Delta^P_{1,t} + \Delta^P_{1,t-1} + \ldots + \Delta^P_{1,t-r+1} \geq k$ *is at most* $c_k(n - \omega_0)(rA/n)^k$.

Proof. By Corollary 8.2 we can assume that $P_{1,t} \leq A$. Then for every i,

$$\Pr[\Delta^P_{1,t} \geq i] = \Pr[B(P_{1,t} - 1, \frac{1}{t}) \geq i] \leq \Pr[B(A, \frac{1}{t}) \geq i]$$

$$= 1 - \sum_{j=1}^{i-1} \binom{A}{j} \left(\frac{1}{t}\right)^j \left(1 - \frac{1}{t}\right)^{A-j}$$

$$\leq \binom{A}{i} (\frac{1}{t})^i$$

The event $\Delta^P_{1,t} + \Delta^P_{1,t-1} + \ldots + \Delta^P_{1,t-r+1} \geq k$ can only happen in a fixed number of circumstances:

- one of the factors is at least k, or
- one of the factors is at least $k - 1$, and another one is at least 1, or
- $\ldots$
- at least k of the factors are at least one.

(a finite number of possibilities). Applying the previous inequality, and taking into account that r, k are fixed immediately proves the lemma. $\qquad\square$

To flesh out the argument outlined before we construct a succession of Markov chains running along $P_{1,t}$, that provide better and better "approximations" to $\overline{Q}_t$. Our use of indices will be slightly nonstandard (to reflect the connection with $P_{1,t}$), in that the sequence of indices starts with $n - \omega_0$ and is decreasing.

DEFINITION 8.9. Let $X_{n-\omega_0} = Y_{n-\omega_0} = Z_{n-\omega_0} = \overline{Q}_{n-\omega_0} = P_{1,n-\omega_0}$ and

$$(8.2) \quad \begin{cases} X_{t-1} = X_t - (p+4)\chi_{p\mathbf{Z}+1}(n - \omega_0 - t) + \Delta^P_{12,t}, \\ Y_{t-1} = Y_t - (p+4)\chi_{p\mathbf{Z}+1}(n - \omega_0 - t) + B(P_{2,n-\omega_1}, 1/t), \\ Z_{t-1} = Z_t - (p+4)\chi_{p\mathbf{Z}+1}(n - \omega_0 - t) + B(P_{2,n-\omega_1}, \frac{1}{n}), \\ \overline{Q}_{t-1} = \overline{Q}_{t-1} - 1 + B(p\lfloor \frac{P_{2,n-\omega_1}}{p+4} \rfloor, \frac{1}{n}). \end{cases}$$

Let $c = \Pr[(\exists t \in [n - \omega_0, n - \omega_1]) : P_{1,t} = 0]$. Note that the amount $p + 4$ is subtracted from X_t, Y_t, Z_t exactly once in every p consecutive steps, so whenever the condition (*) is satisfied it holds that $X_t \leq P_{1,t}$ for every $t \in [n-\omega_0, n-\omega_1]$. By coupling $\Delta^P_{12,t}(= B(P_{2,t}, 1/t))$ with $B(P_{2,n-\omega_1}, 1/t)$ we deduce that we can couple X_t and Y_t so that $Y_t \leq X_t$. We can also couple (using the second coupling idea) Y_t and Z_t such that $Z_t \leq Y_t$. Finally, notice that we can couple $(Z_{n-\omega_0-jp})_{j\geq 0}$ and $(\overline{Q}_{n-\omega_0-j(p+4)})_{j\geq 0}$ such that $\overline{Q}_{n-\omega_0-j(p+4)} \leq Z_{n-\omega_0-jp}$. So an upper bound on c is $\Pr[(\exists t \in [0, n - \omega_0]) : \overline{Q}_t = 0]$. With high probability the Bernoulli distribution in the definition of the chain $\overline{Q}_t$ has the average strictly greater than one, (because the flow from $P_{2,t}$ is approximately Poisson), and $\overline{Q}_{n-\omega_0} = \Omega(\omega_0)$, therefore, by an elementary property of the queuing chain, the probability that $\overline{Q}_t$ hits state 0 is exponentially small. This yields the desired conclusion, that $c = o(1)$.

One word about the way to prove the fact that $\Phi_{n-\omega_1}$ is unsatisfiable (if defined): one can prove that w.h.p. both $P_{1,n-\omega_1}$ and $N_{1,n-\omega_1}$ are $\Omega(\omega_1)$. By the uniformity lemma 7.1 we are left with the following instance of the occupancy problem: there are $P_{1,n-\omega_1}$ white balls, $N_{1,n-\omega_1}$ black balls and $n - \omega_1$ bins. The desired fact now follows from the second part of Lemma 5.8. $\qquad\square$

9. Proof of Theorem 4.4

Theorem 4.4 is proved along lines very similar to the proof of Theorem 4.3. The basis is the following generalization of Lemma 7.1:

LEMMA 9.1. *Suppose that Φ survives up to stage t. Then, conditional on the values $(P_{1,t}, N_{1,t}, \ldots, P_{k,t}, N_{k,t})$, the clauses in $\Phi_{1,t}^P, \Phi_{1,t}^N, \ldots, \Phi_{k,t}^P, \Phi_{k,t}^N$ are chosen uniformly at random and are independent. Also, conditional on the fact that Φ survives stage t as well, the following recurrences hold:*

$$(9.1) \quad \begin{cases} P_{1,t-1} = P_{1,t} - 1 - \Delta_{01,t}^P + \Delta_{12,t}^P, \\ N_{1,t-1} = N_{1,t} + \Delta_{12,t}^N, \\ P_{i,t-1} = P_{i,t} - \Delta_{0i,t}^P - \Delta_{(i-1)i,t}^P + \Delta_{i(i+1),t}^P, \text{ for } i = \overline{2,k}, \\ N_{i,t-1} = N_{i,t} - \Delta_{(i-1)i,t}^N + \Delta_{i(i+1),t}^N, \text{ for } i = \overline{2,k}, \end{cases}$$

where (in distribution)

$$(9.2) \quad \begin{cases} \Delta_{01,t}^P = B(P_{1,t} - 1, 1/t), \\ \Delta_{(i-1)i,t}^P = B(P_{i,t}, (i-1)/t), \\ \Delta_{0i,t}^P = B(P_{i,t} - \Delta_{(i-1)i,t}^P, 1/t), \\ \Delta_{(i-1)i,t}^N = B(N_{i,t}, i/t), \\ \Delta_{k(k+1),t}^P = \Delta_{k(k+1),t}^N = 0. \end{cases}$$

Proof.

The uniformity condition and the justification of the recurrences are absolutely similar to the ones from Lemma 8.1. $\qquad\square$

The additional technical complication in analyzing the recurrences in the previous lemma is that now there is a "positive flow into $P_{2,t}, N_{2,t}$."

LEMMA 9.2. *With high probability it holds that*

$$P_{i,t} = (1 + o(1)) \cdot \frac{c}{n} \cdot \lambda_k \cdot i \cdot \binom{t}{i} \cdot S_{k-i}^{n+1-t},$$

and

$$N_{i,t} = (1 + o(1)) \cdot \frac{c}{n} \cdot \lambda_k \cdot \binom{t}{i} \cdot S_{k-i}^{n+1-t},$$

for every $i \geq 2$, uniformly for $t = n - o(n)$.

Proof.

Let $x_{i,t}, y_{i,t}$ be the expected values of $P_{i,t}, N_{i,t}$. By applying a result of Wormald [29] we deduce that with probability $1 - o(1)$ for all $i \geq 2$ and $t = n - o(n)$

$$P_{i,t} = x_{i,t}[1 + o(1)], \qquad N_{i,t} = y_{i,t}[1 + o(1)].$$

Let us now derive a formula for $x_{i,t}$, $y_{i,t}$, by replacing the binomial distributions in the equations by their expected values.

We have:

(9.3)
$$\begin{cases} x_{i,t-1} = x_{i,t} - \frac{x_{i,t}}{t} - \frac{(i-1)x_{i,t}}{t} + \frac{i x_{i+1,t}}{t}, \text{ for } i = \overline{2,k}, \\ y_{i,t-1} = y_{i,t} - \frac{i y_{i,t}}{t} + \frac{(i+1)y_{(i+1),t}}{t}, \text{ for } i = \overline{2,k}, \end{cases}$$

Rearranging terms the recurrences become

(9.4)
$$\begin{cases} x_{i,t-1} = x_{i,t}(1 - \frac{i}{t}) + x_{i+1,t}\frac{i}{t}, \text{ for } i = \overline{2,k}, \\ y_{i,t-1} = y_{i,t}(1 - \frac{i}{t}) + y_{(i+1),t}\frac{(i+1)}{t}, \text{ for } i = \overline{2,k}. \end{cases}$$

Also,

(9.5)
$$\begin{cases} x_{i,n} = \frac{i\binom{n}{i}}{H_k} \cdot c\lambda_k \cdot \frac{H_k}{n} = \frac{c}{n}\lambda_k \cdot i\binom{n}{i}, \\ y_{i,n} = \frac{\binom{n}{i}}{H_k} \cdot c\lambda_k \cdot \frac{H_k}{n} = \frac{c}{n}\lambda_k \cdot \binom{n}{i}. \end{cases}$$

A simple induction shows that these expected values are $x_{i,t} = \frac{c}{n} \cdot \lambda_k \cdot i \cdot \binom{t}{i} \cdot S_{k-i}^{n+1-t}$, and $y_{i,t} = \frac{c}{n} \cdot \lambda_k \cdot \binom{t}{i} \cdot S_{k-i}^{n+1-t}$.

$\square$

The previous lemma implies that $\Delta_{2,t}^P \sim Po(c \cdot \lambda_k \cdot S_{k-2}^{n+1-t})$ (for $t = n - o(n)$); thus in this range $P_{1,t-1} \sim P_{1,t} - 1 + Po(c \cdot \lambda_k \cdot S_{k-2}^{n+1-t})$. The proof follows exactly the same pattern as in the case $c < 3/2$ for $k = 2$: the conclusion for the stages $[n, n - \omega_0]$ is that the probability that $P_{1.t}$ is zero somewhere in this range differs by $o(1)$ from the corresponding probability for the queuing chain in (4.3). The fact that the stages after $[n, n - \omega_0]$ have a contribution of $o(1)$ to the final accepting probability can be seen by the fact that there is possible to couple the Markov M_1, describing the evolution of PUR on a random k-SAT formula, and M_2 that runs on the 2-CNF component of the formula, such that for every t we have $P_{1,t}^{M_2} \leq P_{1,t}^{M_1}$. Perhaps the most intuitive way to see this coupling is to "paint" the initial clauses of the formula having size at most two in red, and the other clauses in blue. Clauses retain their color as they shrink in size. At every step t $P_{1,t}^{M_2}$ will count only red clauses having unit size at step t, while $P_{1,t}^{M_1}$ will count unit clauses of both colors.

Given this stochastic domination, the desired result follows from the corresponding proof in the case $k = 2$.

$\square$

10. Proof sketches of Theorems 4.5 and 4.6

The proof of Theorem 4.5 is a variation of the one of Theorem 4.2: consider a random *uniform* Horn formula Φ with $\hat{c} \cdot \frac{H_n}{n}$ clauses, and let $\overline{\Phi}$ be its subformula consisting of clauses of size at most k. It is easily seen that the behavior of PUR on the first $k - 1$ steps depends only on the clauses of $\overline{\overline{\Phi}}$, so

$$\Pr[\text{PUR accepts } \Phi \text{ in less than } k \text{ steps}] = \Pr[\text{PUR accepts } \overline{\Phi} \text{ in less than } k \text{ steps}].$$

On the other hand we have

$$0 \leq \Pr[\text{PUR accepts } \Phi \text{ in at least } k \text{ steps}] \leq \Pr[\text{PUR accepts } \overline{\Phi} \text{ in at least } k \text{ steps}].$$

The fact that "$\overline{\Phi}$ is close to a random formula in $\Omega(n, k, c \cdot \frac{H_k}{n})$" (see the discussion in Observation 1) implies that the right-hand side term can be made less than any fixed constant ϵ (for n, k big enough). It follows that

$$|\Pr[\text{PUR accepts } \Phi] - \Pr[\text{PUR accepts } \overline{\Phi}]| \leq 2 \cdot \epsilon,$$

for large enough values of n, k. This immediately implies the desired result.

The proof of theorem 4.6 is only a strenghtened version of the proof of the Theorem 4.3 and is based on an elementary property of the queuing chain Q_t (the expected time that Q_t needs to hit state zero, conditional on actually hitting it is $q(c) + o(1)$).

Indeed, the expected running time of PUR is

$$E[\text{PUR}] = \sum_{i=0}^{n} i \cdot \Pr[\text{PUR accepts at stage } (n - i) \,|\text{PUR accepts.}]$$

The crucial point is to prove that the probability that PUR accepts after stage $n - \omega_0$ is not only $o(1)$, but actually $o(1/n)$, so that the influence of this "bad" event on the average running time is $o(1)$ (clearly the number of steps of PUR is at most n).

This is easy to see: the probabilities that the various steps of the analysis are exponentially small (from Chernoff bound arguments), with the possible exception of the event $\Delta_{1,t}^P + \Delta_{1,t-1}^P + \ldots + \Delta_{1,t-p+1}^P \leq k$ from Lemma 8.8. This one, however is $o(1/n)$.

For stages between n and $n - \omega_0$ we use Lemma 8.6 and the fact that the sum of total variation distances is $o(1/\omega_0)$ and infer the fact that

$$E[\text{PUR}] = \sum_{i=0}^{\omega_0} i \cdot \frac{Pr[Q_t \text{ hits zero at stage } (n - i)]}{Pr[\text{PUR accepts}]} + o(1).$$

Using the facts that $Pr[\text{PUR accepts}] = Pr[Q_t \text{ hits } 0] + o(1)$ and the fact that these two quantities are bounded away from zero the desired conclusion follows.

11. Random Horn satisfiability as a mean-field approximation

What we have shown so far is to prove that (under a suitably rescaled picture) the rescaled probability graphs for random at-most-k Horn satisfiability converge to the graph of random Horn satisfiability. In this section we will show that this latter one can be seen as a result of a mean-field approximation. However the mean-field approximation is *not* the one from [18], and incorporates a correction specific to the properties of random Horn satisfiability.

Let us first see that it is not accurate if no correction is taken into account. Indeed, were it true we would have

$$\lim_{n \to \infty} Pr[\Phi \in \text{HORN-SAT}] = 1 - \lim_{n \to \infty} \prod_{A \in \{0,1\}^n} (1 - \Pr[A \models \Phi]).$$

Since, for an assignment A of Hamming weight i there are exactly $2^i - 1 + (n - i) \cdot 2^i$ Horn clauses that A falsifies, we have

$$\Pr[A \models \Phi] = \left(1 - \frac{2^i - 1 + (n - i) \cdot 2^i}{(n + 2) \cdot 2^n - 1}\right)^{c \cdot 2^n},$$

so the mean-field prediction reads

$$\lim_{n\to\infty} \Pr[\Phi \in \text{HORN-SAT}] = 1 - \lim_{n\to\infty} \prod_{j=0}^{n} \left(1 - \left(1 - \frac{2^j - 1 + (n-j)2^j}{(n+2)\cdot 2^n - 1}\right)^{c\cdot 2^n}\right)^{\binom{n}{j}}$$

All terms in the product are less than 1. Since the term corresponding to $j = 1$ is $\left(1 - \left(1 - \frac{1+2\cdot(n-1)}{(n+2)\cdot 2^n - 1}\right)^{c\cdot 2^n}\right)^n$ has limit 0, the mean-field prediction would imply that $\lim_{n\to\infty}\Pr[\Phi \in \text{HORN-SAT}] = 1$. On the other hand let us observe that, if we do not consider the power $\binom{n}{j}$ in the infinite product we obtain the right result: it is a simple but tedious task to prove that

$$\lim_{n\to\infty} \prod_{j=0}^{n} \left(1 - \left(1 - \frac{2^j - 1 + (n-j)\cdot 2^j}{(n+2)\cdot 2^n - 1}\right)^{c\cdot 2^n}\right) = \prod_{j=0}^{\infty} \left(1 - e^{-c\cdot 2^j}\right).$$

Intuitively this means that "there exist a correction of the mean-field approximation, that only considers a single assignment of each weight, and is accurate." The following simple result gives a precise statement to the above intuition:

LEMMA 11.1. *Suppose Φ is given as a union of formulas $\Phi_1, \ldots, \Phi_n$, where Φ_i contains all clauses of length exactly i. Then there is a set $T = \{T_0, \ldots, T_{n-1}\}$ of assignments, with T_i of Hamming weight exactly i and depending only on $\Phi_1 \cup \ldots \cup \Phi_{i+1}$, such that Φ is satisfiable if and only if it is satisfied by some assignment in T.*

Proof.

Let $\overline{y_1 \ldots y_k}$ denote the assignment that makes $y_1 = \ldots = y_k = 1$, and all the other variables equal to zero.

The set T has two parts: the first is simply the set of assignments implicitly examined by the algorithm PUR in testing satisfiability. That is, if $x_1, \ldots, x_k$ are the variables assigned by PUR in this order, the first part includes the assignments $00000, \overline{x_1}, \ldots, \overline{x_1, \ldots, x_k}$. The second part contains a random assignment for each remaining weight. $\square$

The result has a "mean-field" interpretation: as before, define $f(x_1, \ldots, x_n) = 1 - \prod_{i=1}^{n} x_i$, and the function $g_k[\Phi]$ to be the indicator function for the event "$T_k \not\models \Phi$, given that event $\overline{A}_n \wedge \ldots \wedge \overline{A}_{n-k+1}$ happens," i.e.

$$g_k[\Phi] = \frac{1}{\Pr[\overline{A}_n \wedge \ldots \wedge \overline{A}_{n-k+1}]} \cdot \begin{cases} 1, & \text{if } T_k \not\models \Phi \wedge \overline{A}_n \wedge \ldots \wedge \overline{A}_{n-k+1} \\ 0, & \text{otherwise.} \end{cases}$$

We have

$$E[g_k[\Phi]] = \Pr[\overline{A_{n-k}}|\overline{A}_n \wedge \ldots \wedge \overline{A}_{n-k+1}].$$

Indeed, $g_k[\Phi] \neq 0$ exactly when $R_n \vee \ldots \vee R_{n-k+1}$ or $T_k \not\models \Phi \wedge S_n \wedge \ldots S_{n-k+1}$. The second event is equivalent to $\overline{A_{n-k}} \wedge S_n \wedge \ldots S_{n-k+1}$, hence we have $g_k[\Phi] \neq 0$ exactly when $\overline{A_{n-k}} \wedge \overline{A}_n \wedge \ldots \wedge \overline{A}_{n-k+1}$ holds.

Thus we have, by the discussion in the previous chapter,

$$f(E[g_1[\Phi]], \ldots, E[g_n[\Phi]]) = 1 - \prod_{k=0}^{n} \Pr[\overline{A_{n-k}}|\overline{A}_n \wedge \ldots \wedge \overline{A}_{n-k+1}]$$

$$= \Pr[\Phi \in \text{HORN-SAT}].$$

The above correction seems to be specific to the random model for Horn satisfiability, which allows clauses of varying lengths. To sum up: *the mean-field approximation is true, modulo a correction that takes into account some particular features of the random model for Horn satisfiability.*

12. Discussion

We have characterized the asymptotical satisfiability probability of a random k-Horn formula, and showed that it exhibits very similar behavior to the one uncovered experimentally in [**18**]. We have also displayed an "easy-hard-easy" pattern similar to the ones discussed before in the A.I. literature. In our case the pattern is fully explained by elementary properties of the queuing chain.

As for an explanation of the "convergence to the uniform case", consider an intermediate stage i of PUR and let C_j be the set of clauses of $\Phi_{i,j}^{P}$. It is clear that whether PUR accepts is dependent only on the number of clauses in C_1. The restriction on the clause length acts like a "dampening" perturbation (in that it eliminates the "clause flow into C_k"). The proof of Theorem 4.2 states that when $k(n) \to \infty$, with high probability PUR accepts (if Φ is satisfiable) "before the perturbation reaches C_1", therefore the satisfiability probability is the one from the uniform case. On the other hand, for any constant k, with probability greater than 0 PUR does not halt during the first k iterations (for the exact value see [**16**]), and the dampening has a significant influence. Thus *the explanation for the occurrence (and specific form of) critical behavior is a threshold property for the number of iterations of* PUR *on random satisfiable Horn formulas "in the critical region".*

Acknowledgments

I thank Robert Samal and Zoltan Toroczkai for very useful comments and clarifications concerning the notions from Statistical Mechanics used in this paper.

References

[1] A. San Miguel Aguirre and M. Vardi. Random 3-sat: The plot thickens further. In *Proceedings of the Seventh International Conference on the Principles and Practice of Constraint Programming*, 2001.

[2] N. Alon, P. Erdős, and J. Spencer. *The probabilistic method*. John Wiley and Sons, second edition, 1992.

[3] A. Barbour, L. Holst, and S. Janson. *Poisson Approximation*. Clarendon Press Oxford, 1992.

[4] Baxter. *Exactly solvable models in Statistical Mechanics*. J. Wiley and Sons, 1984.

[5] B. Bollobás, C. Borgs, J.T. Chayes, J. H. Kim, and D. B. Wilson. The scaling window of the 2-SAT transition. Technical report, Los Alamos e-print server, http://xxx.lanl.gov/ps/math.CO/9909031, 1999.

[6] P. Cheeseman, B. Kanefsky, and W. Taylor. Where the really hard problems are. In *Proceedings of the 11th IJCAI*, pages 331–337, 1991.

[7] C. Coarfa, D. Demopoulos, A. San Miguel Aguirre, D. Subramanian, and M. Vardi. Random 3-sat: The plot thickens. In *Proceedings of the Sixth International Conference on the Principles and Practice of Constraint Programming*, 2000.

[8] E. Friedgut. Necessary and sufficient conditions for sharp thresholds of graph properties, and the k-SAT problem. with an appendix by J. Bourgain. *Journal of the A.M.S.*, 12:1017–1054, 1999.

[9] A. Frieze and S. Suen. Analysis of two simple heuristics for random instances of k-SAT. *Journal of Algorithms*, 20:312–355, 1996.

[10] A. Frieze and N. Wormald. Random k-sat: A tight threshold for moderately growing k. In *Proceedings of the Fifth International Symposium on Theory and Applications of Satisfiability Testing*, pages 1–6, 2002.

[11] I. Gent, E. MacIntyre, P. Prosser, and T. Walsh. The scaling of search cost. In *Proceedings of AAAI-97*, pages 315–320, 1997.

[12] N. Goldenfeld. *Lectures on Phase Transitions and the Renormalization Group*. Perseus Publishing, 1992.

[13] B. Hayes. Can't get no satisfaction. *American Scientist*, 85(2):108–112, March–April 1997.

[14] P. Hoel, S. Port, and C. Stone. *Introduction to stochastic processes*. Waveland Press Inc., 1987.

[15] T. Hogg and D. Mammen. A new look at the easy-hard-easy pattern of combinatorial search difficulty. *Journal of Artificial Intelligence Research*, 7:44–66, 1997.

[16] G. Istrate. The phase transition in random Horn satisfiability and its algorithmic implications. *Random Structures and Algorithms*, 4:483–506, 2002.

[17] S. Kirkpatrick, C.D. Gelatt, and M. Vecchi. Optimization by simulated annealing. *Science*, 220(4598):671–680., 1983.

[18] S. Kirkpatrick and B. Selman. Critical behavior in the satisfiability of random boolean expressions. *Science*, 264:1297–1301, 1994.

[19] S. Kirkpatrick and B. Selman. Critical behavior in the computational cost of satisfiability testing. *Artificial Intelligence*, 81, 1996.

[20] T. Lindvall. *Lectures on the coupling method*. John Wiley, 1992.

[21] M. Mézard and G. Parisi. Replicas and optimization. *J. Phys. France*, 46:L771–L778, September 1985.

[22] M. Mézard, G. Parisi, and M. Virasoro. *Spin glass theory and beyond*. World Scientific, 1987.

[23] R. Monasson, R. Zecchina, S. Kirkpatrick, B. Selman, and L. Troyansky. $2 + p$-SAT: Relation of typical-case complexity to the nature of the phase transition. *Random Structures and Algorithms*, 15(3–4):414–435, 1999.

[24] R. Monasson, R. Zecchina, S. Kirkpatrick, B. Selman, and L. Troyansky. Determining computational complexity from characteristic phase transitions. *Nature*, 400(8):133–137, 1999.

[25] B. Møuller, J. Reinhardt, and M.T. Strickland. *Neural Networks: An Introduction (second edition)*. Springer Verlag, 1995.

[26] S. Sheu. The Poisson approximation to the binomial distribution. *The American Statistician*, 38(3):206–207, 1984.

[27] M. Talagrand. Verres de spin et optimisation combinatoire. Seminaire Bourbaki. To appear in Asterisque., 1999.

[28] D. Wilson. The empirical values of the critical k-sat exponents are wrong. Technical Report math.PR/0005136, arxiv.org preprint archive, 2000.

[29] N. Wormald. Differential equations for random processes and random graphs. *Annals of Applied Probability*, 5(4):1217–1235, 1995.

CCS-5, BASIC AND APPLIED SIMULATION SCIENCE, LOS ALAMOS NATIONAL LABORATORY, LOS ALAMOS, NM 87545, MS M 997

E-mail address: istrate@lanl.gov

DIMACS Series in Discrete Mathematics
and Theoretical Computer Science
Volume **63**, 2004

The exchange interaction, spin hamiltonians, and the symmetric group

J. Katriel

ABSTRACT. The exchange interaction model and the Heisenberg model, which coincide for spin-$\frac{1}{2}$ particles, are contrasted for systems consisting of particles with higher elementary spins. The infinite-range version of the former is studied using the representation theory of the symmetric group. For spin-σ particles the relevant irreducible representations are characterized by Young diagrams with up to $2\sigma + 1$ rows, whose lengths give rise to the system's order parameters. The order parameter equations are derived, and their solutions are explored. The solution that corresponds to the thermodynamically stable state is identified. The other non-trivial solutions correspond to saddle points in the free-energy surface, with consecutively increasing indices.

1. Introduction

The exchange-interaction model of ferromagnetism is specified by the many-body hamiltonian

$$(1) \qquad \mathcal{H} = -J \sum_{<ij>} P_{ij} \, ,$$

in which P_{ij} is the transposition of the indices i and j and $<ij>$ stands for a nearest neighbour pair. For two spin-$\frac{1}{2}$ particles Dirac [1] noted that

$$P_{12} = \vec{S}^2 - 1 = 2\vec{s}_1 \cdot \vec{s}_2 + \frac{1}{2} \, ,$$

where $\vec{S} = \vec{s}_1 + \vec{s}_2$. Here, $\vec{s}_j$, $j = 1$, 2, are the single-particle vectorial spin operators, whose components satisfy the commutation relations $[s_{jx}, s_{jy}] = is_{jz}$, etc., and the condition $s_{jx}^2 + s_{jy}^2 + s_{jz}^2 = \frac{3}{4}$. The Dirac identity follows readily, noting that $\vec{S}^2$ has two possible eigenvalues, 2 and 0, corresponding to the symmetric (triplet) states and to the antisymmetric (singlet) state, respectively. Thus, for spin-$\frac{1}{2}$ particles the exchange interaction hamiltonian, eq. 1, and the Heisenberg hamiltonian

$$(2) \qquad \mathcal{H}_H = -J \sum_{<ij>} \vec{s}_i \cdot \vec{s}_j$$

represent the same physical situation.

Generalizing to higher spins, these two points of view suggest diverging outlooks. The spin-σ Heisenberg hamiltonian retains the form presented in eq. 2, with one-particle vectorial spin operators that satisfy $s_{jx}^2 + s_{jy}^2 + s_{jz}^2 = \sigma(\sigma + 1)$. Its infinite-range version amounts to summing over all pairs of particle indices.

Noting that the total N-particle spin operator satisfies

$$\vec{S}^2 = \left(\sum_{i=1}^{N} \vec{s}_i \right)^2 = N\sigma(\sigma + 1) + 2\sum_{i<j} \vec{s}_i \cdot \vec{s}_j$$

the infinite-range Heisenberg hamiltonian can be written in the form

$$\mathcal{H}_H^{(\infty)} = -\frac{J}{2N}\vec{S}^2 .$$

Division by N is required to maintain the "size consistency" of the hamiltonian, i.e., have a total ground state energy which is proportional to the number of particles.

To obtain $\Phi_\sigma(N, S)$, the number of multiplets with total spin quantum number S, for a system of N identical spin-σ particles, we note that

$$(x^\sigma + x^{\sigma-1} + \cdots + x^{-\sigma})^N = \sum_{M=-N\sigma}^{N\sigma} C_\sigma(N, M)x^M ,$$

where $C_\sigma(N, M)$ is the number of N-particle states with a total spin z-component M. Each total spin multiplet consists of $2S + 1$ states specified by $M = -S, -S + 1, \cdots , S$. Thus, for $0 \leq S < N\sigma$,

$$\Phi_\sigma(N, S) = C_\sigma(N, S) - C_\sigma(N, S + 1) ,$$

and for $S = N\sigma$

$$\Phi_\sigma(N, N\sigma) = C_\sigma(N, N\sigma) .$$

Obviously, $\sum_{S=0}^{N\sigma}(2S + 1)\Phi_\sigma(N, S) = (2\sigma + 1)^N$.

For $\sigma = \frac{1}{2}$ we obtain $C_{\frac{1}{2}}(N, M) = \binom{N}{\frac{N}{2}+M}$, so that $\Phi_{\frac{1}{2}}(N, S) = \frac{N!(2S+1)}{(\frac{N}{2}+S+1)!(\frac{N}{2}-S)!}$.

For any hamiltonian $\mathcal{H}(S)$, where S is the total spin quantum number, the free energy can be written in the form $A(S) = \mathcal{H}(S) + kT \log\left((2S + 1)\Phi_\sigma(N, S)\right)$, obtaining its extremal values when the total spin quantum number $S = Ns$ satisfies [2]

$$s = \sigma B_\sigma \left[-\beta\sigma\frac{\partial \mathcal{H}}{\partial s} \right] ,$$

where $\beta = \frac{1}{kT}$ and $B_\sigma[x] = \frac{2\sigma+1}{2\sigma} \coth\left(\frac{2\sigma+1}{2\sigma}x\right) - \frac{1}{2\sigma} \coth\left(\frac{x}{2\sigma}\right)$. For $\sigma = \frac{1}{2}$ this expression reduces to $B_{\frac{1}{2}}[x] = \tanh(x)$. In particular, for the infinite-range Heisenberg hamiltonian we obtain

$$(3) \qquad\qquad s = \frac{1}{2} \tanh(\beta Js) ,$$

which is immediately recognizable as the mean-field equation for the nearest neighbor spin-σ Heisenberg hamiltonian.

The equivalence between the mean-field solution of a finite-range hamiltonian and the exact solution of its infinite-range counterpart had been recognized by several authors. A fairly general treatment of a family of infinite range models that contains the presently discussed systems as special cases was presented by Fannes *et al.* [3].

The Dirac identity was generalized by Schroedinger [4] to pairs of identical particles with arbitrary elementary spin σ. For two spin-1 particles one obtains, most directly by noting that the two-particle states with total spin quantum numbers

2 and 0 are symmetric whereas the state with total spin 1 is antisymmetric with respect to transposition of the particle indices, that

$$P_{12} = \frac{1}{4}(\vec{S}^2)^2 - \frac{3}{2}\vec{S}^2 + 1 = (\vec{s}_1 \cdot \vec{s}_2)^2 + (\vec{s}_1 \cdot \vec{s}_2) - 1 \,.$$

The components of $\vec{s}_i$ satisfy the same commutation relations as above but $s_{jx}^2 + s_{jy}^2 + s_{jz}^2 = 2$.

A transposition of a pair of identical spin-σ particles can be written as a polynomial of degree 2σ in $(\vec{s}_i \cdot \vec{s}_j)$. Thus, for a system of such particles Chen *et al.* [5] introduce a set of $4\sigma(\sigma+1)$ order parameters which are the averages of the components of appropriate spin tensor operators. On the basis of reasonable indications that include an earlier treatment of the spin 1 case [6] they conjecture these order parameters to have a common temperature dependence. The principal result of the mean-field approximation is that for $\sigma > \frac{1}{2}$ the system exhibits a single first order phase transition. This conclusion is supported by more sophisticated treatments [7, 8].

Other types of higher elementary-spin lattice models that use the transposition form of the interaction hamiltonian, or variants thereof, have been investigated by Affleck [9] and by Aizenman and Nachtergaele [10].

In the present article we investigate the infinite-range counterpart of the exchange-interaction model. In several related contexts the equivalence of the mean-field and infinite-range treatments had been established. The infinite-range approach involves a well-defined approximation of the hamiltonian, which is then solved exactly. This approach can be easily adapted to the treatment of hamiltonians that involve arbitrarily complicated dependence on the order parameters. Furthermore, it allows the construction of the full free-energy surface, which is needed for the study of time-dependent phenomena [11] and non-equilibrium properties.

Within the presently proposed treatment of the exchange-interaction model there is no need to express the hamiltonian in terms of spin operators. The eigenvalues and degeneracies are related to the characters and dimensions, respectively, of the irreducible representations (irreps) of the symmetric group (Section 2.1). The relevant quantum numbers are the row lengths of the Young diagrams that specify these representations, and the order parameters are the corresponding thermal averages. The maximal number of rows in the allowed Young diagrams is determined by the value of the elementary spin. The equations satisfied by the 2σ order parameters are derived in Section 2.2 and solved in Section 3.1. All the order parameters bifurcate from the trivial solution at a common temperature, which is a highly degenerate extremum of the free energy. A stability analysis shows that one of the solutions, which coincides with the mean-field solution obtained by Chen *et al.* [5], represents the thermodynamically stable ordered state (Section 3.2). The other solutions are saddle points with well-defined indices on the free-energy surface. Further details are presented in ref. [12].

2. Statistical mechanics of the infinite-range exchange model

2.1. Group theoretical preliminaries. The infinite range exchange-interaction hamiltonian is

$$\mathcal{H}^{(\infty)} = -\frac{J}{N}\sum_{i<j}^{N} P_{ij} \,,$$

where N is the number of spins. The operator $\sum_{i<j}^{N} P_{ij}$ is the sum of all the transpositions of two indices, which form a conjugacy class of the symmetric group S_N. Thus, $\sum_{i<j} P_{ij}$ is an element in the center of the group algebra of S_N, commuting with all elements of the group. Its action within any irreducible subspace is, by Schur's lemma, equivalent to multiplication by a scalar multiple of the unit matrix of appropriate dimension. This is true for any conjugacy class-sum. The corresponding scalar, λ_C^Γ, sometimes referred to as a central character, is related to the corresponding irreducible character, χ_C^Γ, via $\lambda_C^\Gamma = \frac{|C|}{|\Gamma|}\chi_C^\Gamma$. Here, C labels the conjugacy-class and Γ stands for the irrep. $|C|$ is the number of group elements in the class C and $|\Gamma|$ is the degree of the irrep Γ. The latter is specified by means of a Young diagram $\Gamma \equiv \{\mu_1, \mu_2, \cdots\}$, where $\mu_1 \geq \mu_2 \geq \cdots$ and $\mu_1 + \mu_2 + \cdots = N$. Spin-$\sigma$ particles give rise to irreps whose Young diagrams contain at most $n = 2\sigma+1$ rows.

The central characters of the symmetric group can be expressed as polynomials in the symmetric power-sums over the "contents" of the Young diagram specifying the irrep; the "content" of a box in a Young diagram being the difference between its column index j and its row index i. The k'th symmetric power-sum is $\sigma_k = \sum_{(i,j)\in\Gamma}(j-i)^k$.

The central characters corresponding to the first five single-cycle class-sums are [13]

$$\lambda_{[(2)]_N}^\Gamma = \sigma_1$$

$$\lambda_{[(3)]_N}^\Gamma = \sigma_2 - \frac{1}{2}N(N-1)$$

$$\lambda_{[(4)]_N}^\Gamma = \sigma_3 - (2N-3)\sigma_1$$

$$\lambda_{[(5)]_N}^\Gamma = \sigma_4 - (3N-10)\sigma_2 - 2\sigma_1^2 + \frac{1}{6}N(N-1)(5N-19)$$

$$\lambda_{[(6)]_N}^\Gamma = \sigma_5 - (4N-25)\sigma_3 - 6\sigma_1\sigma_2 + (6N^2 - 38N + 40)\sigma_1$$

The central character corresponding to the class of transpositions, the top one in this list, can easily be written in the form $\lambda_{[(2)]_N}^\Gamma = \frac{1}{2}\sum_{i=1}^{n}\mu_i(\mu_i - 2i)$. The dimension of the irrep Γ is $\Omega_\Gamma = N!\frac{\prod_{i<j}^{n}(\mu_i-\mu_j+j-i)}{\prod_{i=1}^{n}(\mu_i+n-i)!}$. N identical particles with an elementary spin σ give rise to a total of $(2\sigma+1)^N$ states. The irrep Γ appears in the space spanned by these states

$$g_\sigma(N,\Gamma) = \frac{\prod_{i<j}^{n}(\mu_i - \mu_j + j - i)}{\prod_{i<j}^{n}(j-i)}$$

times. Thus, $\sum_\Gamma g_\sigma(N,\Gamma)\Omega_\Gamma = (2\sigma+1)^N$.

For N spin-$\frac{1}{2}$ particles a bijection holds between the total spin S and the Young diagram $\Gamma = \{\mu_1, \mu_2\}$, $\mu_1 + \mu_2 = N$. Explicitly, $S = \frac{1}{2}(\mu_1 - \mu_2)$. It follows that $g_{\frac{1}{2}} = 2S+1$ and $\Omega_\Gamma = \frac{N!(2S+1)}{(\frac{N}{2}+S+1)!(\frac{N}{2}-S)!}$. Hence, in this case $\Omega_\Gamma = \Phi_{\frac{1}{2}}(N,S)$.

For $\sigma > \frac{1}{2}$ the relation between $\Phi_\sigma(N,S)$ and Ω_Γ is rather involved. One could define subsets of N-particle states that belong to well defined irreps, Γ, of S_N, and total spin S. Let $\Psi_\sigma(\Gamma,S)$ denote the number of states in each such subset. Clearly, $\sum_{\Gamma\in S_N}\sum_S \Psi_\sigma(\Gamma,S) = (2\sigma+1)^N$. We obviously have $\sum_{\Gamma\in S_N}\Psi_\sigma(\Gamma,S) = (2S+1)\Phi_\sigma(N,S)$ and $\sum_S \Psi_\sigma(\Gamma,S) = g_\sigma(N,\Gamma)\Omega_\Gamma$.

2.2. The order parameter equations. To study the thermodynamic limit $(N \to \infty)$ it will be convenient to introduce the reduced row lengths $\lambda_i = \mu_i/N$, satisfying $\sum_{i=1}^{n} \lambda_i = 1$. The energy per particle, e_Γ, is a function of these reduced row lengths. Since the number of states having energy e_Γ is $g_\sigma(N, \Gamma)\Omega_\Gamma$, the entropy per particle becomes $s_\Gamma = -k \sum_{i=1}^{n} \lambda_i \log(\lambda_i)$. The reduced row lengths serve as the order parameters. Differentiating the free-energy per particle

$$(4) \qquad a_\Gamma = e_\Gamma + kT \sum_{i=1}^{n} \lambda_i \log(\lambda_i)$$

with respect to λ_m we obtain

$$(5) \qquad \lambda_m = \frac{1}{q} \exp\left(-\beta \frac{\partial e_\Gamma}{\partial \lambda_m}\right) \quad ; \quad m = 1, 2. \cdots, n,$$

where $q = \sum_{i=1}^{n} \exp(-\beta \frac{\partial e_\Gamma}{\partial \lambda_i})$. For the infinite-range exchange hamiltonian $e_\Gamma = -\frac{J}{2} \sum_{i=1}^{n} \lambda_i^2$, and eq. 5 implies that $\frac{\exp(\beta J \lambda_1)}{\lambda_1} = \frac{\exp(\beta J \lambda_2)}{\lambda_2} = \cdots = \frac{\exp(\beta J \lambda_n)}{\lambda_n}$. Since the function $\exp(x)/x$, $x > 0$, has a single minimum, it follows that the n order parameters $\lambda_1, \lambda_2, \cdots, \lambda_n$ can have at most two different values. Being the row lengths of a Young diagram they must satisfy $\lambda_1 = \lambda_2 = \cdots = \lambda_m \equiv \lambda_-$ and $\lambda_{m+1} = \lambda_{m+2} = \cdots = \lambda_n \equiv \lambda_+$, where $\lambda_- \geq \lambda_+$ (and $m\lambda_- + (n-m)\lambda_+ = 1$). Here, m can obtain any of the values $1, 2, \cdots, n-1$. The m'th solution is specified by the single order parameter $\delta_m = \lambda_- - \lambda_+$ that satisfies the consistency equation

$$(6) \qquad \delta_m = \frac{\exp(\beta J \delta_m) - 1}{m \exp(\beta J \delta_m) + (n - m)} .$$

For hamiltonians that involve multiple-particle exchange a higher order dependence on the order parameters λ_i may allow phases that are specified by more general Young diagrams. Such generalized hamiltonians can give rise to more complex phase diagrams, that may involve chains of phase transitions between different ordered phases and possibly reentrant behavior.

3. The phase diagram

3.1. Solution of the order parameter equations. We note that the trivial solution $\delta_m = 0$ always exists. This solution describes the "isotropic" phase, for which $\lambda_1 = \lambda_2 = \cdots = \lambda_n = \frac{1}{n}$.

A common temperature $T_0 = \frac{J}{kn}$ is found, at which all the non-trivial solutions bifurcate from the trivial solution. For $m < \frac{n}{2}$ the δ_m vs. T curve bifurcates towards higher temperatures, exhibiting a first-order like behaviour. For $m > \frac{n}{2}$ the curve of δ_m vs. T exhibits a monotonic rise upon lowering the temperature. For even n (half-odd S) $\delta_{\frac{n}{2}}$ exhibits a critical behavior, satisfying the mean-field equation $\delta' = \frac{1}{2} \tanh(\beta J' \delta')$, where $J' = \frac{2J}{n}$ and $\delta' = \frac{n\delta}{4}$. The low temperature limit of δ_m is $\frac{1}{m}$.

In particular, the consistency equation for δ_1 coincides with the mean-field equation derived by Chen *et al.* [5]. For $\sigma = \frac{1}{2}$ it coincides with the magnetization equation of the infinite-range (or mean-field) Heisenberg model, eq. 3, for which the isotropic to ferromagnetic phase transition is of second order, and for $\sigma > \frac{1}{2}$ it yields a first order transition.

The temperatures of the first order transitions of the solutions $\delta_m > 0$, $m < \frac{n}{2}$, can be obtained explicitly, solving the transcendental equation

$$(n-m)\left(1 - \frac{m\delta_m}{2}\right)\log\left(\frac{1}{1-m\delta_m}\right) = m\left(1 + \frac{(n-m)\delta_m}{2}\right)\log\left(1 + (n-m)\delta_m\right) .$$

One finds that $\delta_m^c = \frac{n-2m}{m(n-m)}$ and $T_c(m) = J\frac{n-2m}{2m(n-m)\log(\frac{n-m}{m})}$. For $m = 1$ these expressions coincide with those obtained by Chen *et al.* [5]. $T_c(m)$ is a decreasing function of m, which means that the highest temperature transition from the isotropic phase is the transition into the state $\delta_1 > 0$.

For the spin σ Heisenberg hamiltonian, a well-defined classical limit, $\sigma \to \infty$, exists, in which the Brillouin function becomes the Langevin function $L(x) = \coth(x) - \frac{1}{x}$. Taking $n \to \infty$ to mean the classical limit in the exchange-interaction model, we note that in this limit $T_c \to 0$, which suggests that this model is "strictly quantum mechanical", possessing no classical limit.

3.2. Thermodynamic significance of the different solutions.

To elucidate the nature of the different solutions one needs to evaluate the corresponding free energies, using eq. 4, and to characterize them in terms of the indices of the corresponding Hessian matrices. The latter can be evaluated rather explicitly.

For the isotropic solution $\lambda_1 = \lambda_2 = \cdots = \lambda_n = \frac{1}{n}$ the determinant of the r'th minor of the Hessian matrix is positive for all r when $T > T_0$, and alternates in sign, beginning with a negative sign for the lowest minor, when $T < T_0$. The isotropic solution is the absolute minimum for $T > T_c(1)$. In view of the present stability analysis it follows that this solution is a local minimum within the temperature range $T_0 < T < T_c(1)$ and a maximum below T_0.

For the m'th solution ($\delta_m > 0$) in the low temperature limit the first m minors of the Hessian matrix alternate in sign, beginning with a positive sign for the lowest minor. The remaining minors have a constant sign. Thus, this solution is a saddle point of index $m - 1$ (the number of sign alternations of the sequence of determinants of the principal minors). In particular, the solution $\delta_1 > 0$ is a (local) minimum. For the solutions $\delta_m > 0, m < \lfloor \frac{n}{2} \rfloor$ the "spinodal" point, at which the order parameter curves backwards, can be determined by solving $\frac{\partial T}{\partial \delta_m} = 0$ along with the order parameter equation, eq. 6. The determinant of the Hessian matrix vanishes at the "spinodal point", allowing its index to increase by unity at the lower branch of the order parameter curve relative to the higher branch.

In conclusion, the system exhibits only two thermodynamically stable phases. For temperatures higher then $T_c(1)$ the absolute minimum corresponds to the isotropic phase. Below that temperature the phase $\delta_1 > 0$ has the lowest free energy. The free energy surface exhibits a complex temperature dependence. Above the highest spinodal temperature it has a single (isotropic) minimum. Between that temperature and the highly degenerate bifurcation temperature T_0 it develops a rich manifold of extrema that correspond to the solutions $\delta_m > 0, m < \frac{n}{2}$. Finally, at T_0 a further set of extrema emerges, corresponding to $\frac{n}{2} \leq m \leq n$.

The connection between the order parameter δ_1 and the total spin of the system requires further investigation along the lines pointed out at the bottom of section 2.1.

4. Conclusions

A comparison between the present approach, that depends on the group theoretical significance of the exchange hamiltonian, and the spin-operator analysis due to Chen *et al.*, [5], is instructive. For a system consisting of particles with an elementary spin σ, the total number of order parameters in the group theoretical approach is equal to 2σ, the number of independent row lengths in the allowed Young diagrams. The expression in terms of spin tensor operators introduces $4\sigma(\sigma+1)$ order parameters [5]. Moreover, the present formulation allowed the transparent and straightforward investigation of all the solutions of the coupled mean-field equations without recourse to any prior assumptions. In fact, it was established that the different order parameters are fully decoupled in the sense that each solution can be expressed in terms of a single order parameter. The solution corresponding to the lowest free energy coincides with that given by Chen *et al.* [5], exhibiting, for $\sigma > \frac{1}{2}$, a first order transition. The other solutions correspond to saddle points of different indices in the free-energy vs. order-parameter space. The lowest $\lfloor \sigma \rfloor$ solutions exhibit a first-order like temperature dependence. While all of these solutions but the lowest do not correspond to thermodynamically stable phases, they may well play a significant role in the non-equilibrium kinetics of the phase transformations.

Acknowledgement: The referee's helpful suggestions are gratefully acknowledged.

References

[1] P. A. M. Dirac, Proc. R. Soc. London Ser A **123**, 714 (1929).
[2] J. Katriel and G. F. Kventsel, Solid State Comm. **52**, 689 (1984).
[3] M. Fannes, H. Spohn, and A. Verbeure, J. Math. Phys. **21**, 355 (1980).
[4] E. Schroedinger, Proc. R. Irish Acad. Sect. A **47**, 39 (1941).
[5] H. H. Chen, S. C. Gou and Y. C. Chen, Phys. Rev. B **46**, 8323 (1992).
[6] H. H. Chen and P. M. Levy, Phys. Rev. B **7**, 4267 (1973).
[7] H. A. Brown, Phys. Rev. B **31**, 3118 (1985); *ibid.*, **40**, 775 (1989).
[8] Y. C. Chen, Phys. Rev. B **57**, 5009 (1998).
[9] I. Affleck, J. Phys. C: Condens. Matter **2**, 405 (1990).
[10] M. Aizenman and B. Nachtergaele, Commun. Math. Phys. **164**, 17 (1994).
[11] G. F. Kventsel and J. Katriel, Phys. Rev. B **31**, 1559 (1985).
[12] J. Katriel and G. F. Kventsel, Phys. Rev. B **62**, 12350 (2000).
[13] J. Katriel, Discrete Appl. Math. **67**, 149 (1996).

DEPARTMENT OF CHEMISTRY, TECHNION, 32000 HAIFA, ISRAEL
Current address: Department of Chemistry, Rutgers University, Piscataway, NJ 08854, USA
E-mail address: jkatriel@tx.technion.ac.il

DIMACS Series in Discrete Mathematics
and Theoretical Computer Science
Volume **63**, 2004

A Discrete Non-Pfaffian Approach to the Ising Problem

Martin Loebl

ABSTRACT. We describe a method developed in fifties by Kac, Ward, Potts, Feynman and Sherman to solve the 2-dimensional Ising problem using methods of discrete mathematics. This approach is older and not so well known to discrete mathematics community as the Pfaffian approach. Using the results of Sherman and the theory of Pfaffian orientations developed by Galluccio and Loebl we generalise the results to arbitrary (non-planar) graphs.

1. Introduction

Let $G = (V, E)$ be a finite graph. V is the set of *vertices* and E is the set of unordered pairs of vertices called *edges*. We allow multiple edges and loops. Moreover we will assume that a variable x_e is associated with each edge e of G. If $A \subset E$ then we let $x_A = \prod_{e \in A} x_e$. The *degree* of a vertex is the number of edges incident with it. Each loop contributes 2 to the degree.

A (spanning) subgraph $H = (V, E')$ of G is called *even* if each vertex of H has an even degree, possibly zero. The *generating function of even subgraphs* is a polynomial $\mathcal{E}(G, x) = \sum x_A$ over all even subgraphs (V, A) of G.

This polynomial is extensively studied in discrete mathematics in matching theory and also in statistical physics since it is equivalent via a theorem of van der Waerden [**16**] to the Ising problem partition function of G.

Since the solution of the 2-dimensional (planar) Ising problem by Onsager [**9**], the physisists were trying to reproduce his solution by more understandable methods. In fifties and in the beginning of sixties two discrete solutions appeared: the Pfaffian method of Fisher and Kasteleyn [**2**, **7**] and the 'paths method' of Kac, Ward, Potts, Feynman and Sherman [**6**, **10**, **13**, **14**]. The Pfaffian method seems to be better known to discrete mathematicians. Recently it has been further developed in [**3**, **4**, **5**, **8**] and independently in [**15**] and in [**11**, **12**] to express $\mathcal{E}(G, x)$ or, equivalently, the Ising partition function of an arbitrary graph G as a linear combination of Pfaffians of matrices associated with G.

The aim of this article is to describe the path method and to prove an analogous general theorem. All the results apart of Lemma 3.1, Theorem 3.5 and most of the proofs essentially appeared in Sherman's paper [**13**].

2000 *Mathematics Subject Classification.* 05B35, 05C15, 05A15.

Supported by project LN00A056 of the Ministry of Education of the Czech Republic and by Charles University grants No. 158/99 and 159/99.

Section 2 contains a theorem conjectured by Feynman and proved by Sherman. Section 3 contains the main result: a generalisation of the theorem of Feynman and Sherman to arbitrary graphs which is analogous to the theory of Pfaffian orientations. The motivation for the Feynman's conjecture was a work of Kac, Ward and Potts [**6, 10**]. This together with concrete formulas for the 3-dimensional Ising problem will be described in a continuation of this paper.

2. A Theorem of Feynman and Sherman

Let $G = (V, E)$ be a graph and $D = (V, A(G))$ an arbitrary orientation of G. If $e \in E$ then a_e will denote the orientation of e in $A(G)$ and a^{-1} will be the reversed directed edge to a. We let $x_{a_e} = x_{a_e^{-1}} = x_e$. A circular sequence $p = v_1, a_1, v_2, a_2, ..., a_n, v_{n+1} = v_1$ is called *non-periodic closed walk* if the following conditions are satisfied: $a_i \in \{a_e, a_e^{-1} : e \in E\}$, $a_i \neq a_{i+1}^{-1}$ and $(a_1, ..., a_n) \neq Z^m$ for some sequence Z and $m > 1$. We let $X(p) = \prod_{i=1}^{n} x_{a_i}$ and $sign(p) = (-1)^{1+n(p)}$, where $n(p)$ is the *winding number* of p, i.e. the number of integral revolutions of the tangent vector of p. Finally let $W(p) = sign(p)X(p)$.

There is a natural equivalence on non-periodic closed walks: p is equivalent with reversed p. Each equivalence class has two elements and will be denoted by $[p]$. We let $W([p]) = W(p)$ and note that this definition is correct since equivalent walks have the same sign. The following theorem was conjectured by Feynman and proved by Sherman [**13**].

THEOREM 2.1. *If $G = (V, E)$ is a planar graph then*

$$\mathcal{E}(G, x) = \prod (1 + W([p]))$$

where $\prod$ is the formal infinite product over all equivalence classes of non-periodic closed walks of G.

Note that the product is infinite even for a very simple graph consisting of one vertex and two loops. In fact, Sherman proved a generalisation of this theorem which will be used in section 3. In order to state this generalisation we need to introduce the *crossover condition*: assume graph G is properly drawn in the plane and let v be a vertex of degree 4 of G and let p be a non-periodic closed walk of G. We say that p satisfies the *crossover condition* at v if the way p passes through v is consistent with the crossover pairing of the four edges incident with v.

Remark. The following Theorem 2.2 is formulated for planar graphs such that each degree is even and at most four. Hence in order to show that Theorem 2.1 follows from it, we must reduce Theorem 2.1 to the case of planar graphs with each degree even and at most four. This may be done easily as follows: first double each edge and let the variables associated with the new edges equal to zero. This makes each degree even. Then replace each vertex v with incident edges $e_1, ..., e_{2k}$, $k > 2$, listed in a circular order given by a fixed drawing of G in the plane, by a path of $2k - 2$ vertices. Let the variables of the edges of the path equal to one. Next double each edge of the perfect matching of this path and let the variables of these new edges equal to zero. Finally join the edges $e_1, ..., e_{2k}$ to the vertices of the auxiliary path so that the order is preserved along the path and each degree is four: there is a unique way to do that. Observe that Theorem 2.1 holds for G if and only if it holds for the graph obtained from G by the above construction.

THEOREM 2.2. *Let $G = (V, E)$ be a planar graph properly drawn in the plane and such that each degree is even and at most four. Let $U = \{v_1, ..., v_k\}$ be a subset of vertices of G such that each v_i has degree 4. Let $\prod'_{G,U}(1 + W([p])$ denote the formal infinite product over all equivalence classes of non-periodic closed walks of G which satisfy the crossover condition at each v_i, $i = 1, ..., k$. An even subgraph H of G is called acceptable if for each v_i and two edges incident with v_i and joined by the crossover pairing at v_i, if H contains one of them then it contains also the other. If H is acceptable then we let $c(H)$ equal the number of vertices of U such that H contains all four edges incident with it. Then*

$$\prod{}'_{G,U}(1 + W([p]) = \sum(-1)^{c(H)} x_H$$

where the sum is over all acceptable even subgraphs of G.

Proof. We proceed in two steps. First we show that when the infinite product is expanded as a sum of monomials of variables, the coefficient corresponding to x_H, H acceptable subgraph, equals $(-1)^{c(H)}$. In the second step we show that all the remaining coefficients are zero.

Claim 1. Let H be an admissible subgraph of G. If $\prod'_{G,U}(1+W([p])$ is expanded as a sum of monomials of variables then the coefficient of x_H equals $(-1)^{c(H)}$.

Proof of Claim 1. By induction on the number of vertices of non-zero degree in H. If H has just one vertex then it consists of one loop e or two loops e, f and $c(H)$ equals zero or one. If H consists of one loop only then $\prod'_{G,U}(1 + W([p]) = (1 + x_e) \times$ product of terms which cannot influence the coefficient at x_H. If H consists of two loops and $c(H) = 0$ then $\prod'_{G,U}(1 + W([p])$ equals $(1 + x_e)(1 + x_f)(1 + x_e x_f)(1 - x_e x_f) \times$ product of terms which cannot influence the coefficient at x_H.

Next let $c(H) = 1$ and H consist of two loops. $\prod'_{G,U}(1 + W([p])$ equals $(1 - x_e x_f) \times$ product of terms which cannot influence the coefficient at x_H. Hence the base of the induction is verified.

Now assume Claim 1 is true for all subgraphs H with n vertices and we will prove it for the graphs with $n + 1$ vertices. Hence let H be an acceptable subgraph with $n + 1$ vertices. A vertex v of H will be called *free* if it doesnot contribute to $c(H)$, i.e. if it has degree 2 in H or if the crossover condition is not imposed at it. Let $k = n + 1 - c(H)$ be the number of free vertices. We continue by induction on k. First let $k = 0$, i.e. each vertex of H has degree four and there is a crossover condition imposed at it. The crossover conditions cause that there is a unique decomposition of H into non-periodic closed walks $p_1, ..., p_r$ such that $x_H = \prod_{i=1}^{r} X(p_i)$. If $r = 1$ then observe that $sign(p_1) = (-1)^{c(H)}$. If $r > 1$ then $\prod_{i=1}^{r} sign(p_i) = (-1)^{c(H)}$ since any two of the p_i's mutually intersect in an even number of vertices (and each vertex contributs to $c(H)$).

Hence let $k > 0$ and Claim 1 holds for all acceptable subgraphs with less than k free vertices. If all free vertices have degree 2 in H then we may proceed as in the case $k = 0$. Hence let v be a free vertex of H of degree four in H. Denote the edges incident with v by north, east, south and west according to the cyclic order induced by the planar drawing.

Partition the non-periodic closed walks of G which satisfy the crossover conditions at the vertices of U into four classes. Classes I,II,III contain walks that have an edge incident with v and:

class I contains the walks that are consistent with west-north and east-south pairing,

class II contains the walks that are consistent with west-south and east-north pairing,

class III contains the walks that are consistent with north-south and east-west pairing (i.e. consistent with the crossover condition at v), and finally

class IV contains the walks that do not contain any edge incident with v.

Suppose $p \in I$ and $q \in II$. Then the product $W[p]W[q]$ can make no contribution to x_H and the same is true for II, III and I, II. Hence if $\prod'_{G,U}(1 + W([p])$ is expanded as a sum, the coefficient of x_H is the sum of the corresponding coefficients in $I \times IV$, $II \times IV$ and $III \times IV$.

The contribution to $I \times IV$ can be regarded as the coefficient of $x_{H'}$ in $\prod'_{G',U}(1 + W([p])$ where G' and H' are obtained from G and H by deleting vertex v and by identifying the west, north edges into one edge, and the east, south edges into one edge. Analogously we can treat the case $II \times IV$. Hence by the induction assumption the sum of the contributions from $I \times IV$ and $II \times IV$ is $2(-1)^{c(H)}$. The contribution to $III \times IV$ can be regarded as coming from $\prod'_{G,U \cup \{v\}}(1 + W([p])$, i.e. one additional cross-over condition is imposed on vertex v. Using the induction assumption again (this time for k) we get that this contribution equals $(-1)^{c(H)+1}$.

Summarising when the product $\prod'(1 + W([p])$ is expanded as a sum, the coefficient of x_H equals $2(-1)^{c(H)} + (-1)^{c(H)+1}$ which we wanted to show.

End of proof of Claim 1.

To finish the proof of Theorem 2.2 we need to show that the remaining coefficients, i.e. the coefficients corresponding to the products of variables where at least one of the exponents is greater than one, are all equal to zero. To that end, temporarily consider $\prod'_{G,U}(1 + W(p))$ where now the product is over non-periodic closed walks and so it is the square of the product we are considering. Let $a_1 > a_1^{-1} > ... > ...$ be a linear order of orientations of the edges of graph G.

Let A_1 be the set of all non-periodic closed walks p such that a_1 appears in p. Each $p \in A_1$ has a unique factorisation into words $(W_1, ..., W_k)$ each of which starts with a_1 and has no other appearance of a_1. Some of these words contain a_1^{-1} and some do not. We will need a curious lemma on coin arrangements stated below. The Lemma is proved in [13] and Sherman remarks in [13] that another proof has been devised by A. Selberg.

A Lemma on Coin Arrangements. Suppose we have a fixed collection of N objects of which m_1 are of one kind, m_2 are of second kind,...,m_n are of nth kind. Let b_k be the number of exhaustive unordered arrangements of these symbols into k disjoint, nonempty, circularly ordered sets such that no two circular orders are the same and none are periodic. For example let us have 10 coins of which 3 are pennies, 4 are nickles and 3 are quarters. Then $\{(p,n), (n,p), (p,n,n,q,q,q)\}$ is not a correct arrangement since (p,n) and (n,p) represent the same circular order. If $N > 1$ then $\sum_{i=1}^{N}(-1)^{i+1}b_i = 0$.

Proof of the Lemma. The Lemma follows immediately if we expand the LHS of the Witt equality and collect terms where the sums of the exponents of the z_i's are the same.

Witt Identity (see [1]): let $z_1, ..., z_k$ be commuting variables. Then

$$\prod_{m_1,...,m_k \geq 0} (1 - z^{m_1}...z^{m_k})^{M(m_1,...,m_k)} = 1 - z_1 - z_2 - ... - z_k,$$

where $M(m_1, ..., m_k)$ is the number of different nonperiodic sequences of z_i's taken with respect to circular order.

End of Proof of the Lemma.

Claim 2. $\prod_{p \in A_1}(1 + W(p)) = 1 + x_{a_1} d_{11}$ where d_{11} is formal (possibly infinite) sum of monomials none of which has x_{a_1} as a factor.

Proof of Claim 2.

First note the following simple fact: if p_1, p_2 are two non-periodic closed walks such that $p_1 p_2$ is also non-periodic then $sign(p_1 p_2) = -sign(p_1)sign(p_2)$.

Let D be a monomial summand in the expansion of $\prod_{p \in A_1}(1 + W(p))$. Hence D is a product of finitely many $W(p), p \in A_1$.

Each $p \in A_1$ has a unique factorisation into words defined above. Each word may appear several times in the factorisation of p and also in the factorisation of different non-periodic closed walks. Let $B(D)$ be the set-system of all the words (with repetition) appearing in the factorisations of the aperiodic closed walks of D.

It directly follows from the Lemma on Coin Arrangements that the sum of all monomial summands D in the expansion of $\prod_{p \in A_1}(1 + W(p))$, which have the same $B(D)$ of more than one element is zero.Hence the monomial summands D which survive in the expansion of $\prod_{p \in A_1}(1 + W(p)$ all have $B(D)$ consisting of exactly one word. This word may but neednot contain a_1^{-1}. However only the summands with their word NOT containng a_1^{-1} survive since if $b, c_1, ..., c_k$ contain neither a_1 nor a_1^{-1} then

$$W(a_1 b a_1^{-1} c_1 a_1^{-1} c_2...a_1^{-1} c_k) + W(a_1 b^{-1} a_1^{-1} c_1 a_1^{-1} c_2...a_1^{-1} c_k) +$$

$$W(a_1 b a_1^{-1} c_1 a_1^{-1} c_2...a_1^{-1} c_k^{-1}) + W(a_1 b^{-1} a_1^{-1} c_1 a_1^{-1} c_2...a_1^{-1} c_k^{-1}) = 0.$$

End of Proof of Claim 2.

Analogously let A_2 be the set of all non-periodic closed walks p such that a_1^{-1} appears in p. Note that possibly $A_1 \cap A_2 \neq \emptyset$. Each $p \in A_2$ has a unique factorisation into words $(W_1, ..., W_k)$ each of which starts with a_1^{-1} and has no other appearance of a_1^{-1}. Some of these words contain a_1 and some do not. The following claim may be proved in exactly the same way as Claim 2.

Claim 3. $\prod_{p \in A_2}(1 + W(p)) = \prod_{p \in A_1 - A_2}(1 + W(p)) = \prod_{p \in A_2 - A_1}(1 + W(p)) = \prod_{p \in A_1}(1 + W(p)).$

Let B be the set of non-periodic closed walks in which neither a_1 nor a_1^{-1} appear. We may assume (by some minimality assumption) that

Claim 4. $\prod_{p \in B}(1 + W(p)) = (1 + d_{12})^2$, where d_{12} is a formal sum of monomials, none of which has x_{a_1} as a factor.

In $\prod_{p \in A_1}(1 + W(p)) \times \prod_{p \in A_2}(1 + W(p)) = (1 + x_{a_1} d_{11})^2$, the non-periodic closed walks from $A_1 \cap A_2$ have been counted doubly, while the non-periodic closed walks from $A_1 - A_2$ and $A_2 - A_1$ have been counted singly. Hence

$$\left[\prod_{p \in (A_1 \cup A_2)} (1 + W(p)) \right]^2 =$$

$$\prod_{p \in A_1} (1 + W(p)) \times \prod_{p \in A_2} (1 + W(p)) \times \prod_{p \in A_1 - A_2} (1 + W(p)) \times \prod_{p \in A_2 - A_1} (1 + W(p)) =$$

$$(1 + x_{a_1} d_{11})^4.$$

This means that

$$[\prod{}'_{G,U}(1 + W([p]))]^2 = \prod{}'_{G,U}(1 + W(p)) =$$

$$\prod_{p \in (A_1 \cup A_2)} (1 + W(p)) \times \prod_{p \in B} (1 + W(p)) =$$

$$(1 + x_{a_1} d_{11})^2 (1 + d_{12})^2,$$

and

$$\prod{}'_{G,U}(1 + W([p]) = (1 + x_{a_1} d_{11})(1 + d_{12}).$$

Thus, there are no monomial summands having factors $x_{a_1}^n$, $n \geq 2$. Analogous arguments dispose of summands with factors $x_{a_i}^n$, $i \neq 1, n \geq 2$. Hence the proof of the Theorem is finished.

$\square$

3. A Formula For General Graphs

As Sherman have noticed, Theorem 2.2 may be used to express $\mathcal{E}(G, x)$ for general graphs as a linear combination of infinite products. Let us consider toroidal graphs first, and we will again assume that each degree is even and at most four: by the remark before Theorem 2.2 this may be done without loss of generality. Let us take a representation of the torus as a rectangle with identified edges. We will assume that a graph G is drawn there properly and so that all the vertices belong to the interior of the rectangle. If p is a non-periodic closed walk of G then let $h(p)$ denote the number of horisontal rectangle edge crossings of p and let $v(p)$ analogously denote the number of vertical rectangle edge crossings of p. The notation $h(H)$ and $v(H)$ is also used for even subgraphs H of G. Finally we let $W_h(p) = (-1)^{h(p)} W(p)$, $W_v(p) = (-1)^{v(p)} W(p)$ and $W_{h,v}(p) = (-1)^{h(p)+v(p)} W(p)$.

The following Theorem 3.2 and in particular Theorem 3.5 are based on a curious lemma.

LEMMA 3.1. *Let R be the set of all $0, 1$-vectors of length $2n$ and let a be an arbitrary integer vector of length $2n$. Then*

$$2^{-n}(-1)^{\sum_{i=1}^{n} a_{2i-1} a_{2i}} \left[\sum_{r \in R} (-1)^{ra}(-1)^{s(r)} \right] = 1,$$

where $s(r)$ denotes the number of i such that $r_{2i-1} = r_{2i} = 1$.

Proof. We proceed by induction on n. The initial case $n = 1$ may be easily checked by hand. Next assume that Lemma 3.1 is true for n and we want to prove it for $n + 1$. Let R' be the set of all $0, 1$-vectors of length $2(n + 1)$ and let a' be an arbitrary integer vector of length $2(n+1)$. Let a denote the initial part of a' of length $2n$. Then

$$2^{-n-1}(-1)^{\sum_{i=1}^{n+1} a'_{2i-1} a'_{2i}} \left[\sum_{r \in R'} (-1)^{ra'} (-1)^{s(r)} \right] =$$

$$2^{-1}(-1)^{a'_{2n+1} a'_{2n+2}} \alpha [(-1)^{a'_{2n+1}} + (-1)^{a'_{2n+2}} - (-1)^{a'_{2n+1}+a'_{2n+2}} + 1],$$

where

$$\alpha = 2^{-n}(-1)^{\sum_{i=1}^{n} a_{2i-1} a_{2i}} \left[\sum_{r \in R} (-1)^{ra} (-1)^{s(r)} \right].$$

By induction assumption we have that $\alpha = 1$ and applying again the first step of the induction, we get that the lemma holds. $\qquad \square$

THEOREM 3.2. *If $G = (V, E)$ is a toroidal graph where each degree is even and at most four then*

$$\mathcal{E}(G, x) = 1/2 \left[\prod (1 + W_h([p])) + \prod (1 + W_v([p])) + \prod (1 + W_{h,v}([p])) - \prod (1 + W([p])) \right],$$

where $\prod$ is the formal infinite product over all equivalence classes of non-periodic closed walks of G.

Proof. 'Unglue' the edges of the rectangle. Hence each rectangle edge crossing now corresponds to 'leaving' the rectangle and 'coming back' to the rectangle by the oposite rectangle edge. If we draw all this to the plane, we get $h(G)v(G)$ crossings of the curves representing the edges of G. Let G' be the graph obtained from G by introducing a vertex to each such intersection. Note that G' is properly drawn in the plane and each degree of G' is even and at most four. Let us call the new vertices *special* and note that each special vertex has degree four in G'. Further note that each non-periodic closed walk p of G corresponds to the non-periodic closed walk p' of G' which satisfies the crossover condition at each special vertex. Moreover $sign(p') = (-1)^{h(p)+v(p)} sign(p)$ and thus $W([p']) = W_{h,v}([p])$.

Using Theorem 2.2 we get that

$$\prod (1 + W_{h,v}([p])) = {\prod}'(1 + W([p'])) = \sum (-1)^{h(H)v(H)} x_H,$$

where the sum goes over all acceptable subgraphs H of G', i.e. over all even subgraphs of G. Hence also

$$\prod (1 + W_v([p])) = \sum (-1)^{h(H)v(H)+h(H)} x_H,$$

$$\prod (1 + W_h([p])) = \sum (-1)^{h(H)v(H)+v(H)} x_H$$

and

$$\prod (1 + W([p])) = \sum (-1)^{h(H)v(H)+h(H)+v(H)} x_H.$$

Let H be an arbitrary even subgraph of G. Then the coefficient of x_H in

$$1/2 \left[\prod (1 + W_h([p])) + \prod (1 + W_v([p])) + \prod (1 + W_{h,v}([p])) - \prod (1 + W([p])) \right]$$

equals

$$1/2(-1)^{h(H)v(H)}[(-1)^{h(H)} + (-1)^{v(H)} - (-1)^{h(H)+v(H)} + 1].$$

This equals 1 by the previous Lemma 3.1.

$\qquad \square$

Using the machinery of the theory of Pfaffian orientations we can write down a formula for general graphs. The machinery is based on considering graphs embedded on orientable surfaces of arbitrary genus.

DEFINITION 3.3. *A surface S_g consists of a base B_0 and $2g$ bridges B_j^i, $i = 1,...,g$ and $j = 1,2$, where*

i) *B_0 is a convex $4g$-gon with vertices $a_1,...,a_{4g}$ numbered clockwise;*

ii) *B_1^i, $i = 1,...,g$, is a 4-gon with vertices $x_1^i, x_2^i, x_3^i, x_4^i$ numbered clockwise. It is glued with B_0 so that the edge $[x_1^i, x_2^i]$ of B_1^i is identified with the edge $[a_{4(i-1)+1}, a_{4(i-1)+2}]$ of B_0 and the edge $[x_3^i, x_4^i]$ of B_1^i is identified with the edge $[a_{4(i-1)+3}, a_{4(i-1)+4}]$ of B_0;*

iii) *B_2^i, $i = 1,...,g$, is a 4-gon with vertices $y_1^i, y_2^i, y_3^i, y_4^i$ numbered clockwise. It is glued with B_0 so that the edge $[y_1^i, y_2^i]$ of B_2^i is identified with the edge $[a_{4(i-1)+2}, a_{4(i-1)+3}]$ of B_0 and the edge $[y_3^i, y_4^i]$ of B_2^i is identified with the edge $[a_{4(i-1)+4}, a_{4(i-1)+5(mod4g)}]$ of B_0.*

Observe that in Definition 3.3 we denote by $[a, b]$ edges of polygons and not edges of graphs. The usual representation in the space of an orientable surface S of genus g may be then obtained from S_g by the following operation: for each bridge B, glue together the two segments which B shares with the boundary of B_0, and delete B.

DEFINITION 3.4. *A graph G is called a g-graph if it may be embedded on S_g so that all the vertices belong to the base B_0, and if the embedding of an edge uses a bridge, then it crosses the bridge.*

This is analogous to the situation described above for the torus: we can imagine that we contract all the bridges (and get a usual representation of an orientable surface of genus g), draw our graph there, and then split the bridges back. The resulting drawing is a g-graph with its embedding on S_g. From now on, we shall consider g-graphs together with a fixed embedding on S_g. If G is a g-graph and p is a non-periodic closed walk of G then we denote by $a(p)$ the vector of length $2g$ such that $a(p)_{2(i-1)+j}$ equals the number of times p crosses bridge B_j^i, $i = 1,...,g$, $j = 1,2$. Similarly we will use the notation $a(H)$ where H is an even subgraph of G.

Next we present a formula for general g-graphs. Note that any graph G may be embedded as a a g-graph where g is genus of G and that we only need to consider g-graphs that have all degrees even and at most four (by a remark before theorem 2.2).

Let $R(g)$ denote the set of all $0, 1$-vectors of length $2g$. If G is a g-graph (with all degrees even and at most four), p a non-periodic closed walk of G and $r \in R(g)$, then let $W_r([p]) = (-1)^{ra(p)} W([p])$.

THEOREM 3.5. *If $G = (V, E)$ is a g-graph where each degree is even and at most four then*

$$\mathcal{E}(G, x) = 2^{-g} \sum_{r \in R(g)} (-1)^{s(I-r)} \prod (1 + W_r([p])),$$

where $\prod$ is the formal infinite product over all equivalence classes of non-periodic closed walks of G and I denotes the vector of all ones.

Proof. We proceed analogously as in the proof of Theorem 3.2. We consider G embedded in the plane by the projection of the bridges B_j^i outside B_0. We get $\sum_{i=1}^{g} a(G)_{2i-1}a(G)_{2i}$ crossings of the curves representing the edges of G. Let G' be the graph obtained from G by introducing a vertex to each such intersection. Note that G' is properly drawn to the plane, and each degree of G' is even and at most four. Let us call the new vertices *special* and note that each special vertex has degree four in G'. Further note that each non-periodic closed walk p of G corresponds to the non-periodic closed walk p' of G' which satisfies the crossover condition at each special vertex. Moreover $sign(p') = (-1)^{Ia(p)}sign(p)$ and thus $W([p']) = W_I([p])$.

Using Theorem 2.2 we get that

$$\prod(1 + W_I([p])) = \prod{}'(1 + W([p'])) = \sum(-1)^{\sum_{i=1}^{g} a(H)_{2i-1}a(H)_{2i}}x_H,$$

where the sum is over all acceptable subgraphs H of G', i.e. over all even subgraphs of G. Hence for $r \in R(g)$ we have

$$\prod(1 + W_r([p])) = \sum(-1)^{\sum_{i=1}^{g} a(H)_{2i-1}a(H)_{2i}+(I-r)a(H)}x_H,$$

where the sum is over all even subgraphs H of G.

Let H be an arbitrary even subgraph of G. Then the coefficient of x_H in

$$2^{-g} \sum_{r \in R(g)} (-1)^{s(I-r)} \prod(1 + W_r([p]))$$

equals

$$2^{-g}(-1)^{\sum_{i=1}^{g} a(H)_{2i-1}a(H)_{2i}} \sum_{r \in R(g)} (-1)^{s(I-r)}(-1)^{(I-r)a(H)}.$$

This equals 1 by the previous Lemma 3.1, since we can replace r by $I - r$ in the summation.

$\square$

References

[1] H.M. Hall, The theory of groups, *Macmillan*, New York, 1959.

[2] M.E. Fisher, On the dimer solution of planar Ising models, *Journal of Mathematical Physics* 7,10, 1966.

[3] A. Galluccio and M. Loebl, A Theory of Pfaffian Orientations I: Perfect Matchings and Permanents, *Electronic Journal of Combinatorics* 6,1, 1999.

[4] A. Galluccio and M. Loebl, A Theory of Pfaffian Orientations II: T-joins, k-cuts and duality of enumeration, *Electronic Journal of Combinatorics* 6,1, 1999.

[5] A. Galluccio, M. Loebl and J. Vondrak, A new algorithm for the Ising problem, *Physical Review Letters* 84,26, 2000.

[6] M. Kac and J. C. Ward, *Phys. Rev.* 88, 1952.

[7] P. W. Kasteleyn, Graph theory and crystal physics, *in Graph theory and theoretical physics* , 1967, Academic Press, New York.

[8] M. Loebl, On the Dimer Problem and the Ising Problem in 3-dimensional Lattices, manuscript 2001, submitted for publication.

[9] L. Onsager and B. Kaufman, *Phys. Rev.* 76, 1949.

[10] R. B. Potts and J. C. Ward, *Progr.Theoret. Phys.* 13, 1955.

[11] T. Regge and R. Zecchina, Exact solution of the Ising model on group lattices of Genus $g > 1$, *J. Math. Phys.* 37, 1996.

[12] T. Regge and R. Zecchina, Combinatorial and topological approach to the 3D Ising model, *J. Phys. A* 33, 2000.

[13] S. Sherman, Combinatorial Aspects of the Ising Model of Ferromagnetism I, *J. Math. Phys.* 1, 1960.

[14] S.Sherman, Combinatorial Aspects of the Ising Model of Ferromagnetism II, *Bull. Am. Math. Soc.* 68, 225, 1962.
[15] G. Tessler, Matchings in Graphs on Non-oriented Surfaces, *J. Combinatorial Theory B*, 2000.
[16] B.L. van der Waerden, Die lange Reichweite der regelmassigen Atomanordnung in Mischkristallen, *Z.Physik* 118, 1941.

DEPARTMENT OF APPLIED MATHEMATICS AND INSTITUTE FOR THEORETICAL COMPUTER SCIENCE (ITI), CHARLES UNIVERSITY, MALOSTRANSKE NAM. 25, 118 00 PRAHA 1, CZECH REPUBLIC

E-mail address: `loebl@kam.ms.mff.cuni.cz`

DIMACS Series in Discrete Mathematics
and Theoretical Computer Science
Volume **63**, 2004

Survey: Information flow on trees

Elchanan Mossel

ABSTRACT. Consider a tree network T, where each edge acts as an independent copy of a given channel M, and information is propagated from the root. For which T and M does the configuration obtained at level n of T typically contain significant information on the root variable?

This model appeared independently in biology, information theory, and statistical physics. Its analysis uses techniques from the theory of finite markov chains, statistics, statistical physics, information theory, cryptography and noisy computation. In this paper, we survey developments and challenges related to this problem.

1. Introduction

Consider a process on a tree in which information is transmitted from the root of the tree to all the nodes of the tree. Each node inherits information from its parent with some probability of error. The transmission process is assumed to have identical distribution on all the edges, and different edges of the tree are assumed to act independently.

As this process represents propagation of a genetic property from ancestor to its descendants, it was studied in genetics, see e.g. [**7, 39**]. In communication theory, this process represents a communication network on the tree where information is transmitted from the root of the tree. Earlier, the process was studied in statistical physics, see e.g. [**41, 17, 6**].

The basic question we address in this survey is: Does the configuration obtained at level n of T typically contain significant information on the root variable? The theory of finite markov chains implies that if the underlying markov chain is ergodic (i.e. irreducible and aperiodic), then the variable at a single node at level n and the variable at the root are asymptotically independent as $n \to \infty$. However, for the tree process, information is duplicated, so it is conceivable that the configuration at level n contains a significant amount of information on the root variable.

In Section 2 a precise formulation of the problem is given. In Section 3 we discuss the problem for symmetric binary channels (which correspond to Ising models with no external field). This is the family of channels for which the most is known. In particular, is subsection 3.1 we compare various reconstruction algorithms for symmetric binary channels.

Suppose that instead of the *configuration* at level n, we are given the *census* of the configuration at level n. In section 4 we see how the spectral properties of M determine if the census is asymptotically independent of the root. In Section 5 we present bounds for the problem for Potts models, while in Section 6 we present some examples of channels which are related to secret sharing and demonstrate that level n may contain significant information on the root even if the census of the level contains no such information.

In Section 7 we discuss related problems and terminology. Many unsolved problems are presented throughout the paper.

Disclaimer: This paper is a survey: most of the results are presented without a proof; for others only a sketch is given.

2. The reconstruction problem

Definition of the process. Denote the underlying tree by $T = (V, E)$. The information flow on each edge is given by a channel on a finite alphabet $\mathcal{A} = \{1, \ldots, k\}$. Let $\mathbf{M}_{i,j}$ be the transition probability from i to j; M be the random function (or channel) which satisfies for all i and j that $\mathbf{P}[M(i) = j] = \mathbf{M}_{i,j}$, and $\lambda_2(M)$ be the eigenvalue of $\mathbf{M}$ which has the second largest absolute value ($\lambda_2(M)$ is in general a complex number). We assume throughout the paper that M defines an ergodic markov chain (irreducible and aperiodic). At the root ρ one of the symbols of $\mathcal{A}$ is chosen according to some initial distribution. We denote this (random) symbol by σ_ρ. This symbol is then propagated in the tree in the following way. Given that the parent of v, denoted v', has value $\sigma_{v'}$, the probability that σ_v is j is given by $\mathbf{M}_{\sigma_{v'},j}$. More formally, for each vertex v having as a parent v', we let $\sigma_v = M_{v',v}(\sigma_{v'})$, where the $\{M_{v',v}\}$ are independent copies of M. Equivalently, for a vertex v, let v' be the parent of v, and let $\Gamma(v)$ be the set of all vertices which are connected to ρ through paths which do not contain v. Then the process satisfies:

$$\mathbf{P}[\sigma_v = j | (\sigma_w)_{w \in \Gamma(v)}] = \mathbf{P}[\sigma_v = j | \sigma_{v'}] = \mathbf{M}_{\sigma_{v'},j}.$$

Let $d(,)$ denote the graph-metric distance on T, and $L_n = \{v \in V : d(\rho, v) = n\}$ be the n'th level of the tree. For $v \in V$ and $e = (v, w) \in E$ we denote $|v| = d(\rho, v)$ and $|e| = \max\{|v|, |w|\}$. We denote by $\sigma_n = (\sigma(v))_{v \in L_n}$ the symbols at the n'th level of the tree. We let $c_n = (c_n(1), \ldots, c_n(k))$ where

$$c_n(i) = \#\{v \in L_n : \sigma(v) = i\}.$$

In other words, c_n is the *census* of the n'th level. Note that both $(\sigma_n)_{n=1}^\infty$ and $(c_n)_{n=1}^\infty$ are markov chains.

Reconstruction solvability. For distributions P and Q on the same space, the total variation distance between P and Q is

$$(1) \qquad D_V(P, Q) = \frac{1}{2} \sum_\sigma |P(\sigma) - Q(\sigma)|.$$

DEFINITION 2.1. *The reconstruction problem for T and M is* solvable *if there exist $i, j \in \mathcal{A}$ for which*

$$(2) \qquad \lim_{n \to \infty} D_V(\mathbf{P}_n^i, \mathbf{P}_n^j) > 0,$$

where $\mathbf{P}_n^{\ell}$ denotes the conditional distribution of σ_n given that $\sigma_\rho = \ell$. We define census solvability similarly, where the measures $\mathbf{P}_n^{\ell}$ are replaced by measures $\widetilde{\mathbf{P}}_n^{\ell}$, which are conditional distributions of c_n given that $\sigma_\rho = \ell$.

A stronger definition than Definition 2.1 is obtained by replacing "there exists i, j" by "for all i, j". We choose the weaker definition as we are interested to know if some information is propagated from the root to the boundary, see also Proposition 2.1 below.

Equivalent definitions. If the reconstruction problem is solvable, then σ_n contains significant information on the root variable. This may be expressed in several equivalent ways. Assume that the variable at the root, σ_ρ, is chosen according to some initial distribution $(\pi_i)_{i \in \mathcal{A}}$, and let $\mathbf{P}^{\pi}$ denote the corresponding probability measure. The maximum-likelihood algorithm, which is the optimal reconstruction algorithm of σ_ρ given σ_n, is successful with probability

$$\Delta_n(\pi) = \sum_{\sigma} \mathbf{P}^{\pi}[\sigma_n = \sigma] \max_{i \in \mathcal{A}} \mathbf{P}^{\pi}[\sigma_\rho = i | \sigma_n = \sigma]$$

$$\geq \max_{i \in \mathcal{A}} \sum_{\sigma} \mathbf{P}^{\pi}[\sigma_n = \sigma] \mathbf{P}^{\pi}[\sigma_\rho = i | \sigma_n = \sigma] = \max_{i \in \mathcal{A}} \pi_i.$$

Note that it is possible to reconstruct σ_ρ with probability $\max_i \pi_i$ even when σ_n is unknown (using the algorithm which always reconstructs the i which maximizes π_i). It is therefore natural to consider $\Delta_n(\pi) - \max_i \pi_i$ as a measurement for the dependency between σ_n and σ_ρ.

Let H be the entropy function, and let $I(X, Y) = H(X) + H(Y) - H(X, Y)$ be the mutual-information operator (see e.g. [8] for definitions and basic properties).

For a sequence of random variables X_n defined on the same probability space, let F_n be the σ-algebra defined by $(X_m)_{m \geq n}$, i.e., F_n is the minimal σ-algebra such that all the variables $(X_m)_{m \geq n}$ are measurable with respect to F_n. Let $F_\infty = \cap_{n=1}^{\infty} F_n$. We say that the sequence X_n has a trivial tail, if all the measurable sets with respect to F_∞ have probability either 0 or 1. Otherwise, we say the the sequence has a non-trivial tail.

In the theory of markov random fields the notion of tail triviality is closely related to the extremality of the measure, see e.g. [14].

The following equivalence follows from the fact that σ_n is a markov chain (see e.g. [32]):

PROPOSITION 2.1. *Let T be an infinite tree and M a channel. Then the following conditions are equivalent (where π denotes initial distribution for σ_ρ):*

(1) *The reconstruction problem is solvable*

(2) *There exists a π for which $\lim_{n \to \infty} I(\sigma_\rho, \sigma_n) > 0$.*

(3) *If π is the uniform distribution on $\mathcal{A}$, then $\lim_{n \to \infty} I(\sigma_\rho, \sigma_n) > 0$.*

(4) *For any distribution π with $\min_i \pi_i > 0$, it holds that $\lim_{n \to \infty} I(\sigma_\rho, \sigma_n) > 0$.*

(5) *There exists a π for which $\liminf_{n \to \infty} \Delta_n(\pi) > \max_i \pi_i$.*

(6) *If π is the uniform distribution on $\mathcal{A}$, then $\liminf_{n \to \infty} \Delta_n(\pi) > 1/|\mathcal{A}|$.*

(7) *For all π with $\min_i \pi_i > 0$, $(\sigma_n)_{n=1}^{\infty}$ has a non-trivial tail.*

(8) *There exists a π with $\min_i \pi_i > 0$ such that $(\sigma_n)_{n=1}^{\infty}$ has a non-trivial tail.*

The analogous 8 conditions are equivalent for c_n.

3. The Ising model

The only family of channels for which the problem is well understood is the family of symmetric binary channels

$$\mathbf{M} = \begin{pmatrix} 1 - \epsilon & \epsilon \\ \epsilon & 1 - \epsilon \end{pmatrix}, \tag{3}$$

where $\lambda_2(M)$, the second largest (in absolute value) eigen value of M, satisfies $\lambda_2(M) = 1 - 2\epsilon$. Channel (3) corresponds to the Ising model on the tree. The free measure for the Ising model on a finite tree is the probability measure on configurations σ of ± 1, given by

$$\mathbf{P}[\sigma] = \frac{1}{Z} \exp\Big(\sum_{v \sim w} \sigma_v \sigma_w\Big), \tag{4}$$

where Z is a normalizing constant. The correspondence between (3) and (4) is given by $\epsilon = \frac{\exp(-\beta)}{\exp(-\beta) + \exp(\beta)}$, or equivalently, $\lambda_2(M) = \tanh \beta$.

THEOREM 3.1. *The reconstruction problem is solvable for the binary symmetric channel with error probability ϵ (3), and the b-ary tree T_b, if and only if $b\lambda_2^2(M) > 1$. If $b\lambda_2^2(M) > 1$, then the reconstruction problem is also* census *solvable.*

Reconstruction (and census) solvability when $b\lambda_2^2(M) > 1$ was first proved in [17] ([23] is earlier and does much more, but is formulated in the language of multi-type branching processes).

PROOF. We think of the spin values as ± 1. Write $\lambda = 1 - 2\epsilon$, and let S_n be the sum of the $\pm$ variables at level n of the tree. Given that the spin at the root is $+$, the expected value of S_n satisfies

$$\mathbf{E}^+[S_n] = \sum_{v \in L_n} \mathbf{E}^+[\sigma_v] = b^n \lambda^n. \tag{5}$$

Similarly, $\mathbf{E}^-[S_n] = -b^n \lambda^n$. The second moment of S_n satisfies
(6)

$$\mathbf{E}^+[S_n^2] = \mathbf{E}^-[S_n^2] = \sum_{v,w \in L_n} \mathbf{E}[\sigma_v \sigma_w] = b^n \left(1 + \sum_{j=1}^n (b^j - b^{j-1})\lambda^{2j}\right) = \Theta(b^{2n}\lambda^{2n}),$$

where the last equality follows from the fact that $b\lambda^2 > 1$. By Cauchy-Schwartz,

$$\begin{aligned}
\mathbf{E}^+[S_n] - \mathbf{E}^-[S_n] &= \sum_\sigma (\mathbf{P}^+[\sigma] - \mathbf{P}^-[\sigma])S_n(\sigma) \\
&\leq \sqrt{\sum_\sigma \frac{(\mathbf{P}^+[\sigma] - \mathbf{P}^-[\sigma])^2}{\mathbf{P}^+[\sigma] + \mathbf{P}^-[\sigma]}} \sqrt{\sum_\sigma S_n^2(\sigma)(\mathbf{P}^+[\sigma] + \mathbf{P}^-[\sigma])}.
\end{aligned}$$

It now follows by (5) and (6) that when $b\lambda^2 > 1$,

$$\sum_\sigma \frac{(\mathbf{P}^+[\sigma] - \mathbf{P}^-[\sigma])^2}{\mathbf{P}^+[\sigma] + \mathbf{P}^-[\sigma]} = \Theta(1).$$

which implies that $D_V(\mathbf{P}^+, \mathbf{P}^-) = \Theta(1)$. $\square$

The reconstruction solvability result when $b|\lambda_2(M)|^2 > 1$ is extended to general trees [12] and general channels [23, 30], where b is replaced by the *branching number* of the tree and λ_2 is the second largest eigenvalue of the matrix $\mathbf{M}$ (in absolute value).

The proofs of the non-reconstruction result when $b\lambda_2^2(M) \leq 1$ are harder, and do not generalize to other channels. We know of 4 different proofs for this result

- The first proof [6], is based on recursive analysis of the Gibbs measure.
- A proof of non-reconstruction which is based on information inequalities is given in [12] where it is shown that the mutual information between the variable at the root of the tree and the level n variables satisfy $I(\sigma_\rho, \sigma_n) \leq \sum_{v \in L_n} I(\sigma_\rho, \sigma_v)$, as in the case of conditionally independent variables. This proof extends to general trees when $\mathrm{br}(T)\lambda_2^2(M) < 1$, where $\mathrm{br}(T)$ is the branching number of the tree.
- In [18] recursive analysis is used in order to show $\mathbf{E}[\mathbf{E}^2[\sigma_\rho|\sigma_n]]$ tends to zero as $n \to \infty$ for the n-level tree, when $b\lambda_2^2(M) \leq 1$. The proof [19] applies also to general trees when $\mathrm{br}(T)\lambda_2^2(M) < 1$.
- Glauber dynamics is the following reversible Monte-Carlo method for sampling configurations σ according to the distribution (3) or (4). Given the current configuration σ, a vertex v is picked uniformly at random at rate 1, in which case the variable σ_v is replaced by a random variable σ'_v chosen according to the conditional distribution on the rest of the configuration, $(\sigma_w)_{w \neq v}$. In [2] it is shown that Glauber dynamics have spectral gap which is bounded away from zero when $b\lambda_2^2(M) < 1$. Then using a general principle (which is proven in a much more general context) we obtain that the reconstruction problem is unsolvable when $b\lambda_2^2(M) < 1$.

In [37] the critical case for general trees, $\mathrm{br}(T)\lambda_2^2(M) = 1$, is analyzed in detail.

3.1. Reconstruction algorithms. Theorem 3.1 reveals a surprising phenomenon: reconstruction by global majority vote has the same threshold for success as maximum likelihood reconstruction (which is the optimal reconstruction strategy).

The parsimony method is popular in biology. Given a bicoloring of the boundary of a tree T, a *parsimonious* coloring of the internal nodes is any assignment of the two colors to these nodes that minimizes the total number of bicolored edges. A way of finding a parsimonious coloring is the following: Starting from the parents of the boundary nodes, assign recursively to each internal node the color of the majority of its ± 1-colored children. In case of a tie, assign the non-color "?". Then scan the tree from the root downwards and assign all vertices labeled by ? the same label as their parent.

On a fixed finite tree, when $\epsilon > 0$ is small, the maximum likelihood algorithm will reconstruct the same root value as one of the parsimonious colorings given the boundary.

However, this is not the case when ϵ is larger. For the binary tree, it is shown in [42] that the parsimony reconstruction algorithm has success probability bounded away from $1/2$ as $n \to \infty$ if and only if $\epsilon \geq 1/8$. Thus when $\lambda_2(M) = 1 - 2\epsilon \in (2^{-1/2}, 3/4]$ on the binary tree, majority (and maximum likelihood) will have success probability bounded away from $1/2$, while the parsimony success probability tends to $1/2$.

On a tree where each vertex has k children with k odd, the above algorithm for finding a parsimonious coloring reduces to recursive majority; In [31] it is shown that reconstruction via this method succeeds asymptotically if and only if

$$\epsilon < \beta_k := \frac{1}{2} - \frac{2^k}{4k}\left(\frac{k-1}{\frac{k-1}{2}}\right)^{-1}.$$

(it is interesting to note that the proof is based on exactly the same recursion which is analyzed in the context of noisy computation in [**16, 10**]).

In [**31**] more general reconstruction algorithms on regular trees (and more generally, on ℓ-periodic trees) are analyzed. Suppose that in order to determine the color assigned to a node v, the algorithm is allowed to examine the colors of its descendants ℓ generations down. (However, only a single bit can be stored at each node). Then it is shown in [**31**] that recursively applying majority vote of the descendants ℓ generations down is optimal, yet it succeeds asymptotically only for flip probabilities ϵ below a threshold which is strictly lower than the critical value for reconstruction.

Yuval Peres (private communication) conjectured that

CONJECTURE 1. *Consider the Ising model on the regular tree T_b and reconstruction algorithms which are* local: *algorithms that for each vertex are allowed to scan the information stored at its descendants ℓ generations down, and at most r bits of information are allowed to be stored at each node. Then for all ℓ and r the threshold for reconstruction for such algorithms is strictly below the threshold for reconstruction.*

We remark that it is important to require that the algorithm examines only vertices below the vertex which is being updated, as Glauber dynamics are local and are successful in reconstruction of the root whenever $b\lambda_2^2(M) > 1$. A step of Glauber dynamics is performed as follows. Given the current configuration σ, an internal vertex v is picked uniformly at random at rate 1, in which case the variable σ_v is replaced by a random variable σ_v' chosen according to the conditional distribution on the rest of the configuration, $(\sigma_w)_{w\neq v}$.

We emphasize that the fact that recursive algorithms are asymptotically inferior to global majority does not hold for other Potts models or Ising models with external fields([**32**], see also Section 5).

4. Census solvability

The threshold $b\lambda_2^2(M) = 1$ which appeared as the threshold both for reconstruction solvability and for census solvability for the Ising model, turns out to be in general the threshold for census solvability.

THEOREM 4.1. *Let M be a channel corresponding to an ergodic markov chain. Let T_b be the b-ary tree. The reconstruction problem is census-solvable if $b|\lambda_2(M)|^2 > 1$, and is not census solvable if $b|\lambda_2(M)|^2 < 1$. For general trees, the reconstruction problem is solvable when $\mathrm{br}(T)|\lambda_2(M)|^2 > 1$, where $\mathrm{br}(T)$ is the branching number of the tree.*

CONJECTURE 2. *The reconstruction problem is not census solvable when*

$$b|\lambda_2(M)|^2 = 1.$$

$b|\lambda_2(M)|^2 > 1$ **implies census solvability.** [**23**] proves a limit theorem for the variables c_n. In particular it is shown, that if $b|\lambda_2(M)|^2 > 1$ then the distribution of the limiting variable depends on the initial variable at the root. This implies that the problem is census solvable.

A more elementary proof is given in [**30**]. The proof follows the lines of the proof for the Ising model (Theorem 3.1), where S_n is replaced by the scalar product of c_n with any vector nonzero v satisfying $\mathbf{M}v = \lambda_2(M)v$. Note that this proof generalizes the proof for the Ising model, since for the Ising model, $v = \begin{pmatrix} 1 \\ -1 \end{pmatrix}$. This proof also generalizes to general trees, proving that if $\mathrm{br}(T)|\lambda_2(M)|^2 > 1$, then it is possible to reconstruct the root using the scalar product of v with a weighted census $\tilde{c}_n(i) = \{\sum_x \omega(x) : x \in L_n, \sigma_x = i\}$ for some weights $\{\omega(x)\}_{x \in T}$.

$b|\lambda_2(M)|^2 < 1$ **implies no census solvability.** The CLT in [**23**] implies that if $b|\lambda_2(M)|^2 \leq 1$ then the normalized value of c_n ($c_n/b^{n/2}$ if $b|\lambda_2(M)|^2 < 1$) converges to a nonzero random variable which is independent of the variable of the root. However, this result on does not imply that the reconstruction problem is not census solvable. Presumably, it may the case that the first coordinate of c_n is more likely to be even for some value of the root variable than for others. This dependency between the root variable and c_n would not manifest itself in the limiting normalized variables.

In [**30**] we combine the results of [**23**] with the local central limit theorem to demonstrate that this could not happen. The idea of the proof is to use [**23**] in order to couple c_n^i, the value of c_n given that the root variable is i, and c_n^j, the value of c_n give that the root variable is j in such a way that the variables are close (i.e., $|c_n^i - c_n^j|_\infty < \epsilon b^{n/2}$). Then use the local central limit theorem in order to achieve a coupling of $c_{n+\ell}^i$ and $c_{n+\ell}^j$ with high probability. In [**30**] we also verify Conjecture 2 for Potts models and asymmetric Ising models.

5. Potts models

Two of the natural generalizations of binary symmetric channels are asymmetric binary channels (which correspond to Ising models with external field), and q-ary symmetric channels (which correspond to Potts models with no external field):

- Asymmetric binary channels have the state space $\{0, 1\}$ and the matrices:

$$(7) \qquad \mathbf{M} = \begin{pmatrix} 1 - \delta_1 & \delta_1 \\ 1 - \delta_2 & \delta_2 \end{pmatrix},$$

  with $\lambda_2(M) = \delta_2 - \delta_1$.
- Symmetric channels on q symbols have the state space $\{1, \ldots, q\}$ and the matrices:

$$(8) \qquad \mathbf{M} = \begin{pmatrix} 1 - (q-1)\delta & \delta & & \ldots & \delta \\ \delta & 1 - (q-1)\delta & \delta & \ldots & \\ \vdots & & & \ddots & \vdots \\ \delta & \ldots & & \delta & 1 - (q-1)\delta \end{pmatrix},$$

  with $\lambda_2(M) = 1 - q\delta$.

Depending on the sign of $\lambda_2(M)$ we distinguish between *ferromagnetic* Potts models where $\lambda_2(M) > 0$, and *anti-ferromagnetic* models where $\lambda_2(M) < 0$. When $1 -$

$(q-1)\delta = 0$, we obtain the model of proper colorings of the tree:

$$
(9) \qquad \mathbf{M} = \begin{pmatrix}
0 & (q-1)^{-1} & (q-1)^{-1} & \cdots & (q-1)^{-1} \\
(q-1)^{-1} & 0 & (q-1)^{-1} & \cdots & \\
\vdots & \cdots & & \ddots & \vdots \\
(q-1)^{-1} & \cdots & & (q-1)^{-1} & 0
\end{pmatrix} .
$$

PROBLEM 1. *For the 3 symbols Potts model (8) find the values for which the reconstruction problem is solvable on the b-ary tree.*

It may be easier to solve the analogous problem for the Ising model with external field. The analogous problem for colorings was stated in [**3**]. Applying standard coupon-collector estimates recursively, it is easy to see that if $b \geq (1+\delta)q \log q$ and q is large, then the reconstruction problem is solvable for the coloring model.

PROBLEM 2. *For colorings, for which b and q is the reconstruction problem solvable on the b-ary tree?*

Below we discuss several bounds for the reconstruction problem for Potts models.

- **If $b\lambda_2^2(M) > 1$ then the reconstruction problem is solvable.** This follows from Theorem 4.1, and from the fact that census solvability implies solvability.

- **If $b|\lambda_2(M)| \leq 1$, then the reconstruction problem is unsolvable.**

  PROOF. Assume first that M is a ferromagnetic Potts model, i.e. $\lambda_2(M) > 0$. Consider two measures on the tree, one with i as the root variable and one with j as the root variable. We *couple* these measures in the following way: starting at the root if the two measures agree on the variable at v, then we couple in such a way that the measures also agree for all the children of v. If they do not agree at v, then for each of the children of v, use the optimal coupling in order to couple the measures. For each of the children, this has success probability $q\delta$. Thus the non-coupled vertices are a branching process with parameter $1 - q\delta = \lambda_2(M)$. When $b\lambda_2(M) \leq 1$ this process will eventually die; this means that for large n all the vertices at level n will have the same variables with probability going to 1 as $n \to \infty$, as needed. When M is anti-ferromagnetic, the coupling probability is $(q-2)\delta + 2(1-(q-1)\delta) = 2 - q\delta$, therefore the branching process parameter is $1-(2-q\delta) = -\lambda_2(M) = |\lambda_2(M)|$. Similar arguments apply for Ising models with external fields. $\square$

  If $b\lambda_2(M) > 1$ and q is sufficiently large, then the reconstruction problem is solvable. This is the main result of [**32**]. It implies in particular that $b|\lambda_2(M)|^2 = 1$ is not the threshold for the reconstruction problem for Potts models, as it sometimes possible to reconstruct even when $b|\lambda_2(M)|^2 < 1$. An analogous result is proven for the asymmetric binary channel. The idea behind the proof is the following. Channel (8) may be thought of in the following way: at each step the output is identical to the input with probability $\lambda_2(M)$, otherwise, the output is chosen uniformly among the q symbols. In particular if $\lambda_2(M) > 0$ is fixed and q is very large, then if two of the children of a vertex in the b-ary tree T_b have the same label, then with overwhelming probability, this is also the label of their parent. Now suppose that q is large and there exists a copy of $T_2 \subset T_b$ such that all the vertices of T_2 are labeled by i. Using a recursive argument we see that given this event, with large probability, the variable at the root is i. Moreover we show that when $b\lambda_2(M) > C$ for some constant $C > 1$, such a tree exists with positive probability. Therefore, it is possible to reconstruct the root variable based on the existence of such a unicolored tree. In order to obtain the result for $C = 1$, we replace the unicolored T_2 by a diluted unicolored T_2.

If $b\frac{(1-q\delta)^2}{1-(q-2)\delta} \leq 1$ then the reconstruction problem is unsolvable. This and the analogous result for asymmetric binary channels are proven in [**30**], we sketch the main idea of the proof below.

PROOF. In order to show that the reconstruction problem is unsolvable, it suffices to show that given that the root value is 0 or 1 with probability $1/2$ each, it is asymptotically impossible to conclude from the variables at level n, if the root is 0 or 1. Suppose that in addition to the variables at level n, we are also given all the variables at all levels of the tree having variable j with $j \neq \{0, 1\}$. Since we are given more information, it is easier to reconstruct. The model where we are given this extra information is nothing but the symmetric binary channel with matrix

$$\mathbf{M} = \begin{pmatrix} \frac{1-(q-1)\delta}{1-(q-2)\delta} & \frac{\delta}{1-(q-2)\delta} \\ \frac{\delta}{1-(q-2)\delta} & \frac{1-(q-1)\delta}{1-(q-2)\delta} \end{pmatrix},$$

on the random tree which is obtained from the original tree by independently deleting an edge with probability $(q-2)\delta$ and retaining it with probability $1-(q-2)\delta$. The results of [**12**] imply that the binary symmetric channel on a general tree T, the reconstruction problem is unsolvable if $\mathrm{br}(T)\lambda_2^2(M) < 1$. For the branching process on the regular tree T_b, the branching number is "typically" $b(1 - (q-2)\delta)$. The non-reconstruction criterion $\mathrm{br}(T)\lambda_2^2(M) < 1$, now translates to $b\frac{(1-q\delta)^2}{1-(q-2)\delta} < 1$. $\square$

In Figure 1 we draw several of the bounds, where b is a function of $\lambda = \lambda_2(M)$. The area above the top curve, $b\lambda^2 = 1$, is the area where (census) reconstruction is successful for all channels. The top curve is the critical curve for the symmetric binary channel: above it reconstruction is successful and below it, it fails. Below the second curve, $\lambda = \frac{b(1-3\delta)^2}{1-\delta}$, reconstruction fails for the $q = 3$ Potts model. Below the bottom curve $b\lambda = 1$, reconstruction fails for all Potts models. This curve is also the asymptotic critical curve as $q \to \infty$.

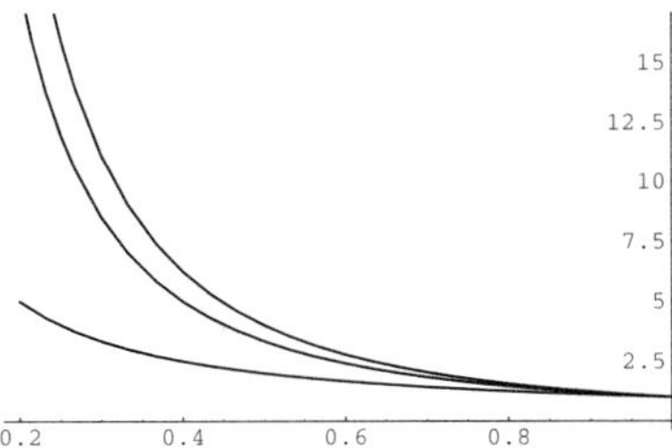

FIGURE 1. bounds for b as a function of $\lambda_2(M)$

5.1. Algorithms. We want to point out that the proofs in [**32**] imply that for Potts models when q is large, reconstruction using recursive schemes has better threshold than any algorithm which uses only the census. We conjecture that the phenomenon that global majority achieves the same threshold for reconstruction as maximum likelihood occurs only for symmetric binary channels.

CONJECTURE 3. *Consider Potts models (8) for $q \geq 3$. Then there exist b and δ such that the reconstruction problem is solvable for the b-ary tree, yet $b\lambda_2^2(M) < 1$.*

Note that this conjecture implies in particular, that any algorithm which uses the census only is inferior to the optimal algorithm for Potts models.

5.2. Monotonicity. For Potts models (8), it is easy to see that if the reconstruction problem is solvable for q and ϵ and $q' < q$, then the reconstruction problem is also solvable for q' and ϵ. We expect that for fixed $\lambda = \lambda_2(M)$, reconstruction is easier when q is larger.

CONJECTURE 4. *Consider two symmetric channels M_1 and M_2 (as in (8)) on q_1 and q_2 symbols respectively, where $q_1 < q_2$. If $0 < \lambda_2(M_1) = \lambda_2(M_2)$ and the reconstruction problem is solvable for M_1, then it is also solvable for M_2.*

This is obvious when q_2 is a multiple of q_1. Using the reconstruction criterion for the binary symmetric channel on 2 symbols, it is easy to prove the conjecture when $q_1 = 2$.

6. General channels

In this section we discuss techniques which apply to the reconstruction problem for general channels.

6.1. Proving solvability. Spectral methods. Theorem 4.1 implies that when $b\lambda_2^2(M) > 1$ the reconstruction problem is census-solvable (and therefore solvable) for the b-ary tree and the channel M.

Recursive methods. Starting at the boundary of the tree, we may try to evaluate recursively the variable at each vertex of the tree. Assuming that the probability of reconstructing the value of a variable at distance n from the boundary is p_n, we obtain recursive bounds on p_{n+1}. Bounding these recursions, we may prove that the reconstruction problem is solvable. Although for symmetric binary channels, these methods always achieve worse thresholds than the spectral method [**31**], in many other cases the recursive methods are superior. The first example which was already discussed in Section 5 is that of the Potts model on q symbols where q is large. It is proven in [**32**] that if $b\lambda_2(M) > 1$ and q is sufficiently large then the reconstruction problem is solvable (the spectral method only applies when $b\lambda_2^2(M) > 1$). In [**30**] we note that a similar argument proves that for the coloring problem for large q, if $b \geq (1 + \epsilon)q \log q$, then the reconstruction problem is solvable. In general this method combined with standard large deviation estimates (see e.g. [**9**]) implies:

THEOREM 6.1. *Let M be a channel, such that for all $i, j \in \mathcal{A}$, there exists an ℓ such that $\mathbf{M}_{i,\ell} \neq \mathbf{M}_{j,\ell}$. Then when b is sufficiently large, the reconstruction problem is solvable for the tree T_b and M.*

This is proven in [**30**] where a general criterion is given to decide: does there exist for the channel M a number b such that the reconstruction problem is solvable for T_b and M.

EXAMPLE 6.2. *Let $\{Z_i\}_{i=1}^{\infty}$ be an i.i.d. sequence of variables such that $\mathbf{P}[Z_i = 0] = 1 - P[Z_i = 1] = p$ where $0 < p < 1$. Let $h \geq 1$ and consider the channel M defined by the markov chain $Y_i = (Z_i, \ldots, Z_{i+h})$. Thus M has state space $\{0,1\}^{h+1}$ with the product $(p, 1-p)$ probability measure. It is easy to see that for the tree process, the variable $(\sigma_v)_{|v| \leq n}$ are independent of $(\sigma_v)_{|v| \geq n+h}$, and therefore the reconstruction problem is not solvable for M. On the other hand, letting*

$$Y_i = \max_{0 \leq j \leq h} \{Z_i = \cdots = Z_{i+j} = 1\}.$$

It is easily seen that Y_i defines a channel M on the space $\{0, \ldots, h\}$. Moreover, it is clear that for all ℓ the variables $\{Z_i\}_{i \geq \ell+h+1}$ and $\{Y_i\}_{i \leq \ell}$ are independent. Therefore $\{Y_i\}_{i \geq \ell+h+1}$ and $\{Y_i\}_{i \leq \ell}$ are independent. It follows that the variables $M^{h+1}(j)$ have the same distribution for all j. Thus $\mathrm{rank}(\mathbf{M}^{h+1}) = 1$, and $\lambda_2(M) = 0$. Writing $\mathbf{M}$:

$$\mathbf{M} = \begin{pmatrix} p & p(1-p) & p(1-p)^2 & \cdots & (1-p)^h \\ 1 & 0 & \cdots & & \cdots & 0 \\ 0 & 1 & 0 & & \cdots & \vdots \\ \cdots & \ddots & & \ddots & & \ddots & \vdots \\ 0 & \cdots & 0 & & p & 1-p \end{pmatrix},$$

if follows from Theorem 6.1 that the reconstruction problem is solvable for M and T_b provided that b is sufficiently large. This is a generalization of a channel appearing in [31]; see also [26].

The example above demonstrates that it may be the case that for the markov chain corresponding to the channel M, the states at times t and $t + h$ are independent, yet, for the tree process, the reconstruction problem is solvable. In fact, a much stronger phenomenon occurs

THEOREM 6.3. *Let $b > 1$ be an integer and T be the 2-level b-ary tree. There exists a channel M such that for any initial distribution, σ_ρ and σ_∂ are independent (where σ_ρ is the root label, and σ_∂ is the configuration at the leaves of the 2-level b-ary tree), yet when B is sufficiently large, the reconstruction problem for the channel M and the infinite B-ary tree T_B is solvable.*

The construction in [30] is motivated by work on secret-sharing protocols [40] and applies Theorem 6.1. We define the channel below. For the proof we refer the reader to [30].

construction. Let $\mathcal{F}$ be a finite field with $q > b + 2$ elements. Let $x_1, \ldots, x_{b+1}$ be a fixed set of non-zero elements of $\mathcal{F}$. We define a channel on the state space

$$\mathcal{F}^b[x] = \{f(x) : f(x) \in \mathcal{F}[x], \deg f \leq b\}.$$

Given f, take I to be a uniform variable in the set $\{1, \ldots, b+1\}$, then take $M(f)$ to be $g \in \mathcal{F}^b[x]$ chosen uniformly among the g's satisfying $g(0) = f(x_I)$. Given the value of f at b of the points $x_1, \ldots, x_{b+1}$, and for all $a \in \mathcal{F}$, there exists a unique polynomial satisfying $f(0) = a$. This implies that b independent copies of the chain at f give no information on f. In [30] it is shown that when the number of copies is sufficiently large, information is retained so that the reconstruction problem is solvable.

6.2. Proving non-solvability. We have a few techniques for proving non-solvability.

Spectral gap of Glauber dynamics. Let $\Lambda(n)$ be the spectral gap of Glauber dynamics for the n-level b-ary tree. In [**2**] we prove a result for general graphs which implies for T_b the following:

THEOREM 6.4. *Suppose that M is a channel such that Glauber dynamics satisfy $\inf_n \Lambda(n) > 0$, then the mutual information between σ_ρ the root variable of T_b, and σ_n, the variables at level n decays exponentially fast: $I(\sigma_\rho, \sigma_n) = O(\exp(-\Omega(n)))$. In particular, the reconstruction problem for the tree T_b is unsolvable.*

PROBLEM 3. *Does it hold for reversible M and the b-ary tree T_b that $I(\sigma_\rho, \sigma_n) = O(\exp(-\Omega(n)))$ if and only if $\inf_n \Lambda(n) > 0$?*

For the Ising model on trees, the answer to the problem is positive, see [**2**].

Recursive analysis of maximum likelihood. A direct approach to the reconstruction problem is to analyze the distribution of the (log) likelihood of the root variable given the boundary variable. This leads to an iteration of random variables. The only case in which this iteration was analyzed is the symmetric binary channel ([**37**]) where this approach yields an exact criterion for reconstruction for general trees. It is an interesting challenge to extended this technique to other channels.

It may be easier to analyze these recursions for "robust" phase transitions which first appeared in [**38**]. Consider the usual reconstruction problem, but suppose that the data at the boundary is given with some additional noise. The proofs that if $b\lambda_2^2(M) > 1$ the reconstruction problem is (census) solvable are immune to this noise. However, this may not be the case for the reconstruction problem. Indeed, we suspect that adding this additional noise (assuming it is fixed but sufficiently strong) will shift the phase transition to the point $b\lambda_2^2(M) = 1$. A similar phenomenon was proven in [**38**] for the phase transition of uniqueness. For n and m, we denote by $\sigma_{n,m}$ the configuration which is obtained from σ_n by applying the random function M^m independently on each of the symbols in σ_n. We denote by $\mathbf{P}^\ell_{n,m}$ the conditional distribution of $\sigma_{n,m}$ given that $\sigma_\rho = \ell$. We then

CONJECTURE 5. *For all M and b, such that $b\lambda_2^2(M) < 1$, there exists m such that for the b-ary tree*

$$\sup_{i,j} \lim_{n \to \infty} D_V(\mathbf{P}^i_{n,m}, \mathbf{P}^j_{n,m}) = 0.$$

7. Terminology and related problems

7.1. Related problems. In this subsection several variants of the reconstruction problem are discussed. Throughout the section we will assume that the variable at the root is chosen uniformly. By Proposition 2.1, reconstruction solvability is equivalent to the fact that there exists $\delta > 0$ such that for all n with *probability at least δ*, the conditional distribution of σ_ρ given σ_n has *total variation distance at least δ* from the uniform distribution.

We may consider the following variants of the problem:

- **non-uniqueness of the Gibbs measure.** The condition here is that for all n there *exists σ_n* such that the distribution of σ_ρ given σ_n has *total variation distance at least $\delta > 0$* from uniform. This is a weaker condition than reconstruction solvability and it was studied in statistical physics for Ising and Potts models. In particular, the phase transition for these models is known, see [15].

- **dismantlable graphs.** Suppose that we require that for all $i \in A$ and for all n there *exists* σ_n such that $\mathbf{P}[\sigma_\rho = i|\sigma_n] = 1$. This requirement clearly fails for all trees if the matrix $\mathbf{M}$ satisfies $\mathbf{M}_{i,j} > 0$ for all i and j. Moreover, this property depends only on which of the entries of $\mathbf{M}$ are non-zero. Define a directed graph G on A such that (i,j) is an edge of G **iff** $\mathbf{M}_{i,j} > 0$. We claim that there exists for sufficiently large b a b-ary tree T for which the requirement holds **iff** for all $i \neq j$ the sets $N(i) = \{\ell : (i,\ell) \in G\}$ and $N(j) = \{\ell : (j,\ell) \in G\}$ satisfy

$$(10) \qquad N(i) \not\subset N(j).$$

  proof. If $N(i) \subset N(j)$ then for all $n \geq 1$ there exists no σ_n such that given σ_n the value of σ_ρ is i with probability 1. On the other hand, if for all $i \neq j$, it holds that $N(i) \not\subset N(j)$, then given i, consider the labeling σ_n of T_b for $b \geq |A|$, which is obtained in the following way: The root satisfies $\sigma_\rho = i$. Given σ_v, label the children of v, denoted $w_1, \ldots, w_b$, in such a way that $\{\sigma_{w_i} : 1 \leq i \leq b\} = \{\ell, (\sigma_v, \ell) \in G\}$. For all n it now holds that given σ_n the value of the root σ_ρ must be i. $\square$

  If (10) holds for all $i \neq j$, then for large b for $\delta > 0$ fraction of the σ_n, it holds that $\mathbf{P}[\sigma_\rho = i|\sigma_n] = 1$ for some i which depends on σ_n. This may be proved using a recursive argument similar to [30, Theorem 2.1].

  We may replace the above requirement by the requirement that there exists $i \in A$ such that for all n there exists σ_n such that given σ_n the variable σ_ρ satisfies $\sigma_\rho \neq i$ (with probability 1). Defining G as in the previous case, the property holds for T_b for sufficiently large b **iff** G is *dismantlable*, see [4].

- **census.** One may ask similar questions about the census.
  - Does there exists a b, such that for the tree T_b, for all n there *exists a census c_n such that the distribution of σ_ρ given c_n has total variation distance at least δ from uniform?* For some models (like ferromagnetic Ising and Potts models) this condition is equivalent to uniqueness of the Gibbs measure. For others (like colorings) it seems that these conditions are not equivalent (we do not know how to demonstrate it for colorings; Example 6.2 is an example of such model).
  - Does there exists a b, such that for the tree T_b and all n there exists a census c_n such that $\mathbf{P}[\sigma_\rho \neq i|c_n] = 1$ for some i? It follows from [30, Lemma 6.2] that such a b does not exist when M is ergodic.

- **phylogeny.** Most of the biological research which is related to reconstruction is devoted to problems in which the underlying tree is unknown and the algorithm is supposed to find both the tree and the variables at the nodes of the tree, see e.g. [7, 13, 15]. These problems seems to be quite hard to analyze; in particular, in some cases there is no well-defined probability space of trees. In a recent work [33, 34] we show that the reconstruction of phylogenetic trees is closely related to the reconstruction problem.

- **Noisy computation.** Von Neumann [43] proposed a model of computation in noisy circuits where each gate computes correctly with probability

$1 - \epsilon$, The analysis of this model in [**43**, **36**, **10**, **11**] has many similarities to the analysis of the reconstruction problem for the symmetric binary channel, in [**6**, **18**, **19**, **12**, **31**]. However, we do not know of any formal relationship between the two models.

7.2. Dictionary. As the reconstruction problem has been studied from different perspectives, different terminology is often used for the same entities. We list below some terms and their translations.

variable. also symbol, state (finite markov chains), label, letter, message (information theory), spin (statistical physics), color (combinatorics), genotype (biology), phenotype (biology).

b-ary trees. also $b + 1$ regular trees (corresponding to the degree as a graph), and Bethe lattice (statistical physics).

channel. the channel (random function) M corresponds to a stochastic matrix $\mathbf{M}$ such that $\mathbf{M}_{i,j} = \mathbf{P}[M(i) = j]$. In the statistical physics literature, when working with Ising and Potts models, it is common to work with the *Hamiltonian*:

$$(11) \qquad H((\sigma_v)_{v \in T}) = \sum_v h_{\sigma_v} + \beta \sum_{(v,w) \text{ edge of } T} \delta_{\{\sigma_v = \sigma_w\}}.$$

The probability of a configuration $\{\sigma_v\}_{v \in T}$ is then

$$(12) \qquad \frac{1}{Z} \exp\left(H((\sigma_v)_{v \in T})\right),$$

where Z is a normalizing constant, known as the *partition function* (it is a function of H and of the tree T). The parameter $1/\beta$ is often referred to as the temperature.

From (11) the matrix $\mathbf{M}$ is given by:

$$\mathbf{M}_{i,j} = \frac{\exp\left(h_j + \beta \delta_{\{i=j\}}\right)}{\sum_\ell \exp\left(h_\ell + \beta \delta_{\{i=\ell\}}\right)}$$

Some families of interest are: The Ising model with no external field, when $|\mathcal{A}| = 2$ and $h_1 = h_2 = 0$; The Ising model with external field, where $|\mathcal{A}| = 2$; and Potts models without external field where $|\mathcal{A}| = q$ and all the h values are 0.

The process on the infinite tree then corresponds to the *Gibbs* measure on that tree with the specification that the root distribution is uniform.

In biology, the matrix $\mathbf{M}$, is related to the *mutation rate* of the process, and it is usually assumed that $\mathbf{M}$ is a perturbation of the identity matrix.

In combinatorics the popular model is proper colorings of the tree, where $\mathbf{M}$ is a $q \times q$ matrix satisfying $\mathbf{M}_{i,j} = \delta_{\{i \neq j\}}(q-1)^{-1}$. This model is also referred to as the zero temperature anti-ferromagnetic Potts model, since $\mathbf{M}$ is obtained as the limit of the corresponding $\mathbf{M}$ for Potts models when $\beta \to -\infty$.

reconstruction solvability. Corresponds in statistical physics to extremality of the above measure. In information theory, it is natural to express reconstruction solvability in term of decay of mutual information (see Proposition 2.1).

8. Very recent results

Since the submitting this survey, a number of new results on reconstruction appeared. In [28] better bounds for the reconstruction problem for Potts and asymmetric binary channels were obtained. Essentially the same bounds for asymmetric binary channels were obtained independently in [27]. [28] provides a comprehensive analysis of the mixing rates of Glauber dynamics for Ising and Potts models under various boundary conditions and gives a positive answer to Problem 3.

In [20] we analyze robust solvability. This may be thought of as reconstruction where the labels at the bottom level are further perturbed. We show that the threshold for robust reconstruction is given by $b|\lambda_2(M)|^2 = 1$ as conjectured in [30].

Finally, we note that the crucial role of the reconstruction problem in Phylogeny was recently demonstrated in [34] and [35] extending the results of [33].

Acknowledgments: I learned about the reconstruction problem from Yuval Peres. I want to thank him for many fruitful conversations about the problem. I thank Olle Häggström, Claire Kenyon, Làszlò Lovàsz, Jeff Steif and Peter Winkler for helpful discussions, and the referee for many helpful comments.

References

[1] Athreya, K. B. and Ney, P. E. (1972) *Branching Processes*, Springer-Verlag.

[2] N. Berger, C. Kenyon, E. Mossel, and Y. Peres. Glauber dynamics on trees and hyperbolic graphs. Submitted. Extended abstract by Kenyon, Mossel and Peres appeared in [22], 2003.

[3] Brightwell, G. and Winkler, P. (2001). Random colorings of a Cayley tree, Contemporary Combinatorics, B. Bollobas, ed., Bolyai Society Mathematical Studies.

[4] Brightwell, G. and Winkler, P. (2000). Gibbs measures and dismantlable graphs, *J. Comb. Theory (Series B)* **78**, 141–169.

[5] Brightwell, G. and Winkler, P. (1999). Graph homomorphisms and phase transitions, *J. Comb. Theory (Series B)*, **77**, 415–435.

[6] Bleher, P. M., Ruiz, J. and Zagrebnov V. A. (1995) On the purity of limiting Gibbs state for the Ising model on the Bethe lattice, *J. Stat. Phys* **79**, 473–482.

[7] Cavender, J. (1978). Taxonomy with confidence. *Math. BioSci.* **40**, 271–280.

[8] Cover, T. M. and Thomas, J. A. (1991) *Elements of Information Theory*, John Wiley and Sons.

[9] Dembo, A. and Zeitouni O. (1997) *Large Deviations, Techniques and Applications*, Springer.

[10] Evans, W.(1994). *Information Theory and Noisy Computation*. PhD thesis, Dept. of Computer Science, University of California at Berkeley.

[11] Evans, W. and Schulman, L. J. (1993). Signal propagation, with application to a lower bound on the depth of noisy formulas. In *Proceedings of the 34th Annual Symposium on Foundations of Computer Science*, 594–603.

[12] Evans, W., Kenyon, C., Peres, Y. and Schulman L. J. (2000) Broadcasting on trees and the Ising Model, *Ann. Appl. Prob.*, **10 no. 2**, 410–433.

[13] Fitch, W. M. (1971). Toward defining the course of evolution: minimum change for a specific tree topology. *Syst. Zool.* **20**, 406–416.

[14] H. O. Georgii (1988). *Gibbs measures and phase transitions.* de Gruyter Studies in Mathematics, 9.

[15] Häggström, O. (1996). The random-cluster model on a homogeneous tree. *Probab. Theory Related Fields* **104 no. 2**, 231–253.

[16] Hajek, B. and Weller, T. (1991). On the maximum tolerable noise for reliable computation by formulas. *IEEE Trans. on Information Theory* **37**(2), 388–391.

[17] Higuchi, Y. (1977). Remarks on the limiting Gibbs state on a (d+1)-tree. *Publ. RIMS Kyoto Univ.* **13**, 335–348.

[18] Ioffe, D. (1996a). A note on the extremality of the disordered state for the Ising model on the Bethe lattice. *Lett. Math. Phys.* **37**, 137–143.

[19] Ioffe, D. (1996b). A note on the extremality of the disordered state for the Ising model on the Bethe lattice. In *Trees*, B. Chauvin, S. Cohen, A. Roualt (Editor).

[20] Janson, S. and Mossel, E. (2003). Robust reconstruction on trees is determined by the second eigenvalue, to appear in *Ann. Probab.*

[21] Kenyon, C., Mossel, E. and Peres, Y. (2001). Glauber dynamics on trees and hyperbolic graphs, *Preprint*.

[22] C. Kenyon, E. Mossel, and Y. Peres. Glauber dynamics on trees and hyperbolic graphs. In *Proceedings of the Forty-Second Annual Symposium on on Foundations of Computer Science*, pages 568–578, 2001.

[23] Kesten, H. and Stigum, B. P. (1966) Additional limit theorem for indecomposable multidimensional Galton-Watson processes, *Ann. Math. Statist.* **37**, 1463–1481.

[24] Lyons, R. (1990) Random walks and percolation on trees. *Ann. Probab.* **18**, 931–958.

[25] Lyons, R. and Pemantle R. (1992) Random walk in a random environment and first-passage percolation on trees. *Ann. Probab.* **20,1** 125–136.

[26] Lovàsz, L. and Winkler P. (1998) Mixing times, in *DIMACS Series in Discrete Mathematics and Theoretical Computer Science* **41**, 85–133.

[27] Martin, J. (2003) Reconstruction thresholds on regular trees. In C. Banderier and C. Kratten-thaler, editors, *Discrete Random Walks*, Discrete Math. Theoret. Comput. Sci., pages 191–204. 2003. Availible at `http://dmtcs.loria.fr/proceedings/dmACind.html`.

[28] Martinelli, F., Sinclair, A. and Weitz, D. (2003). The ising model on trees: Boundary conditions and mixing time. Submitted to communication in mathematical physics. Extended abstarct appeared in [**29**].

[29] Martinelli, F., Sinclair, A. and Weitz, D. (2003). The ising model on trees: Boundary conditions and mixing time. In *Proceedings of the Forty Fourth Annual Symposium on Foundations of Computer Science*, pages 628–639.

[30] Mossel, E. and Peres, Y (2003) Information flow on trees. *Ann. Appl. Probab.*, 13(3):817–844.

[31] Mossel, E. (1998) Recursive reconstruction on periodic trees, *Random Structures and algorithms* **13,1** 81–97

[32] Mossel, E. (2001) Reconstruction on trees: Beating the second eigenvalue, *Ann. Appl. Probab.*, **11**, 285–300.

[33] Mossel, E (2003). Phase transitions in phylogeny. To appear in Trans. of AMS.

[34] Mossel, E (2003). On the impossibility of reconstructing ancestral data and phylogenies. *Jour. Comput. Bio.*, 10(5):669–678.

[35] Mossel, E. and Steel, M. (2003). A phase transition for a random cluster model on phylogenetic trees. Submitted to Mathematical Biosciences.

[36] Pippenger, N. (1988). Reliable computation by formulas in the presence of noise. *IEEE Transactions on Information Theory* **34**, 194–197.

[37] R. Pemantle and Y. Peres (1995). Recursions on trees and the Ising model at critical temperatures. *Unpublished manuscript.*

[38] Pemantle, R. and Steif, J. E. (1999). Robust phase transitions for Heisenberg and other models on general trees. *Ann. Probab.* **27 no. 2**, 876–912

[39] M. Steel and M. Charleston (1995). Five surprising properties of parsimoniously colored trees. *Bull. Math. Biology* **57**, 367–375.

[40] Shamir, A. (1979). How to share a secret? Communications of the ACM **22** , 612–613.

[41] Spitzer, F. (1975). Markov random fields on an infinite tree. *Ann. Probab.* **3**, 387–394.

[42] Steel, M. (1989). Distributions in bicolored evolutionary trees. *Ph.D. Thesis*, Massey University, Palmerston North, New Zealand.

[43] von Neumann, J. (1956). Probabilistic logics and the synthesis of reliable organisms from unreliable components. In *Automata Studies*, C. E. Shannon and J. McCarthy (Editors), 43–98. Princeton University Press, New Jersey.

STATISTICS, UNIVERSITY OF CALIFORNIA, BERKELEY, CA 94720
E-mail address: `mossel@stat.berkeley.edu`

DIMACS Series in Discrete Mathematics
and Theoretical Computer Science
Volume **63**, 2004

Chromatic numbers of products of tournaments: Fractional aspects of Hedetniemi's conjecture

Claude Tardif

ABSTRACT. The chromatic number of the categorical product of two n-tournaments can be strictly smaller than n. We show that
$$\min\{\chi(S \times T) : S \text{ and } T \text{ are } n\text{-tournaments}\}$$
is asymptotically equal to λn, where $\frac{1}{2} \leq \lambda \leq \frac{2}{3}$.

1. Introduction

A *tournament* is an orientation of a complete graph. We will write $u \to v$ to indicate the presence of an arc from u to v in a directed graph. The *categorical product* $G \times H$ of two directed graphs G and H is the directed graph with vertex set $V(G) \times V(H)$ and arcs $(u_1, u_2) \to (v_1, v_2)$ for all $u_1 \to v_1$ in G and $u_2 \to v_2$ in H.

A *n-colouring* of a directed graph G is a map $\phi : V(G) \mapsto \{0, 1, \ldots, n-1\}$ such that $u \to v$ implies $\phi(u) \neq \phi(v)$. The *chromatic number* $\chi(G)$ of G is the least n such that G admits a n-colouring. Note that colourings and chromatic number are independent of the orientation of the edges of a graph. However the orientation of the factors do affect the structure of a product, and its chromatic number. In this paper we will investigate bounds on chromatic numbers of products of directed graphs in general, and of tournaments in particular.

For any two directed graphs G and H, we have
$$\chi(G \times H) \leq \min\{\chi(G), \chi(H)\},$$
since a colouring of $G \times H$ can be derived from a colouring of either factor. As pointed out in [**4**], the inequality can be strict, even in the case of tournaments. We consider two functions $g, t : \mathbb{N} \mapsto \mathbb{N}$ defined as follows:
$$g(n) = \min\{\chi(G \times H) : \chi(G) = \chi(H) = n\},$$
$$t(n) = \min\{\chi(G \times H) : G \text{ and } H \text{ are } n\text{-tournaments}\}.$$

The function g was introduced by Poljak and Rödl [**4**], who showed that $g(n) = n$ for $n \leq 3$ and $g(n) < n$ for $n \geq 4$. Little more is known about the behaviour of g. Clearly, g is nondecreasing, and it is generally thought that g should grow without bounds, but a proof of this seems to be out of reach at the moment. Ironically,

2000 *Mathematics Subject Classification.* 05C15.
Key words and phrases. Hedetniemi's conjecture, chromatic number, tournaments.

if g turns out to be bounded, then it is completely characterised by the following result:

THEOREM 1.1 ([**3, 8**]). *Either* $g(n) = \min\{3, n\}$ *for all* n, *or* $\lim_{n\to\infty} g(n) = \infty$.

By restricting the factors to tournaments instead of general n-chromatic directed graphs, we define the function t such that $g \leq t$. In this paper we prove that t is essentially linear:

THEOREM 1.2. *The sequence* $(t(n)/n)_{n\geq 1}$ *converges to a limit* λ, *where*

$$\frac{1}{2} \leq \lambda \leq \frac{2}{3}.$$

The proof of this result is the object of the next two sections. We will then show how this result is related to fractional approaches to Hedetniemi's conjecture on the chromatic number of the product of two undirected graphs.

2. Lexicographic products of tournaments

The *lexicographic product* $G[H]$ of two directed graphs G and H is the directed graph with vertex set $V(G[H]) = V(G) \times V(H)$, and arcs $(u, v) \to (u', v')$ for all $u \to u'$ in G, and $(u, v) \to (u, v')$ for all $v \to v'$ in H. Thus, $G[H]$ is obtained from G by replacing every vertex by a copy of H. Note that the lexicographic product of two tournaments is again a tournament. Let T_n denote the *transitive tournament* on n vertices, defined by $V(T_n) = \{0, 1, \ldots, n-1\}$ and $i \to j$ if $i < j$.

LEMMA 2.1. *Let* G, H *be directed graphs. Then*

$$\chi(G[T_n] \times H[T_n]) \leq n\chi(G \times H).$$

Proof. Since $V(G \times H)$ can be covered with $\chi(G \times H)$ independent sets, it suffices to show that for any independent set I of $G \times H$, the subgraph X of $G \times H$ induced by

$$V(X) = \{((u, i), (v, j)) \in V(G[T_n] \times H[T_n]) : (u, v) \in I\}$$

is n-colourable. We put

$$J \;=\; \{(u, v) \in I : \text{there exists } (u', v) \in I \text{ such that } u' \to u\},$$
$$J' \;=\; \{(u, v) \in I : \text{there exists } (u, v') \in I \text{ such that } v' \to v\},$$

and we split I into the four sets $I_{\max} = J \cap J'$, $I_{\mathrm{pr}_1} = J' \setminus J$, $I_{\mathrm{pr}_2} = J \setminus J'$ and $I_{\min} = I \setminus (J \cup J')$. Now we define $\phi : X \mapsto \{0, 1, \ldots, n-1\}$ by

$$\phi((u, i), (v, j)) = \begin{cases} \max\{i, j\} & \text{if } (u, v) \in I_{\max}, \\ i & \text{if } (u, v) \in I_{\mathrm{pr}_1}, \\ j & \text{if } (u, v) \in I_{\mathrm{pr}_2}, \\ \min\{i, j\} & \text{if } (u, v) \in I_{\min}. \end{cases}$$

We show that ϕ is a n-colouring of X. For $((u', i'), (v', j')) \to ((u, i), (v, j))$ in X, we have $(u', i') \to (u, i)$ in $G[T_n]$ and $(v', j') \to (v, j)$ in $H[T_n]$. If $u' \to u$, then we must have $v' = v$ because I is an independent set and contains (u', v') and (u, v). This implies $j' < j \leq \phi((u, i), (v, j))$. We must then have $\phi((u', i'), (v', j')) \leq j'$, for otherwise I would contain a vertex (u', v'') such that $v'' \to v' = v$ whence $(u', v'') \to (u, v)$, a contradiction to the independence of I. Therefore $u' \to u$ implies

$$\phi((u', i'), (v', j')) \leq j' < j \leq \phi((u, i), (v, j)).$$

Similarly, $v' \to v$ implies

$$\phi((u',i'),(v',j')) \leq i' < i \leq \phi((u,i),(v,j)).$$

The remaining possibility is $u' = u, v' = v$. We then have $i' < i$ and $j' < j$. Since the functions max, min and the projections pr_1, pr_2 are all proper n-colourings of $T_n \times T_n$, we again have $\phi((u',i'),(v',j')) \neq \phi((u,i),(v,j))$. $\square$

COROLLARY 2.2. $(t(n)/n)_{n\geq 1}$ *converges to* $\lambda = \inf\{t(n)/n : n \geq 1\}$. $\square$

Poljak and Rödl [**4**] provide examples of 4-tournaments S and T such that $\chi(S \times T) = 3$; therefore $\lambda \leq \frac{3}{4}$. We do not know any n-tournaments S and T such that $\chi(S \times T) \leq \frac{2}{3}n$. In the next section the bound $\lambda \leq \frac{2}{3}$ is achieved asymptotically.

3. Bounds on λ

Let S_{2n+1} be the tournament on the vertex set $V(S_{2n+1}) = \{0, 1, \ldots, 2n\}$ with arcs $i \to (i+k)$ for all $i \in V(S_{2n+1})$ and $k \in \{1, \ldots, n\}$ (where addition is carried modulo k). Recall that T_{2n+1} is the transitive tournament on $\{0, 1, \ldots, 2n\}$.

LEMMA 3.1. $\chi(T_{2n+1} \times S_{2n+1}) \leq 2\left\lceil \frac{2n+2}{3} \right\rceil$.

Sketch of proof. For $n = 5$ we have $2n + 1 = 11$ and $2\left\lceil \frac{2n+2}{3} \right\rceil = 8$. A typical 8-colouring of $T_{11} \times S_{11}$ is depicted below (arcs not shown). Colour class 1 (in bold) is a "staircase" of height $n+1$ ($=6$), where the normal step has width 3. The other colour classes are successively layered below in a similar fashion.

$$T_{11} \times S_{11}$$

S_{11}											
10	**1**	**1**	**1**	8	7	7	7	7	6	6	6
9	2	2	**1**	**1**	**1**	8	8	7	7	7	6
8	3	2	2	2	**1**	**1**	**1**	8	8	7	7
7	3	3	3	2	2	2	**1**	**1**	**1**	8	8
6	4	4	3	3	3	2	2	2	**1**	**1**	**1**
5	5	4	4	4	3	3	3	2	2	2	**1**
4	5	5	5	4	4	4	3	3	3	2	2
3	6	6	5	5	5	4	4	4	3	3	3
2	7	6	6	6	5	5	5	4	4	4	4
1	7	7	7	6	6	6	5	5	5	5	4
0	8	8	7	7	7	6	6	6	6	5	5
	0	1	2	3	4	5	6	7	8	9	10

$$T_{11}$$

It is an elementary (though lengthy) exercise to show that similar "staircase colourings" achieve the specified bound for all values of n. $\square$

COROLLARY 3.2. $\lim\limits_{n\to\infty} t(n)/n \leq \dfrac{2}{3}$. $\square$

The next result provides an upper bound on the independence number of a product of a tournaments, from which the lower bound on its chromatic number is easily derived.

LEMMA 3.3. *Let S, T be tournaments. Then* $\alpha(S \times T) \leq |V(S)| + |V(T)| - 1$.

Proof. Let I an independent set of $S \times T$. We consider the bipartite graph B whose vertex set is the disjoint union of $V(S)$ and $V(T)$ whose edges are the pairs $\{s, t\}$ such that $(s, t) \in I$. It is easily seen that for any path $s_1, t_1, s_2, t_2, \ldots, s_n, t_n$ of B, the sets $\{s_1, \ldots, s_n\}$ and $\{t_1, \ldots, t_n\}$ induce transitive tournaments in S and T respectively, with one labelled in increasing order and the other one in decreasing order. ¿From this follows that B cannot contain any cycles, whence B is a forest and $|I| \leq |V(S)| + |V(T)| - 1$. $\square$

By Lemma 3.3, we have $t(n) > n/2$. With Corollaries 2.2 and 3.2, this completes the proof of Theorem 1.2. We will see in the next section that the lower bound derived from Lemma 3.3 is just a special case of a much more general result, which motivated the present work.

4. Fractional aspects of Hedetniemi's conjecture

Let $\mathcal{I}(G)$ denote the family of all independent sets of a directed or undirected graph G. A function $\mu : \mathcal{I}(G) \mapsto [0, 1]$ is called a *fractional colouring* of G if we have $\sum_{u \in I} \mu(I) \geq 1$ for all $u \in V(G)$. The value $\sum_{I \in \mathcal{I}(G)} \mu(I)$ is called the *weight* of μ. The *fractional chromatic number* $\chi_f(G)$ of G is the minimum weight of a fractional colouring of G. In [**6**], Scheinermann and Ullman show how similar linear relaxations of integer-valued graph parameters lead to new insight into old graph-theoretic problems.

One such problem involves a formula for the chromatic number of the categorical product of undirected graphs. According to Hedetniemi's conjecture, the equality

$$(1) \qquad \chi(G \times H) = \min\{\chi(G), \chi(H)\}$$

should hold for all undirected graphs G and H. By now the conjecture has remained open for more than thirty years, in spite of the efforts of many people (see [**5**, **8**] for surveys).

Now for any graph X, we have $\chi_f(X) \leq \chi(X)$. Thus the equality (1) implies

$$(2) \qquad \chi(G \times H) \geq \min\{\chi_f(G), \chi_f(H)\}.$$

This is a weaker form of Hedetniemi's conjecture, and it is yet unproved. However it appears to be more tractable than the original conjecture. In particular, the following bound is given in [**7**]:

$$(3) \qquad \chi(G \times H) \geq \frac{1}{2} \min\{\chi_f(G), \chi_f(H)\}.$$

Now, as mentioned in [**7**], the inequality (3) is valid in the case of directed graphs as well, though neither of (1) and (2) hold in that case. Therefore it would be interesting to know to which extent can (3) be improved in the case of undirected graphs:

PROBLEM 4.1. *Does there exist a constant $c > \frac{1}{2}$ such that the bound*

$$(4) \qquad \chi(G \times H) \geq c \min\{\chi_f(G), \chi_f(H)\}$$

holds for all undirected graphs G and H?

The results of the present paper indicate that the bound given by (3) is probably best possible in the case of directed graphs, given that the class of potential counterexamples to (4) is much larger than the class of tournaments inspected here.

Therefore a positive answer to Problem 4.1 would probably allow to separate the case of undirected graphs from that of directed graphs.

In the long run, it would be desirable to eliminate the recourse to fractional parameters altogether. For instance, the following problem is still open:

PROBLEM 4.2. *Suppose that $\chi(G) \geq 1000$ and $\chi(G \times H) = 4$. Then does the inequality $\chi(H) \leq 1000$ necessarily hold?*

In fact, even the weaker assertion $\alpha(H) \geq |V(H)|/1000$ is not yet proved. For the moment, the best construction of independent sets in H is the basis of the inequality (3); so an improvement of that bound could yield better constructions of independent sets.

Acknowledgements The idea to use some graph product to obtain an upper bound on $t(n)/n$ is essentially due to Béla Bollobás and was communicated to the author during the DIMACS-DIMATIA Workshop on Graphs, Morphisms and Statistical Physics (DIMACS Center, Rutgers University, March 19-21 2001). The 5-colouring of $S_7 \times T_7$ was originally found using Michael Trick's program in the Stony Brook Algorithm Repository
(http://www.cs.sunysb.edu/ algorith/)

References

[1] M.El-Zahar, N.Sauer, The chromatic number of the product of two 4-chromatic graphs is 4, *Combinatorica* **5** (1985), 121–126.

[2] S.H.Hedetniemi, Homomorphisms of graphs and automata, University of Michigan Technical Report 03105-44-T, 1966.

[3] S.Poljak, Coloring digraphs by iterated antichains, *Comment. Math. Univ. Carolin.* **32** (1991), 209–212.

[4] S.Poljak, V.Rödl, On the arc-chromatic number of a digraph, *J. Combin. Theory Ser. B* **31** (1981), 339–350.

[5] N.Sauer, Hedetniemi's conjecture—a survey, *Discrete Math.* **229** (2001), 261–292.

[6] E.R.Scheinerman, D.H.Ullman, Fractional Graph Theory, Wiley-Interscience Series in Discrete Mathematics and Optimization, John Wiley & Sons, New York, 1997, xviii+211 pp.

[7] C.Tardif, The chromatic number of the product of two graphs is at least half the minimum of the fractional chromatic numbers of the factors, *Comment. Math. Univ. Carolin.* **42** (2001), 353–355.

[8] X.Zhu, A survey on Hedetniemi's conjecture, *Taiwanese J. Math.* **2** (1998), 1–24.

DEPARTMENT OF MATHEMATICS AND COMPUTER SCIENCE, ROYAL MILITARY COLLEGE OF CANADA

E-mail address: `tardif@math.uregina.ca`

DIMACS Series in Discrete Mathematics
and Theoretical Computer Science
Volume **63**, 2004

Perfect graphs for generalized colouring – circular perfect graphs

Xuding Zhu

ABSTRACT. Suppose $\mathcal{F}$ is a family of graphs linearly ordered by homomorphisms. A graph G is perfect with respect to $\mathcal{F}$ if for every induced subgraph H of G, there exists a graph $K \in \mathcal{F}$ such that H and K are homomorphically equivalent. If $\mathcal{F} = \{K_1, K_2, K_3, \cdots\}$ consists of the complete graphs, then perfect graphs with respect to $\mathcal{F}$ are the perfect graphs defined by Berge. In this paper, we consider circular perfect graphs, which are perfect graphs with respect to the graph family $\mathcal{F} = \{K_{k/d} : k \geq d \geq 1\}$. We present a sufficient condition for a triangle free graph to be circular perfect. This sufficient condition is used in a companion paper [**10**] to prove an analogue of Hajós theorem for the circular chromatic number and for graphs with $2 < \chi_c(G) < 3$.

1. Introduction

Suppose G and H are graphs. A *homomorphism* from G to H is a mapping f from $V(G)$ to $V(H)$ such that $f(x)f(y) \in E(H)$ whenever $xy \in E(G)$. It is easy to see that a vertex colouring of a graph G with n-colours is equivalent to a homomorphism from G to K_n. Thus homomorphism can be viewed as a generalization of graph colouring. We write $G \preceq H$ if there exists a homomorphism from G to H. Then "$\preceq$" defines a partial order on the set of graphs. Two graphs G and G' are *homomorphically equivalent*, written as $G \leftrightarrow G'$, if $G \preceq G'$ and $G' \preceq G$.

Suppose $\mathcal{F}$ is a family of graphs linearly ordered by homomorphisms, i.e., under order relation "$\preceq$". A graph G is *perfect with respect to* $\mathcal{F}$ if for every induced subgraph H of G, there exists a graph $K \in \mathcal{F}$ such that $H \leftrightarrow K$. The set of complete graphs $\{K_1, K_2, K_3, \cdots\}$ forms an infinite increasing chain under "$\preceq$". The chromatic number $\chi(G)$ of a graph G is equal to the least k such that $G \preceq K_k$. The clique number $\omega(G)$ of G is the maximum k such that $K_k \preceq G$. Denote by $\mathcal{Z}_G$ the set of complete graphs. Then a graph G is perfect with respect to $\mathcal{Z}_G$ if and only if every induced subgraph H of G has $\chi(H) = \omega(H)$. Therefore a graph G is perfect with respect to $\mathcal{Z}_G$ if and only if G is perfect (in the usual sense).

We may view $\mathcal{Z}_G$ as a scale that measures a dimension of graphs, where each complete graph corresponds to a natural number. Just as the set of natural numbers can be extended to rational numbers, we can extend the set of complete graphs to

---

This research was partially supported by the National Science Council under grant NSC 89-2115-M-110-003.

a larger set of graphs. For a fraction k/d with $(k, d) = 1$ and $k \geq 2d$, let $K_{k/d}$ be the graph with vertex set $\{0, 1, \cdots, k-1\}$ in which ij is an edge if and only if $d \leq |i - j| \leq k - d$. Denote by $\mathcal{Q}_G$ the set $\{K_{k/d} : (k, d) = 1$ and $k \geq 2d\} \cup \{K_1\}$. Note that $K_{k/1} = K_k$, and hence $\mathcal{Q}_G$ is indeed an extension of $\mathcal{Z}_G$. Moreover, the set $\mathcal{Q}_G$ is also linearly ordered. It was shown in [**2, 4**] that for any two fractions k'/d', $k/d \geq 2$, $k'/d' \leq k/d$ if and only if $K_{k'/d'} \preceq K_{k/d}$. Thus the set $\mathcal{Q}_G$ together with the order $\preceq$ may be viewed as a representation of those rationals $r \geq 2$ or $r = 1$.

The *circular chromatic number* $\chi_c(G)$ of a graph G is defined as the infimum of those k/d for which G admits a homomorphism to $K_{k/d}$. It is known [**8**] that "infimum" in the definition is always attained and hence can be replaced by "minimum". Since $\mathcal{Q}_G$ is an extension of $\mathcal{Z}_G$ and $\mathcal{Q}_G$ is also linearly ordered, it follows that for any graph G, we have

$$\chi(G) - 1 < \chi_c(G) \leq \chi(G).$$

Therefore $\chi_c(G)$ is a refinement of $\chi(G)$ and it contains more information about the structure of G than $\chi(G)$ does. The parameter of circular chromatic number is a natural generalization of the concept of chromatic number from many different points of view, and has been studied extensively in the past decade. Readers are referred to [**8**] for a comprehensive survey on this subject.

The *circular clique number* of a graph G, introduced in [**11**], is defined as the supremum of those k/d for which $K_{k/d}$ admits a homomorphism to G. It was proved in [**11**] that the circular clique number of a finite graph G is equal to the maximum of those k/d for which $K_{k/d}$ is an induced subgraph of G.

DEFINITION 1.1. *A graph G is called* circular perfect *if G is perfect with respect to $\mathcal{Q}_G$.*

In other words, a graph G is circular perfect if for every induced subgraph H of G we have $\chi_c(H) = \omega_c(H)$. In [**11**], some basic properties of the circular clique number of a graph were proved. Some necessary conditions for a graph to be circular perfect were also given in [**11**]. It follows from the definition that every perfect graph is circular perfect. However, circular perfect graphs could be non-perfect. In particular, it was proved in [**11**] that for any $k \geq 2d$, the graph $K_{k/d}$ is circular perfect. If k/d is not an integer, then certainly $K_{k/d}$ is not perfect. The main result of [**11**] is the following sufficient condition for a graph to be circular perfect.

THEOREM 1.1. [**11**] *If for each vertex x of G, $N_G[x]$ induces a perfect graph and $G - N_G[x]$ induces a bipartite graph which contains no induced P_4, then G is a circular perfect graph.*

This result was shown to be relatively tight [**11**] and useful [**9**]. However, for graphs G with $2 < \chi_c(G) < 3$, its application is very much restricted. Indeed, for $k/d < 3$, the fact that $K_{k/d}$ is a circular perfect graph does not follow from Theorem 1.1.

This paper presents a sufficient condition for a triangle free graph to be circular perfect. We shall prove the following theorem:

THEOREM 1.2. *Suppose G is a triangle free graph such that for every vertex x of G, $G - N[x]$ is a bipartite graph with no induced C_n for $n \geq 6$, and any induced path of $G - N[x]$ is well-linked. Then G is circular perfect.*

The definition of well-linked paths is a little bit technical, and we defer it to Section 2. We remark that the graphs $K_{k/d}$ for $k/d < 3$ satisfy the condition of Theorem 1.2, and hence the fact that $K_{k/d}$ is circular perfect (which was proved in [11]) follows from this theorem.

Theorem 1.1 was used in [9] to prove an analogue of Hajós theorem for $k/d \geq 3$. In [10], we shall use Theorem 1.2 to prove an analogue of Hajós Theorem for $2 < k/d < 3$. Namely, we shall design a few graph operations and prove that for any $2 < k/d < 3$, starting from the graph $K_{k/d}$, one can construct all graphs of circular chromatic number at least k/d by repeatedly applying these graph operations. We remark that general Hajós type theorem for homomorphisms was given in [3], where some graph operations are designed to construct all graphs that do not admit a homomorphism to H. The result of [3] applies to $H = K_{k/d}$ as well. However, there are some genuine differences between the Hajós Theorem of [9, 10] and the Hajós type theorem of [3]. First of all, when applied to $H = K_{k/d}$, the operations in [3] constuct all graphs G with $\chi_c(G) > k/d$, where the operations in [9, 10] constuct all graphs G with $\chi_c(G) \geq k/d$. Secondly, all constructions need some graphs to start with. In [3], a finite set of graphs is needed to start the construction, but the finite set is not explicitly given, and the set is usually very large. In [9, 10], one starts with a single graph $K_{k/d}$.

2. Some notation

The remainder of this paper is devoted to the proof of Theorem 1.2. First we introduce some notation.

We write $x \sim_G y$ to mean that x is adjacent to y in G. If the graph G is clear from the context, we write $x \sim y$ instead of $x \sim_G y$. For a vertex x of G, $N_G(x) = \{y \in V(G) : x \sim y\}$ denotes the set of neighbours of x, and $N_G[x] = N_G(x) \cup \{x\}$. We also use $N_G(x)$ and $N_G[x]$ to denote the subgraphs induced by these sets. When the graph G is clear from the context, we write $N(x)$ and $N[x]$ for $N_G(x)$ and $N_G[x]$. For a subset X of V, let $N(X) = \cup_{x \in X} N(x)$. We denote by $G - N[x]$ the subgraph of G induced by $V(G) - N[x]$.

For two sets A, B, we write $A \subseteq B$ to mean that A is a subset of B, and write $A \subset B$ to mean that A is a proper subset of B.

Given a graph G and a proper subgraph H of G. A homomorphism from G to H which fixes every vertex of H is called a *retraction*. If there is a retraction from G to H, then we say G *retracts* to H. A graph G is a *core* if G does not retract to any of its proper subgraphs (or equivalently, G admits no homomorphism to any of its proper subgraphs). A core of a graph G is a subgraph H of G such that H is a core and G retracts to H. It is well-known that each finite graph has a unique core (up to isomorphism). If H is the core of G then G and H have the same circular chromatic number and the same circular clique number. Thus for the calculation of circular chromatic number and circular clique number, we need only consider core graphs. It is easy to see that if G is a core graph, and x, y are two vertices x, y of G, then none of $N(x)$ and $N(y)$ is a subset of the other. For two vertices $u, v \in V - N[x]$, we frequently need to compare the neighbourhood of u, v in $N[x]$. We shall use the following notation:

- $u \leq^x v$ means $N(u) \cap N(x) \subseteq N(v) \cap N(x)$;

- $u <^x v$ means $N(u) \cap N(x) \subset N(v) \cap N(x)$;

- $u =^x v$ means $N(u) \cap N(x) = N(v) \cap N(x)$.

DEFINITION 2.1. *Given an induced path $P_n = (p_0, p_1, \cdots, p_n)$ of $G - N[x]$, we say P_n is* badly-linked *with respect to x if one of the following holds:*

   (1) *There are three indices $i < j < k$ of the same parity such that $p_i \not\leq^x p_j$ and $p_k \not\leq^x p_j$.*

   (2) *There are three indices $i < j < k$ of the same parity such that $p_j \not\leq^x p_i$ and $p_j \not\leq^x p_k$.*

   (3) *There are two even indices $i < j$ and two odd indices $i' < j'$ such that $p_i \not\leq^x p_j$ and $p_{i'} \not\leq^x p_{j'}$.*

An induced path P of $G - N[x]$ is called well-linked *with respect to x if it is not badly-linked with respect to x.*

It is easy to see that if G satisfies the condition of Theorem 1.2, then any induced subgraph of G satisfies that condition. Therefore to prove Theorem 1.2, it suffices to prove the following:

THEOREM 2.1. *Suppose G is a triangle free graph. If for every vertex x of G, $V - N[x]$ is a bipartite graph which contains no induced C_n for $n \geq 6$, and every induced path of $G - N[x]$ is well-linked, then $\chi_c(G) = \omega_c(G)$.*

3. Proof of Theorem 2.1

Assume that $G = (V, E)$ is a connected core graph which is a counterexample to Theorem 2.1. We shall derive a contradiction. For each $x \in V$ let H_x be the bipartite graph $G - N[x]$. The main part of the proof is to analyse the structure of H_x. The argument is a little complicated and divided into a few subsections.

3.1. Connectness of H_x. In this subsection, we prove that the bipartite graph H_x is connected. Note that H_x contains no isolated vertices because such vertices could be retracted to x, contradicting the fact that G is a core.

LEMMA 3.1. *Assume H_x is not connected and $Q = A \cup B$ is a connected component of H_x. Then either $N(A) \cap N(x) = \varnothing$ or $N(B) \cap N(x) = \varnothing$.*

PROOF. Assume to the contrary that there exist $a \in A$ and $b \in B$ such that $N(a) \cap N(x) \neq \varnothing$ and $N(b) \cap N(x) \neq \varnothing$. Let P be a path of Q connecting a and b. Let

$$u \in N(a) \cap N(x), \quad v \in N(b) \cap N(x).$$

Then $P \cup \{u, v, x\}$ induces a nonbipartite subgraph of G.

Let $Q' = A' \cup B'$ be another connected component of H_x. Then any element w of Q' is adjacent to at least one of u, v, for otherwise, H_w contains $P \cup \{u, v, x\}$ and hence is nonbipartite. On the other hand, w cannot be adjacent to both u and v, for otherwise, let w' be a neighbour of w in Q', then one of u, v is adjacent to both w, w', contrary to the fact that G contains no triangle. In particular, this shows that $u \neq v$.

Let $a'b'$ be an edge of Q', where $a' \in A'$ and $b' \in B'$. Assume a' is adjacent to u and b' is adjacent to v. (Note that a', b' cannot be adjacent to a same vertex as

G is triangle free.) For any neighbour c' of b' in Q', c' must be adjacent to every neighbour u' of a, (because c' must be adjacent to one of u', v, and c' cannot be adjacent to v). Repeat this argument we conclude that

$$\forall a'' \in A', \quad N(a) \cap N(x) \subseteq N(a'') \cap N(x),$$

and similarly,

$$\forall b'' \in B', \quad N(b) \cap N(x) \subseteq N(b'') \cap N(x).$$

Let a', b' play the roles of a, b, we conclude that

$$\forall c \in A \cup A', \quad N(c) \cap N(x) = N(a') \cap N(x) = N(a) \cap N(x)$$

and

$$\forall d \in B \cup B', \quad N(d) \cap N(x) = N(b') \cap N(x) = N(b) \cap N(x).$$

So the mapping that sends all vertices of $A \cup A'$ to a' and all vertices of $B \cup B'$ to b' and fixes every other vertex is a retraction, contrary to the assumption that G is a core. $\qquad\square$

LEMMA 3.2. *For each $x \in V$, H_x is connected.*

PROOF. Assume to the contrary that H_x is not connected and $Q = A \cup B$ is a connected component of H_x. By Lemma 3.1, we may assume that $N(A) \cap N(x) = \emptyset$. Then the map which sends B to x and sends A to a neighbour of x is a retraction, contrary to the assumption that G is a core. $\qquad\square$

Since H_x is connected, there is a unique partition of H_x into two parts, which we denote by A_x and B_x. A vertex u of H_x is called a $\leq^x$-*maximum vertex* if $v \leq^x u$ for every other vertex v in the same part of H_x.

3.2. Ordering of A_x and B_x. In this subsection, we prove that all the vertices of A_x (respectively of B_x) are $\leq^x$-comparable. An induced path P of H_x is called a *maximal induced path* if P is not a proper subpath of an induced path P' of H_x. We denote by $\ell(P)$ the length (i.e., number of edges) of P.

LEMMA 3.3. *If $P = (p_0, p_1, p_2, \cdots, p_{k-2}, p_{k-1}, p_k)$ is a maximal induced path of H_x, then $N_{H_x}(p_0) \subseteq N_{H_x}(p_2)$ and $N_{H_x}(p_k) \subseteq N_{H_x}(p_{k-2})$.*

PROOF. Since P is a maximal induced path of H_x, $N_{H_x}(p_0) \subseteq \cup_{i=1}^k N_{H_x}(p_i)$. Assume to the contrary that there is a vertex $w \in N_{H_x}(p_0) - N_{H_x}(p_2)$. Let i be the minimum index such that $w \in N_{H_x}(p_i)$. As H_x is bipartite, i is even. Therefore $i \geq 4$. Now $(p_0, p_1, \cdots, p_i, w)$ is an induced cycle of H_x of length at least 6, contrary to our assumption. Similarly we can prove that $N_{H_x}(p_k) \subseteq N_{H_x}(p_{k-2})$. $\qquad\square$

LEMMA 3.4. *If P is an induced path of H_x which contains two $\leq^x$-incomparable vertices of A_x (or B_x), then $\ell(P) \leq 2$.*

PROOF. Assume to the contrary that P is an induced path of H_x which contains two $\leq^x$-incomparable vertices of A_x (or B_x), and has length at least 3. As each induced path of H_x is contained in a maximal induced path, we may assume that P is a maximal induced path of H_x.

Suppose $P = (p_0, p_1, \cdots, p_k)$. By Lemma 3.3, $N_{H_x}(p_0) \subseteq N_{H_x}(p_2)$ and $N_{H_x}(p_k) \subseteq N_{H_x}(p_{k-2})$. As G is a core, we conclude that $p_0 \not\leq^x p_2$ and $p_k \not\leq^x p_{k-2}$. As P is well-linked with respect to x and $k \geq 3$, we conclude that k is odd, and

$$p_1 \leq^x p_3 \leq^x \cdots \leq^x p_{k-2} <^x p_k,$$

$$p_{k-1} \leq^x p_{k-3} \leq^x \cdots \leq^x p_2 <^x p_0.$$

This is in contrary to the assumption that P contains two $\leq^x$-incomparable vertices of A_x (or B_x). $\qquad\square$

COROLLARY 3.1. *If* $a, a' \in A_x$ *are* $\leq^x$-*incomparable, then* $N_{H_x}(a) = N_{H_x}(a')$.

PROOF. For otherwise H_x contains an induced path of length at least 3 which contains both a and a', contrary to Lemma 3.4. $\qquad\square$

Suppose $x \in V$, we now define a graph Q with vertex set A_x as follows: $a \sim_Q a'$ if and only if a and a' are $\leq^x$-incomparable.

COROLLARY 3.2. *Let* A *be a connected component of* Q. *Let* $B = N_{H_x}(A)$. *Then the subgraph* H *of* H_x *induced by* $A \cup B$ *is a complete bipartite graph.*

LEMMA 3.5. *Assume* H_x *is not complete. Then the vertices of* A_x *(respectively of* B_x*) are pairwise* $\leq^x$-*comparable.*

PROOF. Assume to the contrary that H_x is not complete, and A_x contains two $\leq^x$-incomparable vertices. Then the graph Q defined above has a component A which contains at least two vertices. Let $B = N_{H_x}(A)$. By Corollary 3.2, the union $A \cup B$ induce a complete bipartite subgraph H of H_x. Since H_x is not complete, each edge of H_x is contained in an induced path of length at least 3. Let $a \in A$ and $b \in B$ and P be a longest induced path containing the edge ab. Assume

$$P = (\cdots, a_0, b_0, a, b, a_1, b_2 \cdots).$$

Note that a or b_0 could be the initial vertex of P. In this case, a_0 does not exist. Also b could be the terminal vertex of P. In this case, a_1 does not exist. However, since P has length at least 3, at least one of a_0, a_1 exists and does not belong to A. (Recall that all the vertices of A has the same neighbourhood in H_x.) Without loss of generality, we assume that a_1 exists and does not belong to A. Then a_1 is $\leq^x$-comparable to every vertex of A.

Assume $a, a' \in A$ are two $\leq^x$-incomparable vertices. Let

$$u \in (N(a) - N(a')) \cap N(x), \quad v \in (N(a') - N(a)) \cap N(x).$$

Fig. 1 below is a depiction of the named vertices (not all edges are drawn).

If $a_1 \leq^x a$ then $a_1 \leq^x a'$. Then $a_1 \not\sim u, a_1 \not\sim v$, hence a, u, x, v, a' induce a P_4 in H_{a_1}. Since

$$b \in N(a_1) \cap N(a) \cap N(a') \text{ and } b \notin N(x),$$

we conclude that $a \not\leq^{a_1} x$ and $a' \not\leq^{a_1} x$. Hence the path induced by a, u, x, v, a' is not well-linked with respect to a_1, contrary to our assumption.

Thus we assume that $a <^x a_1$ and $a' <^x a_1$. As G is a core, we know that $N(a) \not\subseteq N(a_1)$, hence $N_{H_x}(a) \not\subseteq N_{H_x}(a_1)$. By Lemma 3.3, P contains the vertex b_0. Because P is well-linked with respect to x, and $a <^x a_1$, we conclude that $b \leq^x b_0$.

Assume first that $b <^x b_0$. Let $w \in (N(b_0) - N(b)) \cap N(x)$. If w is not adjacent to a_1, then a', v, a_1, u, a is an induced path in H_w (recall that $N(x)$ is an independent set, because G is triangle free). Since

$$b_0 \in N(w) \cap N(a) \cap N(a') \text{ and } b_0 \notin N(a_1),$$

we have $a' \not\leq^w a_1$ and $a \not\leq^w a_1$. Hence the path induced by a', v, a_1, u, a is not well-linked with respect to w, contrary to our assumption. Thus we assume that w is adjacent to a_1. Let $w' \in N(x) - N(a_1)$ (w' exists because G is a core). If w' is

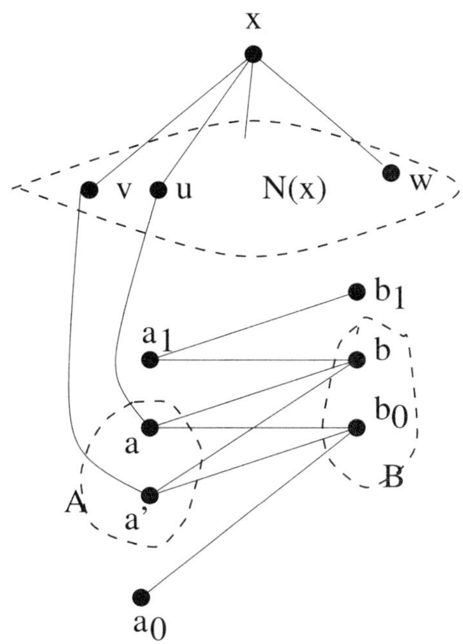

FIGURE 1. A depiction of $x, N(x), H_x$ for the proof of Lemma 3.5

adjacent to b_0, then a', v, a_1, u, a is an induced path in $H_{w'}$, which is not well-linked (for the same reason as above). If w' is not adjacent to b_0, then w, b_0, a, b, a_1 induce a pentagon in $H_{w'}$, again contrary to our assumption.

So we must have $b =^x b_0$, i.e., $N(b) \cap N(x) = N(b_0) \cap N(x)$. Then because $N(b_0) \not\subseteq N(b)$ (as G is a core). By Lemma 3.3, b_0 is not the initial vertex of the induced path P, i.e., P contains the vertex a_0 (note that if a_0 is adjacent to b_1, then H_x contains an induced C_6, contrary to our assumption). Because P is well-linked, we have $a_0 \leq^x a$, and hence $a_0 \leq^x a'$. This is symmetric to the case that $a_1 \leq^x a$ and $a_1 \leq^x a'$ which we discussed above. Indeed, in this case, it can be verified that a, u, x, v, a' is an induced path of H_{a_0} which is not well-linked with respect to a_0. $\qquad\square$

3.3. A new ordering on H_x. This subsection introduces a new ordering $\leq_N^x$ on A_x and B_x. This ordering is a refinement of the ordering $\leq^x$. Then we prove that each pair of vertices of A_x (as well as of B_x) are comparable with respect to $\leq_N^x$.

First we discuss for which $a, a' \in A_x$ we could have $a =^x a'$.

LEMMA 3.6. *If $a, a' \in A_x$ and $a =^x a'$ then $N(a) \cap N(x) = N(a') \cap N(x) = \emptyset$.*

PROOF. Assume to the contrary that $a, a' \in A_x$ are vertices such that $N(a) \cap N(x) = N(a') \cap N(x) \neq \emptyset$. Let

$$u \in N(a) \cap N(x) = N(a') \cap N(x).$$

Since G is a core, both sets $N_{H_x}(a) - N_{H_x}(a')$ and $N_{H_x}(a') - N_{H_x}(a)$ are nonempty. Let

$$b \in N_{H_x}(a) - N_{H_x}(a'), \quad b' \in N_{H_x}(a') - N_{H_x}(a).$$

(Since G is a core, b, b' exists). By applying Lemma 3.5, b and b' are $\leq^x$ comparable. Without loss of generality, we assume that $b \leq^x b'$. The named vertices are as depicted in Fig. 2 below.

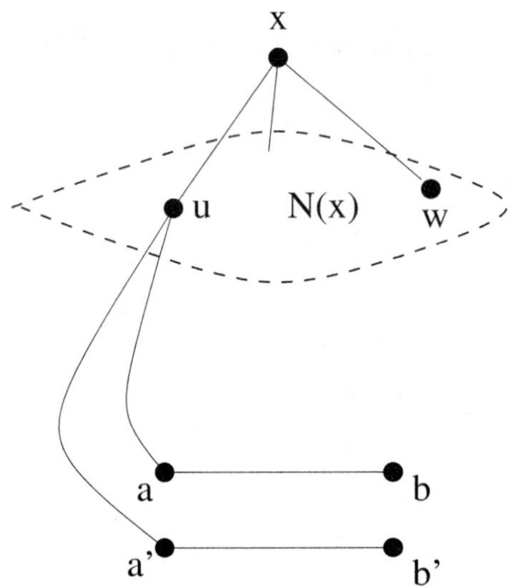

FIGURE 2. For the proof of Lemma 3.6

If $N(b') \cap N(x) = \emptyset$, then let $w \in N(x) - N(a) = N(x) - N(a')$. If $b \sim w$, then a, b, w, x, u induce a pentagon in H_w. If $b \not\sim w$, then $P = (b, a, u, a', b')$ is an induced path of H_w which is not well-linked with respect to w, because $x \in N(w) \cap N(u)$ but $x \not\sim b, b'$, i.e., $u \not\leq^w b$ and $u \not\leq^w b'$.

Assume $N(b') \cap N(x) \neq \emptyset$, and let $w \in N(b') \cap N(x)$. If $w \not\sim b$, then $P = (b, a, u, a')$ is an induced path of H_w. As $x \in N(w) \cap N(u)$ but $x \not\sim b$, so $u \not\leq^w b$; $b' \in N(w) \cap N(a')$ but $b' \not\sim a$, so $a' \not\leq^w a$. Therefore P is not well-linked with respect to w.

If $w \sim b$ then $P = (x, u, a, b)$ is an induced path in $H_{b'}$, which is not well-linked with respect to b' (because $a' \in N(b') \cap N(u)$ but $a' \not\sim b$ which implies that $u \not\leq^{b'} b$, and $w \in N(x) \cap N(b')$ and $w \not\sim a$, which implies that $x \not\leq^{b'} a$). $\qquad\square$

For $x \in V$, we let

$$A^*_x = \{a \in A_x : N(a) \cap N(x) \neq \emptyset\},$$

$$B^*_x = \{b \in B_x : N(b) \cap N(x) \neq \emptyset\}.$$

Lemma 3.6 says that if H_x is not a complete bipartite graph, then every two vertices of A^*_x (respectively of B^*_x) are strictly $\leq^x$-comparable (i.e., no two of them are equal with respect to $\leq^x$. However, all the vertices of $A_x - A^*_x$ are equal with respect to the order relation $\leq^x$. We shall introduce a refined order relation that distinguishes the vertices of $A_x - A^*_x$. Before introducing the refined order relation, we first prove that A^*_x and B^*_x are not empty.

LEMMA 3.7. *If H_x is not complete then $A^*_x \neq \emptyset$ and $B^*_x \neq \emptyset$.*

PROOF. If $N(a) \cap N(x) = \emptyset$ for all $a \in A_x$, then the mapping f which sends every vertex of B_x to x, and sends every vertex of A_x to a neighbour of x is a retraction, contrary to the assumption that G is a core. $\square$

COROLLARY 3.3. *If H_x is not complete then each of A_x and B_x has a unique $\leq^x$-maximum vertex, belonging to A_x^* and B_x^*, respectively.*

PROOF. Assume to the contrary that A_x has two $\leq^x$-maximum vertices a, a'. Then $a =^x a'$. By Lemma 3.6, we must have $N(a) \cap N(x) = N(a') \cap N(x) = \emptyset$, contrary to Lemma 3.7. $\square$

For $a \in A_x - A_x^*$, let $\phi(x, a)$ be the distance in H_x from a to A_x^*. By Lemma 3.7, the function ϕ is well-defined, and $\phi(x, a) \geq 2$ for any $a \in A_x - A_x^*$. For $a \in A_x - A_x^*$, let $A_x^*(a)$ be the set of vertices $w \in A_x^*$ for which there is a path P in H_x of length $\phi(x, a)$ connecting a and w. Now we define a new ordering $\leq_N^x$ on the vertices of A_x. For $a, a' \in A_x$, we write $a <_N^x a'$ if and only if one of the following holds:

- $a <^x a'$;

- $a =^x a'$ and $\phi(x, a) > \phi(x, a')$ (note the reverse of the inequality, which is not a typo):

- $a =^x a'$, $\phi(x, a) = \phi(x, a')$ and $A_x^*(a) \subset A_x^*(a')$.

We write $a =_N^x a'$ if $a =^x a'$, $\phi(x, a) = \phi(x, a')$ and $A_x^*(a) = A_x^*(a')$; and write $a \leq_N^x a'$ if either $a <_N^x a'$ or $a =_N^x a'$. Similarly we define the orderings $<_N^x$, $\leq_N^x$ on B_x.

We shall prove in the following that every pair of vertices of A_x is $\leq_N^x$-comparable. In other words, if $a =^x a'$ and $\phi(x, a) = \phi(x, a')$ then one of the two sets $A_x^*(a)$, $A_x^*(a')$ contains the other.

Suppose P is a path of G and $u, v \in P$. Then we denote by $P[u, v]$ the subpath of P connecting u to v.

LEMMA 3.8. *If $u, u' \in A_x$ and $u =^x u'$, then one of the two sets $A_x^*(u)$ and $A_x^*(u')$ contains the other. Similarly for any $v, v' \in B_x$, if $v =^x v'$ then one of the sets $B_x^*(v)$ and $B_x^*(v')$ contains the other.*

PROOF. Suppose $u, u' \in A_x$ and $u =^x u'$. By Lemma 3.6, $u, u' \in A_x - A_x^*$. Assume to the contrary that there exist

$$w \in A_x^*(u) - A_x^*(u'), \quad w' \in A_x^*(u') - A_x^*(u).$$

Let P, P' be shortest paths of H_x connecting u to w, and u' to w' respectively. First we show that no vertex of P is adjacent to a vertex of P'. Assume to the contrary that $s \in P$ is adjacent to $s' \in P'$.

If $\ell(P[s, w]) \geq \ell(P'[s', w']) + 1$, then the concatenation of $P[u, s], ss', P'[s', w']$ is a path connecting u to w' which has length at most $\phi(x, u)$, contrary to our assumption that $w' \notin A_x^*(u)$. . So $\ell(P[s, w]) \leq \ell(P'[s', w'])$. By symmetry we have $\ell(P[s, w]) \geq \ell(P'[s', w'])$. Therefore $\ell(P[s, w]) = \ell(P'[s', w'])$. However, this is impossible because $\ell(P[s, w])$ and $\ell(P'[s', w'])$ have different parity (note that both $w, w' \in B_x$ and $s \sim s'$, so one of the s, s' is in B_x and the other is in A_x).

Now let P'' be a shortest path (and hence an induced path) of H_x connecting u to u'. Since P'' is well-linked with respect to x, we conclude that $P'' \cap A_x^* = \emptyset$ (as the two ends of P'' are vertices in $A_x - A_x^*$). Now the union of P, P', P'' contains an induced path P^* connecting w and w'. If P^* contains any vertex, say w'', of A_x other than w, w', then because $w'' \in A_x - A_x^*$ (observe that $(P \cup P' \cup P'') \cap A_x^* = \{w, w'\}$), P'' is not well-linked with respect to x. Thus P^* contains no vertex of A_x, i.e., $P^* = (w, b, w')$ for some $b \in B_x$. If b is adjacent to any vertex of P other than w, say $b \sim s$ for $s \in P$, then the concatenation of $P[u, s]$ and (s, b, w) would be a shortest path of H_x connecting u to w. But this path contains a vertex b which is adjacent to a vertex w' of P', contrary to the previous paragraph. Similarly b is not adjacent to any vertex of P' other than w'. Hence the concatenation of $P[u, w], (w, b, w'), P[w', u']$ is an induced path of H_x. However, this path is not well-linked with respect to x, because $N(w) \cap N(x) \neq \emptyset$, and yet $N(u) \cap N(x)$ and $N(u') \cap N(x)$ are both emptyset. $\square$

By Lemma 3.8, each pair of vertices of A_x are $\leq_N^x$-comparable, and similarly each pair of vertices of B_x are $\leq_N^x$-comparable.

3.4. Ordering B_x and B_a. Suppose a is a $\leq^x$-maximum vertex of A_x. This subsection and the next subsection discuss the relation between the two orderings $\leq_N^x$ and $\leq_N^a$. In this subsection, we prove that if a is the $\leq^x$-maximum vertex of A_x, then in the set $B_x \cap B_a$ which is contained in their common domain, these two ordering relations coincide with each other. Moreover, $B_a - B_x = \{x\}$ and x is the $\leq^x$-maximum vertex of B_a.

LEMMA 3.9. *Suppose H_x is not complete and a is the $\leq^x$-maximum vertex of A_x. Then the two parts of H_a are $A_a = (A_x - \{a\}) \cup (N(x) - N(a))$ and $B_a = (B_x - N(a)) \cup \{x\}$. Moreover, the following are true:*

(1) H_a is also not complete;

(2) x is the $\leq^a$-maximum vertex of B_a;

(3) all $y \in A_a - A_x = N(x) - N(a)$ and $y' \in A_a \cap A_x$, then $y \leq^a y'$;

(4) if $y, y' \in A_a \cap A_x$ then $y <^x y'$ implies $y <^a y'$;

(5) if $y, y' \in B_a \cap B_x$, then $y <^a y'$ implies $y <^x y'$.

PROOF. Obviously we have

$$V(H_a) = (A_x - \{a\}) \cup (N(x) - N(a)) \cup (B_x - N(a)) \cup \{x\}.$$

It can be easily proved by induction on the length of the paths that any path in H_a starting from x with an odd length ends at a vertex of $(A_x - \{a\}) \cup (N(x) - N(a))$, and any path starting from x with an even length ends at a vertex of $(B_x - N(a)) \cup \{x\}$. As H_a is connected, so the two parts of H_a are as described in the lemma. Of course, the names A_a and B_a can be interchanged. However, in the following we shall always name A_a and B_a as described above.

Since x is not adjacent to any vertex of $A_x \cap A_a = A_x - \{a\}$, we know that H_a is not complete. (Note that $A_x - \{a\} \neq \emptyset$, for otherwise H_x is complete.) It is obvious that for any $b \in B_x$, $N(b) \cap N(a) \subseteq N(x)$. Therefore x is the $\leq^a$-maximum vertex of B_a. Next we prove that if $y \in A_a - A_x = N(x) - N(a)$ and $y' \in A_a \cap A_x$, then $y \leq^a y'$. Assume to the contrary that $w \in N(y) \cap N(a) - N(a') \cap N(a)$ for some $a' \in A_a \cap A_x$. Because a is $\leq^x$-maximum, by Lemmas 3.6

and 3.7, there exists $w' \in N(a) \cap N(x) - N(y') \cap N(x)$. Now w, y, x, w', a induce a nonbipartite graph in $H_{y'}$, contrary to our assumption. It remains to show that $\leq^a$ coincides with $\leq^x$ on their common domain. If $y, y' \in A_a \cap A_x$ and $y <^x y'$, then $N(y) \cap N(x) \subset N(y') \cap N(x)$. Since a is $\leq^x$-maximum, $N(y) \cap N(x) \subset N(a)$ and $N(y') \cap N(x) \subset N(a)$. Therefore $N(y) \cap N(a) \subset N(y') \cap N(a)$. If $y, y' \in B_a \cap B_x$ and $y <^a y'$, then $N(y) \cap N(a) \subset N(y') \cap N(a)$. Since $N(y) \cap N(a) \subset N(x)$ and $N(y') \cap N(a) \subset N(x)$. Therefore $N(y) \cap N(x) \subset N(y') \cap N(x)$. $\qquad\square$

Lemma 3.9 basically says that $\leq^x$ and $\leq^a$ coincide in their common domain. Now we shall compare the refined orderings $\leq^x_N$ and $\leq^a_N$. First, for $u \in (B_a - B_a^*) \cap (B_x - B_x^*)$, we want to study the relation between $\phi(x, u)$ and $\phi(a, u)$.

LEMMA 3.10. *Assume $a \in A_x$ is a $\leq^x$-maximum vertex of A_x. If $u \in B_x - (B_x^* \cup N(a))$ then for any shortest path P of H_x connecting u to B_x^*, $P \cap N(a) = \emptyset$.*

PROOF. Assume $P = (u, a_1, b_1, \cdots, b_k)$ and b_k is adjacent to $w \in N(x)$. First we observe that $a \notin P$. Otherwise, say $a_j = a$ then $a_1 <^x a_j$, $u <^x b_k$ implies that P is not well-linked with respect to x. Assume now that $P \cap N(a) \neq \emptyset$. Let j be the least index such that $a \sim b_j$. If $j = k$ then $P' = (u, a_1, b_1, \cdots, b_k, a)$ is an induced path of H_x which is not well-linked (as $u <^x b_k$ and $a_1 <^x a$).

Assume $j < k$. Then $a \not\sim b_{j-1}$ and $a \not\sim b_{j+1}$, for otherwise we may replace a_j or a_{j+1} by a and obtain a shortest path in H_x connecting u to B_x^* which contains a, contrary to the conclusion above.

Let $w' \in N(a) \cap N(x) - (\cup_{i+1}^k N(a_i))$ (since a is the $\leq^x$-maximum vertex of A_x and every two vertices of A_x are $\leq^x$-comparable, the vertex w' exists). Consider the path

$$P'' = (b_{j-1}, a_j, b_j, a_{j+1}, b_{j+1}),$$

where b_{j-1} could be u. We must have $b_{j+1} \not\sim w'$, for otherwise $w, a, b_j, a_{j+1}, b_{j+1}$ induce a nonbipartite graph in $H_{b_{j-1}}$. So P'' is an induced path in $H_{w'}$ (note that $w' \not\sim b_j, b_{j-1}$, because $b_j, b_{j-1} \in B_x - B_x^*$). As $a \in N(w') \cap N(b_j)$ and $a \not\sim b_{j-1}$ $a \not\sim b_{j+1}$ which imply that $b_j \not\leq^x b_{j-1}$ and $b_j \not\leq^x b_{j+1}$, we conclude that P'' is not well-linked with respect to w'. $\qquad\square$

COROLLARY 3.4. *Assume $a \in A_x$ is a $\leq^x$-maximum vertex of A_x. If $u \in B_x - (B_x^* \cup N(a))$ and $u \notin B_a^*$, then either*

- $B_x^*(u) \cap B_a^* \neq \emptyset$ *and $\phi(a, u) = \phi(x, u)$, or*

- $B_x^*(u) \cap B_a^* = \emptyset$ *and $\phi(a, u) = \phi(x, u) + 2$.*

PROOF. Observe that $B_a^* \subseteq B_x^* \cup \{x\}$. Assume P is a shortest path of H_a connecting u to B_a^*. If P does not contain any vertex of $N[x]$, then P is also a path of H_x connecting u to B_x^*. If P does contain a vertex, say w, of $N(x)$, then let b' be the vertex of P preceding w, then $P[u, b']$ is a path of H_x connecting u to B_x^*. Therefore $\phi(a, u) \geq \phi(x, u)$.

If $B_x^*(u) \cap B_a^* \neq \emptyset$ then let P be a shortest path in H_x connecting u to a vertex $w \in B_x^*(u) \cap B_a^*$. By Lemma 3.10, P is also a path in H_a. Therefore $\phi(a, u) = \phi(x, u)$.

If $B_x^*(u) \cap B_a^* = \emptyset$ then any shortest path in H_x connecting u to B_x^* is not a path connecting u to B_a^*. Therefore $\phi(a, u) > \phi(x, u)$ and hence $\phi(a, u) \geq \phi(x, u) + 2$. Let P be a shortest path of H_x connecting u to $w \in B_x^*$, and let $v \in N(w) \cap N(x)$, then the concatenation of P and (w, v, x) is a path connecting u to x. As $x \in B_a^*$, we have $\phi(a, u) \leq \phi(x, u) + 2$ and hence $\phi(a, u) = \phi(x, u) + 2$. $\qquad\square$

Now we are ready to study the ordering relation between two vertices $u, u' \in B_a \cap B_x$ with respect to the two orderings $\leq_N^x$ and $\leq_N^a$, where a is the $\leq^x$-maximum vertex of A_x. We shall eventually prove that $u \leq_N^x u'$ implies that $u \leq_N^a u'$. First we consider the case that $u \in B_x^*$.

LEMMA 3.11. *Suppose $a \in A_x$ is a $\leq^x$-maximum vertex of A_x. Assume $u, u' \in B_x \cap B_a$, $u \neq u'$ and $u \leq^x u'$. If $u \in B_x^*$ then $u' \in B_a^*$.*

PROOF. Since $u \in B_x^*$, $u \neq u'$ and $u \leq^x u'$, by Lemma 3.6), $u <^x u'$ and

$$\varnothing \neq N(u) \cap N(x) \subset N(u') \cap N(x).$$

Let

$$w \in N(u) \cap N(x), \quad w' \in N(u') \cap N(x) - N(u) \cap N(x).$$

Assume to the contrary of this lemma that $u' \notin B_a^*$. Then $w, w' \notin N(a)$. Therefore $w, w' \notin N(a')$ for any $a' \in A_x$. Since G is a core, $N(w') \not\subseteq N(w)$. Let $u'' \in N(w') - N(w)$. Then $u'' \in B_x$. But then none of $N(u'') \cap N(x)$ and $N(u) \cap N(x)$ contains the other, contrary to Lemma 3.5. $\qquad\square$

Now we are ready to prove that $u \leq_N^x u'$ implies $u \leq_N^a u'$.

LEMMA 3.12. *Suppose H_x is not complete and a is the $\leq^x$-maximum vertex of A_x. If $u, u' \in B_x \cap B_a$ and $u \leq_N^x u'$ then $u \leq_N^a u'$.*

PROOF. Observe that

$$N(u) \cap N(a) \subseteq N(u) \cap N(x)$$

and

$$N(u') \cap N(a) \subseteq N(u') \cap N(x).$$

Since $u \leq_N^x u'$, we have $N(u) \cap N(x) \subseteq N(u') \cap N(x)$, which implies (by applying Lemma 3.5) that $N(u) \cap N(a) \subseteq N(u') \cap N(a)$.

If $N(u) \cap N(a) \subset N(u') \cap N(a)$ then $u <^a u'$ and hence $u <_N^a u'$, we are done. Assume now that $N(u) \cap N(a) = N(u') \cap N(a)$. By Lemma 3.6,

$$N(u) \cap N(a) = N(u') \cap N(a) = \varnothing.$$

By Lemma 3.11, $u \notin B_x^*$, i.e., $N(u) \cap N(x) = \varnothing$. (For otherwise, we would have $u' \in B_a^*$, i.e., $N(u') \cap N(a) \neq \varnothing$.)

If $N(u') \cap N(x) \neq \varnothing$, then let $w \in N(u') \cap N(x)$. Then $P = (u', w, x)$ is a path from u to B_a^* (as $x \in B_a^*$). Therefore $\phi(a, u') = 2 \leq \phi(a, u)$. If $\phi(a, u) > 2$ then $u \leq_N^a u'$ and we are done. Assume $\phi(a, u) = 2$. We need to show that $B_a^*(u) \subseteq B_a^*(u')$. Since $N(u) \cap N(x) = \varnothing$, there is no path of length 2 in H_a connecting u to x, i.e., $x \notin B_a^*(u)$. Therefore, by applying Lemma 3.8, $B_a^*(u) \subset B_a^*(u')$, hence $u <_N^a u'$.

Assume now that $N(u') \cap N(x) = \varnothing$. Then $u =^x u'$, and $\phi(x, u) \geq \phi(x, u')$. If $\phi(x, u) > \phi(x, u')$, then $\phi(x, u) \geq \phi(x, u') + 2$. By Corollary 3.4,

$$\phi(a, u) \geq \phi(x, u) \geq \phi(x, u') + 2 \geq \phi(a, u').$$

If $\phi(a, u) > \phi(a, u')$ then $u <_N^a u'$ and we are done. Assume that $\phi(a, u) = \phi(a, u')$. Then $x \in B_a^*(u') - B_a^*(u)$. By applying Lemma 3.8, we conclude that $B_a^*(u) \subset B_a^*(u')$. So $u <_N^a u'$.

Assume now that $\phi(x,u) = \phi(x,u')$. As $u \leq_N^x u'$, we know that $B_x^*(u) \subseteq B_x^*(u')$. By Corollary 3.4, either $B_x^*(u) \cap B_a^* \neq \varnothing$ and $\phi(a,u) = \phi(x,u)$ or $B_x^*(u) \cap B_a^* = \varnothing$ and $\phi(a,u) = \phi(x,u) + 2$.

In the former case, we have $B_x^*(u') \cap B_a^* \neq \varnothing$ (as $B_x^*(u) \subseteq B_x^*(u')$) and $\phi(a,u') = \phi(x,u')$. So $\phi(a,u') = \phi(a,u)$. As $B_x^*(u) \subseteq B_x^*(u')$, and $B_a^*(u) = B_x^*(u) \cap B_a^*$ and $B_a^*(u') = B_x^*(u') \cap B_a^*$ we conclude that $B_a^*(u) \subseteq B_a^*(u')$. So $u \leq_N^a u'$.

In the latter case, if $\phi(a,u') = \phi(x,u')$ then we have $\phi(a,u') < \phi(a,u)$ and hence $u \leq_N^a u'$; if $\phi(a,u') = \phi(x,u') + 2$ then we need to prove that $B_a^*(u) \subseteq B_a^*(u')$. Assume to the contrary that there is a vertex $b \in B_a^*(u) - B_a^*(u')$. Let

$$P = (u, \cdots, v, a', b)$$

be a shortest path of H_a connecting u to b. Since $b \notin B_a^*(u')$ we know that $v \notin B_x^*(u')$ and hence $v \notin B_x^*(u)$, because $B_x^*(u) \subseteq B_x^*(u')$. Therefore $N(v) \cap N(x) = \varnothing$. Let $v' \in B_x^*(u')$. Then $\phi(a,v) = \phi(a,v') = 2$. Note that $N(v') \cap N(b) = \varnothing$, for otherwise we would have $b \in B_a^*(u')$. Therefore $x \in B_a^*(v') - B_a^*(v)$ and $b \in B_a^*(v) - B_a^*(v')$, i.e., none of the sets $B_a^*(v')$ and $B_a^*(v)$ contains the other. This is in contrary to Lemma 3.8. This completes the proof. $\qquad\square$

3.5. Ordering A_x and A_a. In this subsection, we compare the two orderings $\leq_N^x$ and $\leq_N^a$. We shall prove that if $v' \in A_x \cap A_a$ and $v \in A_a - A_x$ then $v \leq_N^a v'$. If $v, v' \in A_x \cap A_a$ and $v \leq_N^a v'$ then $v \leq_N^x v'$.

LEMMA 3.13. *Suppose H_x is not complete and a is the $\leq^x$-maximum vertex of A_x. Then for any $v' \in A_x \cap A_a$ and $v \in A_a - A_x$ we have $v \leq_N^a v'$.*

PROOF. Since a is the $\leq^x$-maximum vertex of A_x, $N(v') \cap N(x) = N(v') \cap N(a) \cap N(x)$. If $N(v') \cap N(x) \neq \varnothing$ then since $N(v) \cap N(x) = \varnothing$ (as G is triangle free), we have $N(v') \cap N(a) \not\subseteq N(v) \cap N(a)$. By Lemma 3.6, $N(v) \cap N(a) \subset N(v') \cap N(a)$. This implies that $v <^a v'$ and hence $v <_N^a v'$.

Assume now that $N(v') \cap N(x) = \varnothing$. First we show that $N(v) \cap N(a) \subseteq N(v') \cap N(a)$. Assume to the contrary that $b \in N(v') \cap N(a) - N(v) \cap N(a)$. By Lemma 3.3, there is a vertex $w' \in N(a) \cap N(x) - N(v')$. Then v, x, w', a, b induce a pentagon in $H_{v'}$, contrary to our assumption that $H_{v'}$ is bipartite.

If $N(v) \cap N(a) \subset N(v') \cap N(a)$, then $v <_N^a v'$ and we are done. Assume now that $N(v) \cap N(a) = N(v') \cap N(a)$. By Lemma 3.6,

$$N(v) \cap N(a) = N(v') \cap N(a) = \varnothing.$$

Let w be any vertex of $A_a^*(v)$ and let P be a shortest path of H_a connecting v to a vertex w. Let $w' \in N(a)$ be a vertex adjacent to w, and let $w'' \in N(x) \cap N(a)$. If v' is not adjacent to any vertex of the path P, then the concatenation of P and (w, w', a, w'', x, v) is an odd closed walk in H_v, contrary to our assumption. Thus v' is adjacent to some vertex of the path P. Therefore $\phi(a,v') \leq \phi(a,v)$. Moreover, if $\phi(a,v') = \phi(a,v)$ then $w \in A_a^*(v')$. As w is an arbitrary vertex of $A_a^*(v)$ we have $A_a^*(v) \subseteq A_a^*(v')$. So in any case we have $v \leq_N^a v'$. $\qquad\square$

LEMMA 3.14. *Suppose H_x is not complete and a is the $\leq^x$-maximum vertex of A_x. If $v, v' \in A_x \cap A_a$ and $v <^a v'$ then $v \leq_N^x v'$.*

PROOF. If $v <^a v'$, then $N(v) \cap N(a) \subset N(v') \cap N(a)$. Let

$$w \in N(v') \cap N(a) - N(v) \cap N(a).$$

If $N(v') \cap N(x) \neq \emptyset$ then by Lemma 3.6 we know that $N(v) \cap N(x) \subset N(v') \cap N(x)$, hence $v <_N^x v'$ and we are done.

Assume

$$N(v') \cap N(x) = N(v) \cap N(x) = \emptyset.$$

So $v, v' \in A_x - A_x^*$. By Lemma 3.6, $v <^a v'$ implies that $N(v') \cap N(a) \neq \emptyset$. Let $w \in N(v') \cap N(a)$. Then (v', w, a) is a path of H_x connecting v' to A_x^*. Therefore $\phi(x, v') = 2$ (because $a \in A_x^*$).

If $\phi(x, v) > 2$ then $v <_N^x v'$ and we are done. Thus we assume that $\phi(x, v) = 2$. If $A_x^*(v) \subseteq A_x^*(v')$ then $v \leq_N^x v'$ and we are done. Thus we assume that there is a vertex $a' \in A_x^*(v) - A_x^*(v')$. Let (v, w', a') be a path in H_x. Observe that (v', w, a) is a path of H_x as well.

We must have $a' \not\sim w$ and $v' \not\sim w'$ for otherwise $a' \in A_a^*(v')$.

Let P be a shortest path of H_x connecting v to v'. Since the two end vertices of P are not in A_x^*, P does not contain any vertex of A_x^*, for otherwise P is not well-linked with respect to x. Now the concatenation of (a', w', v), P and (v', w, a) is a walk W in H_x connecting a' to a. It contains an induced path P' of H_x connecting a' and a. If P' contains any other vertex of A_x other than a, a', then P' is not well-linked with respect to x, because all the other vertices of $W \cap A_x$ is in $A_x - A_x^*$. Thus P' is of the form (a, b, a') for some $b \in P \cap B_x$. We have $b \not\sim v'$, for otherwise $a' \in A_x^*(v')$. But then the concatenation of $P[v, b]$ and (b, a, w, v') is a v, v'-path of H_x whose length is not longer than P. Hence P^* is a shortest v, v'-path, and hence is an induced path. But P^* is not well-linked with respect to x, because the two end vertices of P^* are not in A_x^*, and it contains the vertex a which is in A_x^*. This completes the proof of Lemma 3.14. $\square$

LEMMA 3.15. *Suppose H_x is not complete and a is the $\leq^x$-maximum vertex of A_x. If $v, v' \in A_x \cap A_a$ and $v \leq_N^a v'$ then $v \leq_N^x v'$.*

PROOF. By Lemma 3.14, it suffices to consider the case that $v =^a v'$, i.e., $N(v) \cap N(a) = N(v') \cap N(a)$. By Lemma 3.6, $N(v) \cap N(a) = N(v') \cap N(a) = \emptyset$, i.e., $v, v' \in A_a - A_a^*$.

Claim *For $u \in A_a - A_a^*$, if $A_a^*(u)$ contains a vertex w such that $N(w) \cap N(x) \neq \emptyset$, then $\phi(a, u) = \phi(x, u)$. Otherwise $\phi(a, u) = \phi(x, u) - 2$.*

The proof of this claim is similar to that of Corollary 3.4, and we omit the details.

If $\phi(a, v') < \phi(a, v)$, then $\phi(a, v') \leq \phi(a, v) - 2$ and hence

$$\phi(x, v') \leq \phi(a, v') + 2 \leq \phi(a, v) \leq \phi(x, v).$$

Moreover, in case $\phi(x, v') = \phi(x, v)$, we have $\phi(x, v') = \phi(a, v') + 2$ which implies that $a \in A_x^*(v')$ and $\phi(x, v) = \phi(a, v)$ which implies that $a \notin A_x^*(v)$. So in this case $A_x^*(v) \subset A_x^*(v')$, and hence $v <_N^x v'$.

Assume $\phi(a, v') = \phi(a, v)$. As $v \leq_N^a v'$, we have $A_a^*(v) \subseteq A_a^*(v')$. By the claim we have $\phi(x, v') \leq \phi(x, v)$. If $\phi(x, v') < \phi(x, v)$, then $v <_N^x v'$ and we are done. Assume $\phi(x, v') = \phi(x, v)$. By the claim, either

- $\phi(x, v') = \phi(a, v')$ and $\phi(x, v) = \phi(a, v)$, or
- $\phi(x, v') = \phi(a, v') + 2$ and $\phi(x, v) = \phi(a, v) + 2$.

In the former case, since a is the $\leq^x$-maximum vertex of A_x which implies that $A_x^* - \{a\} \subseteq A_a^*$, we have $A_x^*(v') = A_a^*(v') \cap A_x^*$ and $A_x^*(v) = A_a^*(v) \cap A_x^*$. Therefore, $A_x^*(v) \subseteq A_x^*(v')$, hence $v \leq_N^x v'$ and we are done.

Assume now that

$$\phi(x, v') = \phi(a, v') + 2, \quad \phi(x, v) = \phi(a, v) + 2.$$

We need to prove that $A_x^*(v) \subseteq A_x^*(v')$. Assume to the contrary that there is a vertex $a' \in A_x^*(v) - A_x^*(v')$. Let

$$P = (v, \cdots, b, w, b', a')$$

be a path of H_x of length $\phi(x, v)$ connecting v to a'. Then $w \notin A_a^*$, for otherwise $w \in A_a^*(v) \subseteq A_a^*(v')$ and hence $a' \in A_x^*(v')$. Therefore $b' \nsim a$.

Let $u \in A_a^*(v) \subseteq A_a^*(v')$, and let $b'' \in N(u) \cap N(a)$. Let

$$P' = (v, \cdots, u, b'', a)$$

be a path of H_x of length $\phi(x, v)$ connecting v to a. Now $b'' \nsim w$, for otherwise we would have $w \in A_a^*$. Since

$$N(a') \cap N(x) \neq \varnothing$$

and

$$N(a') \cap N(x) \subseteq N(a) \cap N(x),$$

$a' \in A_a^*$, and there is a vertex $z \in N(a') \cap N(a) \cap N(x)$. Because $z \notin N(u)$ (for otherwise we would have $u \in A_x^*(v)$ and $\phi(x, v) = \phi(a, v)$), by Lemma 3.5, we know that $N(u) \cap N(a) \subset N(a') \cap N(a)$. Hence $b'' \sim a'$. So the concatenation of $P[v, a']$ and $P[v, b'']$ is a closed walk Z of H_x. The segment of this closed walk "near" b'' is as follows:

$$(\cdots, b, w, b', a', b'', u, \cdots).$$

Moreover, each of these six vertices occurs in the walk just once (as the paths P, P' are shortest paths from v to a', a). Note that $b \nsim a'$, for otherwise there is a path of H_x of length $\phi(a, v) - 2$ connecting v to a', contrary to the definition of $\phi(x, v)$. We observed before that $b'' \nsim w$. These imply that Z contains an induced cycle of length at least 6, contrary to our assumption. $\square$

3.6. The final contradiction. Now we shall complete the proof of Theorem 2.1.

Let $V' = \{x \in V : H_x \text{ is not complete}\}$. If $V' = \varnothing$ then for each $x \in V$, $V - N[x]$ contains no induced P_4, hence G is circular perfect by Theorem 1.1.

Thus we may assume that $V' \neq \varnothing$. Define a graph Q with vertex set V', in which $x \sim y$ if and only if y is a $\leq^x$-maximum vertex of A_x or B_x. It follows from Lemma 3.9 and Corollary 3.3 that Q is a graph in which each vertex has degree 2. Let $C = (x_0, x_1, x_2, \cdots, x_{k-1})$ be a cycle which is a connected component of Q. Arbitrarily assign a direction of traversal to the cycle C, say $x_0 \to x_1 \to x_2 \to \cdots \to x_{k-1}$. For any x_i, the two neighbours x_{i-1} and x_{i+1} are the two $\leq^x$-maximum vertices of x_i. It follows from Lemma 3.9 that we can properly label the two parts of H_{x_i} so that $x_{i+1} \in A_{x_i}$ and $x_{i-1} \in B_{x_i}$, for all i (where summation and subtraction in the index are modulo k). By the moreover part of Lemma 3.9 and Lemma 3.15, for each i, there is a $d_i \geq 2$ such that $A_{x_i} = \{x_{i+1}, x_{i+2}, \cdots, x_{i+d_i-1}\}$. Indeed, x_{i+1} is the $\leq^x$-maximum vertex of A_{x_i}. Now x_{i+2} is the $\leq^x$-maximum vertex of $A_{x_{i+1}}$ and by Lemma 3.15, x_{i+2} is the $\leq^x$-maximum vertex of $A_{x_i} - \{x_{i+1}\}$, provided $A_{x_i} - \{x_{i+1}\} \neq \varnothing$. In general, if

$$A_{x_i} - \{x_{i+1}, x_{i+2}, \cdots, x_{i+t}\} \neq \varnothing,$$

then by Lemma 3.15, x_{i+t+1} is the $\leq^x$-maximum vertex of $A_{x_{i+t}}$ which is the $\leq^x$-maximum vertex of

$$A_{x_i} - \{x_{i+1}, x_{i+2}, \cdots, x_{i+t}\}.$$

By Lemma 3.13 and Lemma 3.15, for each i, there is a $d_i' \geq 2$ such that $B_{x_i} = \{x_{i-1}, x_{i-2}, \cdots, x_{i-d_i'-1}\}$.

By Lemma 3.9, $A_{x_{i+1}} = (A_{x_i} - \{x_{i+1}\}) \cup R$, where $R = N(x_i) - N(x_{i+1})$. Since $R \neq \varnothing$ (for otherwise $N(x_{i+1}) \subseteq N(x_i)$, contrary to the assumption that G is a core), we conclude that $|A_{x_{i+1}}| \geq |A_{x_i}|$ for all i (again summation in the index is modulo k). So

$$|A_{x_0}| \leq |A_{x_1}| \leq |A_{x_2}| \leq \cdots \leq |A_{x_{k-1}}| \leq |A_{x_0}|.$$

Therefore $|A_{x_i}| = |A_{x_j}|$ for all $0 \leq i, j \leq k - 1$. So $d_i = d_j$ for all $0 \leq i, j \leq k - 1$. Let $d = d_i$.

Similarly, by Lemma 3.9, $B_{x_{i+1}} = (B_{x_i} - N(x_{i+1})) \cup \{x_i\}$. As $N(x_{i+1}) \cap B_{x_i} \neq \varnothing$, we conclude that $|B_{x_{i+1}}| \leq |B_{x_i}|$ for all i (again summation in the index is modulo k). So

$$|B_{x_0}| \geq |B_{x_1}| \geq |B_{x_2}| \geq \cdots \geq |B_{x_{k-1}}| \geq |B_{x_0}|.$$

Therefore $|B_{x_i}| = |B_{x_j}|$ for all $0 \leq i, j \leq k - 1$. So $d_i' = d_j'$ for all $0 \leq i, j \leq k - 1$. Because $\Sigma_{i=0}^{k-1} d_i = \Sigma_{i=0}^{k-1} d_i'$, which is equal to the number of nonedges of the subgraph of G induced by $x_0, x_1, \cdots, x_{k-1}$, we conclude that $d_i' = d$ for all i.

Now for any $x_i \in C$, any vertex x of G not adjacent to x either belong to A_{x_i} or belong to B_{x_i}. So $x = x_j \in C$ for some j, and $d \leq |i - j| \leq k - d$. Therefore, the subgraph of G generated by $x_0, x_1, \cdots, x_{k-1}$ is isomorphic to $K_{k/d}$. Moreover, if $y \in V(G) - C$ then y is adjacent to every vertex of C.

If $V(G) \neq C$, then for any $y \in V(G) - C$, we have $C \subset N[y]$, contrary to the assumption that G is triangle free (observe that C contains at least one edge because H_x is connected). If $V(G) = C$, then $G = K_{k/d}$, contrary to our assumption that $\omega_c(G) \neq \chi_c(G)$. This completes the proof of Theorem 2.1.

References

[1] B. Bauslaugh and X. Zhu, *Circular colouring of infinite graphs*, Bulletin of The Institute of Combinatorics and its Applications, 24(1998), 79-80.

[2] J. A. Bondy and P. Hell, *A note on the star chromatic number*, J. Graph Theory 14 (1990), 479-482.

[3] J. Nešetřil, *Homomorphism structure of classes of graphs*, Combinatorics, Probability and computing 8 (1999), 177-184.

[4] A. Vince, *Star chromatic number*, J. Graph Theory 12 (1988), 551-559.

[5] X. Zhu, *Star chromatic numbers and products of graphs*, J. Graph Theory 16 (1992), 557-569.

[6] X. Zhu, *Circular colouring and graph homomorphisms*, Bulletin of Australian Mathematical Society, 59(1999), 83-97.

[7] X. Zhu, *Circular chromatic number and graph minor*, Taiwanese Journal of Mathematics, 4(2000), 643-660.

[8] X. Zhu, *The circular chromatic number, a survey*, Discrete Mathematics, Vol. 229 (1-3) (2001), 371-410.

[9] X. Zhu, *An analogue of Hajós theorem for the circular chromatic number*, The Proceeding of the American Mathematical Society, 129(2001), 2845-2852.

[10] X. Zhu, *An analogue of Hajós theorem for the circular chromatic number (II)*, Graphs and Combinatorics, to appear.

[11] X. Zhu, *Circular perfect graphs (I)*, manuscript, 1999.

Department of Applied Mathematics, National Sun Yat-sen University, Kaohsiung, Taiwan 80424

E-mail address: zhu@math.nsysu.edu.tw

Titles in This Series

For a complete list of titles in this series, visit the
AMS Bookstore at **www.ams.org/bookstore/**.